Advances in Biology

This is a numbered series focused on the latest research in biology.

nova
science publishers
www.novapublishers.com

Advances in Biology

Advances in Biology. Volume 1
Charles D. Grant (Editor)
2022. ISBN: 979-8-88697-201-6 (Hardcover)
2022. ISBN: 979-8-88697-218-4 (eBook)

More information about this series can be found at
https://novapublishers.com/product-category/series/advances-in-biology/

Charles D. Grant
Editor

Advances in Biology

Volume 2

nova
science publishers
www.novapublishers.com

Library of Congress Cataloging-in-Publication Data

ISBN: 979-8-88697-422-5
ISSN: 2833-1753

Published by Nova Science Publishers, Inc. † New York

Contents

Preface

This volume contains eight selected chapters on recent advances in the science of biology. Chapter One explores a combined DFT and electrostatics approach to coupling biological electron and proton transfer reactions in Cytochrome c oxidase. Chapter Two reviews a new process for the detection of amyloid fibrils in the brain using unsymmetric cationic cyanine dyes. Chapter Three discusses the membrane effects of amyloid fibrils in the body via fluorescence study and computer modeling. Chapter Four updates the model of the curvature of graphitic materials through an analysis of molecular modeling studies on the spiral structure of graphene and eumelanin models. Chapter Five reviews the generation of antigen specific CD8+ T-cells by B-cells and Dendritic cells. Chapter Six provides a categorization of prophage genes in *Bacillus subtilis* 168 and assesses their relative impotence through RNA-seq gene expression analysis. Chapter Seven reviews the optimization of naturally occurring compounds through the click chemistry approach, giving researchers a new powerful synthetic toolbox. Chapter Eight is an overview of studying asafoetida phytoconstituents as biological inhibitors.

Chapter 1 - Cytochrome c oxidase (CcO) is the terminal enzyme in the respiratory electron transport chain of aerobic organisms. It is a redox-driven proton pump that converts atmospheric oxygen to water and couples the oxygen reduction with generating a membrane proton gradient that subsequently drives ATP synthesis. The function of CcO as a biomolecular nanomachine that transforms the energy of redox reaction into protonmotive force across a biological membrane has been the subject of intense research, debate, and controversy. For a long time, the molecular mechanism of electron transfer coupled to proton pumping in CcO has been one of the central unsolved problems in biochemistry, molecular biology, and bioenergetics. The enzyme structure has been solved for several organisms; however, details of its molecular mechanism of proton pumping still remain elusive. Mainly, the nature and position of the proton-loading site, the key element of the mechanism, are under dispute. However, nowadays, many essential details and principles have emerged with the accelerating progress in this field. Recent calculations indicate that one of the histidine ligands of the CuB center, His291, may play the role of the pumping element. In this chapter, the authors review the first principles calculation used to study models of the catalytic center of CcO and calculate the pKa of the His291 residue for both the reduced and oxidized states of the enzyme catalytic center. The combined density functional theory (DFT) and continuum electrostatic calculations (QM/CE method) are employed to explore the coupling of the conformational changes of Glu242 residue, the primary proton donor of both chemical and pump protons, to its pKa, and the pKa of His291, a putative proton-loading site of the pumping model. The pKa values of His291 and Glu242, the two crucial residues of the model, are calculated for different redox states of the enzyme, and the influence of different factors on the pKas is analyzed in detail. To understand how different factors affect the apparent pKa values of His291 and Glu242, the authors have considered several computational QM/CE models of the membrane–enzyme–cavities– solvent system and tested different dielectric properties of the water-filled cavities. In addition, the DFT is applied to two different sizes of quantum-mechanical (QM) systems of interest. The authors also review the structure and function of CcO, describe the proposed mechanism of proton pumping, identify the fundamental problems that can be addressed computationally in this area and describe the methods for their solution. The outstanding theoretical and computational challenges of the area are also discussed.

Chapter 2 - Amyloid fibrils, β-sheet-rich linear protein aggregates formed by a large variety of proteins and peptides, have received much interest in the recent decades due to their involvement in a range of human disorders among which are neurodegenerative diseases, type II diabetes, systemic amyloidoses, etc. One of the simplest and inexpensive methods for the identification and characterization of amyloid fibrils in vitro and the visualization of intracellular and intercellular amyloid inclusions in vivo is based on the use of fluorescent dyes sensitive to the beta-pleated motifs. Several classes of fluorescent compounds, such as near-infrared fluorescent DANIR probes, benzanthrone dyes, curcumine-based fluorescence imaging probes, cyanine, BODIPY and styryl derivatives have been successfully employed as amyloid sensors along with classical amyloid markers thioflavin T and Congo Red. This contribution is focused on the application of cyanine dyes to differentiating between the fibrillar and monomeric states of protein molecules. More specifically, a series of novel unsymmetric trimethine cyanine dyes have been tested for their ability to identify and characterize amyloid fibrils of insulin and lysozyme.

The spectral properties of the trimethines were examined in a free state and in the presence of non-fibrillized and fibrillar proteins. The fluorescence studies showed that the examined trimethines have negligible emission in the buffer solution as well as in the presence of the non-fibrillized insulin and lysozyme due to their H-type self-association. However, the binding of the examined cyanines to amyloid fibrils was followed by a significant increase of their fluorescence, with the magnitude of this effect being dependent on the dye chemical structure. The quantitative parameters of the trimethine association with fibrillar proteins have been derived from the results of fluorimetric titration. The spectral properties and the binding characteristics of cyanines proved to be different for the lysozyme and insulin amyloid fibrils, highlighting the sensitivity of these dyes to the distinctions in fibril morphology. The molecular docking studies revealed the possible binding sites, as well as the nature of interactions involved in the dye–protein complexation in the monomeric protein and amyloid assemblies. The possibility of differentiating between the non-fibrillized and fibrillar states of insulin using the Förster resonance energy transfer between Thioflavin T as a donor and one of a series of trimethine cyanine dyes as an acceptor has been demonstrated. By comparing the dye preference to the fibrillar protein aggregates, the most prospective amyloid probes have been selected. Overall, the authors' findings demonstrate that the novel usymmetric trimethine cyanine dyes are prospective amyloid tracers capable of operating in both fluorescence and absorption modes.

Chapter 3 - Amyloid fibrils, highly ordered elongated protein aggregates with a characteristic cross-β-sheet molecular architecture, have received growing research attention throughout the past decades. These nanoassemblies are explored from three main perspectives concerning the pathological, functional and artificial amyloids. Pathological amyloids are proved to be associated with a range of human diseases, among which are neurodegenerative disorders, type II diabetes, systemic amyloidoses, etc. Ongoing research efforts are directed at elucidating the nature of protein misfolding and amyloid conversion, search for therapeutic molecules capable of preventing these processes, increasing the sensitivity and specificity of amyloid detection. Functional amyloids, found in bacteria, fungi, plants and mammals, are involved in diverse physiological activities, such as biofilm formation, storage of peptide and proteins, information carriage, signaling, etc. Artificial amyloids are scrutinized as building blocks for bioinspired functional nanomaterials with versatile applications, including tissue engineering, drug delivery, biosensing, nanowire production, etc. Major research efforts are currently focused on pathological amyloid with the emphases on the molecular-level details of cell damage by pathogenic protein assemblies. Accumulating evidence supports the idea that cytotoxic action of amyloid fibrils and their oligomeric intermediates is primarily targeted at cellular membranes. The cytotoxicity of oligomeric aggregates is displayed in the formation of non-specific ion channels, the alterations in calcium homeostasis and activity of membrane enzymes, and the loss of membrane integrity, while the mature fibrils are capable of producing the increase in membrane permeability, lipid bilayer thinning, membrane fragmentation and the uptake of lipids, etc. The membrane effects of aggregated proteins essentially depend on the basic physicochemical characteristics of a lipid bilayer, such as chemical nature of phospholipid headgroups, electrostatic surface potential, degree of hydration, lipid packing density, phase state, bilayer curvature, etc. which, in turn, are determined by the exact lipid composition. This emphasizes the importance of systematic investigation of the model protein-lipid systems with controllable structural and physicochemical properties. The present contribution is a quintessence of the authors' research efforts focused on elucidating the membrane effects of amyloid fibrils formed from structurally different proteins – lysozyme, insulin, N-terminal fragment of apolipoprotein A-I and Aβ-peptide. The lipid component of the model protein-lipid systems was represented by the unilamellar liposomes composed of zwitterionic (phosphatidylcholine, PC), anionic (cardiolipin, CL) phospholipids and cholesterol (Chol). The following main questions have been addressed: i) whether the membranotropic action of the examined fibrillar proteins depends on the electrostatic potential of lipid bilayer; ii) how the amyloid-induced structural alterations of the model membranes are modulated by cholesterol in the two- (PC/Chol) and three-component (PC/CL/Chol) lipid systems; iii) what the membrane disposition of amyloid fibrils is. To answer these questions, several fluorescence-based approaches have been used, involving the fluorescent probes located in the polar (Laurdan) and nonpolar (pyrene, diphenylhexatriene) regions of a lipid bilayer, the intrinsic protein fluorescence and its quenching by a polar quencher acrylamide. To uncover the preferable modes of the fibril-membrane binding, the computer modeling tools, such as CreateFibril, AQUASOL, PPM and molecular docking have been employed. The results obtained provide evidence for i) different modes of interactions of monomeric and fibrillar proteins with the model membranes; ii) more pronounced influence of fibrillar proteins on the degree of hydration and free volume of the lipid bilayer; iii) essential roles of electrostatic, hydrophobic protein-lipid interactions and amyloid twist in determining the bilayer location of amyloid fibrils; iv) strong dependence of the membrane-modifying effects of fibrillar proteins on the lipid bilayer composition. Overall, the authors' findings support the notion that the development of effective anti-amyloid strategies should involve the search for compounds capable of preventing the changes in membrane structure and physicochemical properties.

Chapter 4 - In the last few years, molecular modeling has become a relevant tool for research and teaching purposes in chemistry and molecular biology, allowing the design and development of new molecular structures and materials. In the present chapter, the focus of molecular modeling methodology is addressed towards the mechanistic origin of spiral organization of graphitic materials. A recently proposed catechol-porphycene (CPo) model for eumelanin, and both "arm-chair" (A) and "zig-zag" (Z) single layer graphene spirals were chosen as case studies to be analyzed by molecular modeling after MM+, PM3, and extended Hückel geometry optimization using HyperChem v7 software. The poly-CPo eumelanin model corresponds to a highly conjugated graphite-like planar structure based on the polymerization of 5,6-dihydroxy-indole monomers, using all the reactive carbon radicals. Formation of ether bridges between catechol groups of the polymer induces a constrained curvature due to the newly formed furan rings, which in turn results in a spiral organization of poly-CPoe. Interestingly, published molecular modeling studies on graphene properties show a spiral structure, which generates improved opto-electronic properties. However, it is not obvious why a graphene layer should be curved lacking any chemical modification of the original planar polymer. Regarding oxidation, it is tempting to speculate that the possible attack of graphene edges by oxidant agents (e.g., singlet oxygen), could introduce hydroxyls along the reactive A and Z edges, which after ether bridging would result in curved and then spiral graphenes just as it is revealed by molecular modeling. Likewise, in addition to furan, cyclopenta-dienyl and pyrrole rings, oxo and peroxide bridges along graphene edges do result in a spiral structure. In all cases, oligomer stacking generates fused LUMOs at low energies. Stacked poly-CPoe and graphene spirals show compact and very close HOMO (valence band, VB) and LUMO (conduction band, CB) levels, fused LUMOs and strongly reduced energy gap (Eg). It is worth to note the clear similarity between the fused LUMOs of a spiral graphene and the poly-CPoe eumelanin model, which is related to the striking semiconductor characteristics of these amorphous graphitic compounds. Likewise, molecular modeling methods allow the design of new bioinspired materials (e.g., planar and spiral poly-catechol-imidacene) and explain the massive fusion of longitudinal and transversal LUMOs continuum from stacked, intercalated and bilayered graphitic spirals yielding compact and close VB and CB.

Chapter 5 - The role of CD8+ T cells in protecting the host from the intracellular pathogens and cancer is well established. APCs (antigen presenting cells) present the antigens to CD8+ T cells in the form of MHC-I/peptide complex. Following the antigen specific stimulation, CD8+ T cells proliferate and differentiate; the cytotoxic T cells generated kill the target cells (pathogen infected cells or cancer cells), thus clear the infection or regress the tumor. The initial interaction between the antigen bearing APCs and CD8+ T cells occurs in the peripheral lymphoid organs. Dendritic cells are the most efficient APCs and known to initiate CD8+ T cell response. However, emerging evidence suggests the role of other APCs such as B cells in activation of CD8+ T cells through a mechanism known as antigen cross-presentation. Since CD8+ T cells play a significant role during antiviral immune response, various efforts are being made to design vaccines that can generate specific antiviral CD8+ T cells. The present chapter discusses the generation of CD8+ T cells during a natural infection or vaccination. It also highlights the mechanisms behind CD8+ T cell generation and provides insights to design better vaccines.

Chapter 6 - Bacteriophage evolves to control the population of fast-growing bacterial cells, without which explosion in bacterial population may induce unimaginable harm to diverse ecosystems. But, bacteriophage also “hide” in bacterial genomes when nutritional and environmental circumstances are unfavourable. This involves the integration of phage genome into the host genome at appropriate genomic loci in a process known as lysogeny. This work sought to delineate the prophages present in the annotated genome of *Bacillus subtilis* 168, and assess their relative importance through RNA-seq expression analysis. Firstly, examination of the annotated genome of the model Gram-positive bacterium revealed five distinct prophage regions: SPBeta, prophage 6, PBSX, prophage 3 region, and prophage 1 region. All prophage regions contain host genes, which suggests that host transposase activity have swapped in host genes for phage genes in the prophage genome. Given the significant number of phage genes that have been swapped into each of the prophage genome, all prophage regions are deemed to be defective. BLAST analysis further highlighted that many of the prophages in *B. subtilis* are extinct given that they do not have ancestral or daughter brethren. However, RNA-seq transcriptome analysis of *B. subtilis* turned out an interesting paradox indicative of the important role that host transposase have in swapping in host promoters for prophage genes. Specifically, a significant number of prophage genes are highly expressed, which is implausible given that phage genes should be transcriptionally silent. The result and phenomenon further suggests the relative facile nature in which host promoters could be swapped in for phage genes, which is indicative of presence of genomic motifs in prophage genome recognizable by host transposase. Existence of such sequence motifs is thus indicative of possible co-evolution of transposase and phages where transposases were originally a part of the phage genome, which latter "jumped out" into the host genome to aid the swapping in of host genes into the prophage genome for augmenting prophage genetic repertoire in the face of changing environmental conditions. Overall, it is not uncommon for bacterial species to harbour multiple prophages. But, lysogeny may not be a viable option for long-term preservation of prophage genetic repertoire given that host transposase would inevitable swap in host genes at random locations in the prophage genome.

Chapter 7 - In this rapidly evolving era, naturally occurring product-driven discovery of hybrid molecules by click chemistry have recently revolutionized promising results. Molecular optimization depending on modification of pharmacophoric moiety is often necessary for improvement on target specificity, toxicity, potency and stability. Doubtlessly, click chemistry is one the most efficient synthetic approaches which gives rise to molecular diversity. This review focuses on optimization of naturally occurring products by click chemistry protocol which is a powerful synthetic toolbox and their promising mode of actions compared to parental molecules.

Chapter 8 - Asafoetida, an oleo gum resin obtained from *Ferula asafoetida* of medicinal and nutritional importance. It has been traditionally used for the treatment of various ailments as spice and medicine from centuries. A number of phytoconstituents are present in *Ferula asafoetida* which are responsible for pharmacological properties. To evaluate to potential of *Ferula asafoetida* in the treatment of microbial infections, molecular modeling technique was utilized. 74 phytoconstituents of asafoetida were created from ACD Chem sketch software and PDBs for 14α-demethylase and transpeptidase were retrieved from RCSB PDB database. Docking studies were carried out using Mol Dock score software. All the prepared ligands were evaluated for drug likeliness properties using Swiss ADME. Among 74 compounds, Asacoumarin A was found to be good inhibitor of enzymes under investigation and can be a potential lead molecule for treatment of bacterial and fungal infections.

Chapter 1

Combined DFT/Electrostatics Approach: Exploring the Coupling of Biological Electron and Proton Transfer Reactions in Cytochrome *c* Oxidase

Dragan M. Popović*

Department of Chemistry, Institute of Chemistry, Technology and Metallurgy, National Institute of the Republic of Serbia, University of Belgrade, Belgrade, Serbia

Abstract

Cytochrome *c* oxidase (C*c*O) is the terminal enzyme in the respiratory electron transport chain of aerobic organisms. It is a redox-driven proton pump that converts atmospheric oxygen to water and couples the oxygen reduction with generating a membrane proton gradient that subsequently drives ATP synthesis. The function of C*c*O as a *biomolecular nanomachine* that transforms the energy of redox reaction into protonmotive force across a biological membrane has been the subject of intense research, debate, and controversy. For a long time, the molecular mechanism of electron transfer coupled to proton pumping in C*c*O has been one of the central unsolved problems in biochemistry, molecular biology, and bioenergetics. The enzyme structure has been solved for several organisms; however, details of its molecular mechanism of proton pumping still remain elusive. Mainly, the nature and position of the proton-loading site, the key element of the mechanism, are under dispute. However, nowadays, many essential details and principles have emerged with the accelerating progress in this field. Recent calculations

* Corresponding Author's Email: dragan.popovic@ihtm.bg.ac.rs.

In: Advances in Biology. Volume 2
Editor: Charles D. Grant
ISBN: 979-8-88697-422-5

indicate that one of the histidine ligands of the Cu_B center, His291, may play the role of the pumping element. In this chapter, we review the first principles calculation used to study models of the catalytic center of C*c*O and calculate the pK_a of the His291 residue for both the reduced and oxidized states of the enzyme catalytic center. The combined density functional theory (DFT) and continuum electrostatic calculations (QM/CE method) are employed to explore the coupling of the conformational changes of Glu242 residue, the primary proton donor of both chemical and pump protons, to its pK_a, and the pK_a of His291, a putative proton-loading site of the pumping model. The pK_a values of His291 and Glu242, the two crucial residues of the model, are calculated for different redox states of the enzyme, and the influence of different factors on the pK_as is analyzed in detail. To understand how different factors affect the apparent pK_a values of His291 and Glu242, we have considered several computational QM/CE models of the membrane–enzyme–cavities– solvent system and tested different dielectric properties of the water-filled cavities. In addition, the DFT is applied to two different sizes of quantum-mechanical (QM) systems of interest. We also review the structure and function of C*c*O, describe the proposed mechanism of proton pumping, identify the fundamental problems that can be addressed computationally in this area and describe the methods for their solution. The outstanding theoretical and computational challenges of the area are also discussed.

Keywords: bioenergetics, heme-copper oxidases, nitric-oxide reductase, copper center, heme, binuclear complex, histidine ligand, bovine, proton pumping, molecular mechanism, redox-coupled pK_a, pK_a calculations, DFT, reaction and protein field

Abbreviations

BNC	binuclear (complex) center
C*c*O	cytochrome *c* oxidase
Cyt*c*	cytochrome *c*
DFT	density functional theory
ET	electron transfer
ETC	electron transport chain
MD	molecular dynamics
MeIm	methylimidazole
PB	Poisson-Boltzmann
PT	proton transfer (translocation)

PLS	proton-loading site
pmf	protonmotive force
PRA*a*, PRD*a*	propionates A/D of heme *a*
PRAa_3, PRDa_3	propionates A/D of heme a_3
QM/CE	quantum-mechanical/continuum electrostatic (method)
SCRF	self-consistent reaction field

Introduction

Cellular respiration, a process of oxidizing nutrient molecules to carbon dioxide and water, involves the transfer of electrons through a sequence of membrane-bound protein complexes in the respiratory electron transport chain (ETC). The oxidation of nutrient molecules and electron transport are coupled processes. The initial electron donors are the nutrient molecules, such as monosaccharides/carbohydrates, fatty acids, or amino acids; the final acceptor is atmospheric molecular oxygen bound to the catalytic center of cytochrome *c* oxidase (C*c*O). Thus, one can say that oxygen in the binuclear center (BNC) is the *ultimate sink* for the electrons from the respiratory electron transport chain. In the process, oxygen is reduced to form water, and protons are pumped across the membrane (Antonini et al. 1970; Wikström 1977). This process generates an electrochemical proton gradient across the membrane, which drives the synthesis of ATP that provides energy for all energy-consuming processes in living organisms (Mitchell 1961, 1976). Generating the electrochemical proton gradient across the organelle membrane is how living cells temporarily store the energy and use it later when needed. The overall process is the central mechanism of biological energy transduction in cells.

Cytochrome *c* oxidase (C*c*O), the terminal enzyme of the respiratory electron transport chain, is situated in the plasma membrane of aerobic prokaryotes or the inner mitochondrial membrane of eukaryotes. Since heme-copper oxidases catalyze most of the biological oxygen consumption (Brzezinski and Gennis 2008), their importance for cellular respiration and energy supply in aerobic organisms is essential. Mitochondrial C*c*O plays a crucial role in aerobic cellular respiration, reducing O_2 to H_2O in a process coupled with the pumping of protons across the inner mitochondrial membrane (Babcock and Wikström 1992; Ferguson-Miller and Babcock 1996; Wikström 1998; Gennis 1998b; Michel et al. 1998), leading to a net positive charge movement to the intermembrane space and generating in part the membrane proton gradient that provides the *protonmotive force* required to drive ATP synthesis (Babcock and Wikström 1992; Ferguson-Miller and Babcock 1996; Gennis 1998b).

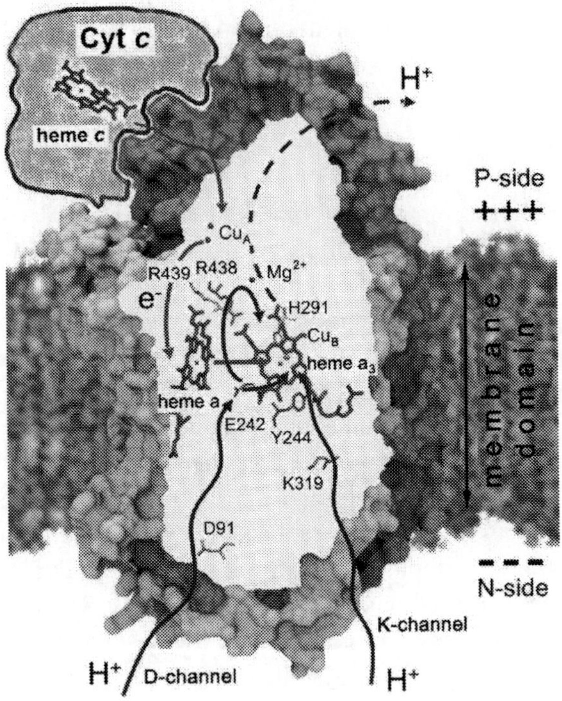

Figure 1. Structure of the bovine heart cytochrome *c* oxidase (Yoshikawa et al. 1998; Tsukihara et al. 2003) placed in the inner mitochondrial membrane. Subunits I and II are depicted in yellow and green along with the four redox centers – Cu_A complex, heme *a*, heme a_3, and Cu_B complex, forming the electron transfer pathway (red arrows). The catalytic center is treated with density functional theory. The rest of the protein is represented by partial atomic charges embedded in the inhomogeneous dielectric of the protein-membrane-solvent system and treated by electrostatic calculation. Cytochrome *c*, schematically shown at the *P*-side of the membrane, shuttles and provides electrons to C*c*O owing to their surfaces' electrostatic and structural complementarity. The proton transport pathway, including the proton entrance D- and K-channels and the proton exit channel, is displayed by blue arrows. In addition, several essential residues are also highlighted.

C*c*O contains four redox-active metal centers: Cu_A, a low-spin heme *a* (Fe_a), and the binuclear catalytic center (BNC) consisting of a high-spin heme a_3 (Fe_{a3}) and a Cu_B. Electrons are shuttled and delivered, one at a time, on the

P-side by the water-soluble and membrane peripheral protein cytochrome *c* (Cyt*c*) to the Cu_A center. The electrostatic and surface complementarity between Cyt*c* and the globular domain of subunit II of *Cc*O enables the binding and facilitates the ET between heme *c* and Cu_A cofactors (Figure 1). Electrons supplied to Cu_A by reduced Cyt*c* are sequentially transferred through heme *a* (Fe_a) to the active site (BNC), where oxygen reduction occurs (Ferguson-Miller and Babcock 1996; Konstantinov et al. 1997; Ostermeier et al. 1997; Hosler, Ferguson-Miller, and Mills 2006). During each cycle (an O_2 molecule reduced to water), eight protons are taken up from the matrix space, four protons are used for water formation (*chemical* or *substrate* protons), and four protons are pumped across the membrane (*vectorial* or *pumped* protons). The overall chemical reaction catalyzed by the *Cc*O enzyme is deceptively simple and can be expressed as follows:

$$O_2 + 4e^-{}_{(P)} + 8H^+{}_{(N)} \rightarrow 2H_2O + 4H^+{}_{(P)}, \qquad (1)$$

where (*N*-) and (*P*-) indicate two sides of the mitochondrial membrane: the inner (negatively charged) matrix side and the outer (positively charged) intermembrane space, respectively. Notice that the electrons and protons come from the opposite sides of the membrane. The chemical and pumped protons are transferred from inside the matrix space through the inner mitochondrial membrane via two hydrogen-bond networks – the entrance proton-conducting D- and K-channels. The majority of 6-7 protons are delivered along the D-channel, which provides all four pumped protons (Gennis 1998c; Qin et al. 2009). As established experimentally, Glu242 residue is presumably the major proton donor in the enzyme (Hellwig et al. 1998; Pawate et al. 2002; Mills et al. 2003), while Lys319 plays a role in the reductive halve of the catalytic cycle providing up to two chemical protons (Konstantinov et al. 1997; Dürr et al. 2008).

In the past decades, the structure of *Cc*O has been solved for several organisms (Ostermeier et al. 1997; Yoshikawa et al. 1998; Svensson-Ek et al. 2002; Tsukihara et al. 2003; Dürr et al. 2008; Qin et al. 2009; Yano et al. 2016); however, details of the molecular mechanism of proton pumping still remain a subject of intense debate. In particular, the identity of the proton-pumping site, the essential element of the mechanism, is under dispute. Based on the experimental data, computer simulations, and theoretical studies, various models have been proposed to explain the proton pumping mechanism of this complex protein system, see e.g., (Riistama et al. 1997; Michel 1998; Das et al. 1999; Wikström 2000; Tsukihara et al. 2003; Wikström,

Verkhovsky, and Hummer 2003; Popović and Stuchebrukhov 2004a, b; Bloch et al. 2004; Siletsky et al. 2004; Soares et al. 2004; Faxen et al. 2005; Popović and Stuchebrukhov 2006; Brzezinski and Ädelroth 2006; Song, Michonova-Alexova, and Gunner 2006; Olsson and Warshel 2006; Siegbahn and Blomberg 2007; Sharpe and Ferguson-Miller 2008; Fee, Case, and Noodleman 2008; Kaila et al. 2008; Fadda, Yu, and Pomès 2008; Siletsky and Konstantinov 2012; Yano et al. 2016; Blomberg 2021). However, there is no agreement on significant issues, such as where the pumping site is located and how the enzyme pumps protons. It is also an incredibly challenging experimental task since the mutagenesis studies and the kinetic measurements of proton translocation cannot uniquely pinpoint a single residue as a pumping site (Siletsky et al. 2004; Bloch et al. 2004; Belevich et al. 2007; Siletsky and Konstantinov 2012; Sugitani, Medvedev, and Stuchebrukhov 2008). Similar problems are encountered with other advanced experimental techniques due to the strong mutual coupling and proximity between potential PLS sites, which significantly affect each other.

Although a molecular mechanism of proton pumping of C*c*O remains a subject of intense debate, many common features and principles appear in the most recently proposed models, and they will be briefly reviewed in this chapter. See the following literature for a detailed review of the enzyme structure, function, and kinetics of coupled electron and proton transfer reactions, e.g., (Han, Takahashi, and Rousseau 2000; Wikström 2004; Brzezinski 2004; Gennis 2004; Belevich, Verkhovsky, and Wikström 2006; Hosler, Ferguson-Miller, and Mills 2006; Belevich et al. 2007; Brzezinski and Gennis 2008; Dürr et al. 2008; Sharpe and Ferguson-Miller 2008; Qin et al. 2009; Siletsky and Konstantinov 2012; Sarti et al. 2012; Popović 2013; Wikström, Krab, and Sharma 2018; Du, Götz, and Noodleman 2018).

The correlated motion of electrons and protons coming from the opposite sides of the membrane is required for the proper function of the enzyme (Konstantinov et al. 1997; Jasaitis et al. 1999; Ruitenberg et al. 2002; Stuchebrukhov 2003a; Bloch et al. 2004; Siletsky et al. 2004; Quenneville, Popović, and Stuchebrukhov 2006; Belevich et al. 2007; Kaila et al. 2009; Hammes-Schiffer and Stuchebrukhov 2010; Popović and Stuchebrukhov 2012). The electrostatic interactions between electrons and protons, which result in the dependence of the protonation state of the enzyme on its redox state, are of fundamental importance. Computer simulations can estimate such interactions and thereby get insights into possible mechanisms of the coupled reactions.

In redox metalloenzymes, the theoretical problem is multifaceted; it involves questions about pure electron transfer, with all complexities of the subject, proton transfer, no lesser problem in protein medium, and finally, their coupling. Electrons are transferred between metal centers via thermally assisted quantum tunneling, whereas protons, much heavier particles, need water intermediates to move between protonatable groups in the protein. Electron tunneling and internal water in the proteins are both rich subjects that the computer simulations can address. The overall kinetics of the coupled reactions in proteins can be quite complicated and involve multiple possibilities, a topic that only recently gets attention in detailed experimental and theoretical studies (Olsson et al. 2007; Siegbahn and Blomberg 2007; Pisliakov et al. 2008; Kaila et al. 2009; Blomberg and Siegbahn 2010; Johansson et al. 2011; Blomberg and Siegbahn 2012).

One of the most fundamental questions one needs to address in any discussion of this rich subject is the nature and free energy of the initial and final states of the system. Once the initial and final states are well defined and the free energy of these states is known, one can further address the kinetics questions about transition state complexes and kinetic barriers, i.e., how the system goes from one state to another during the catalytic cycle of the enzyme. In this review, we will focus solely on thermodynamic issues. An interested reader is referred to review other subjects from this area published elsewhere (Stuchebrukhov 2003a, b; Hammes-Schiffer and Stuchebrukhov 2010).

More precisely, the problem that we are interested in is formulated as follows: for a given redox state of the metal centers of C*c*O (Cu_A, heme *a*, heme a_3, and Cu_B), find the equilibrium protonation state of the enzyme, and evaluate the energy of possible non-equilibrium states of interest that might be involved in the cycle of the enzyme. The equilibrium protonation states at given ambient conditions can be described in terms of pK_as, protonation energies, and probabilities of the protonatable groups in the protein, which are redox-dependent in our molecular system. Since the details of the functioning mechanism are not well known, the essence of this type of calculation is to estimate the feasibility of the proposed hypothesis and make specific predictions that could be later tested experimentally. One of the problems of understanding the proton transport in C*c*O is that the key residues involved in redox-coupled proton activity still remain unidentified (Wikström and Verkhovsky 2007; Brzezinski and Gennis 2008; Popović 2013; Johansson et al. 2013). Here we will show how computations allow one to provide crucial information about the enzyme.

The organization of the chapter is as follows. We first discuss the necessary details of the structure and function of C*c*O, describe the proposed mechanism of proton pumping, and identify the fundamental problems that can be addressed computationally. Next, we describe the methods for the solution of the formulated problems and then present the results of calculations with a discussion of the assumptions of the computational models, emphasizing the important factors in such calculations and the theoretical and computational challenges of the area.

Structure and Function

The Heme–Copper Superfamily

The mitochondrial cytochrome *c* oxidase belongs to a heme–copper superfamily. These enzymes include O_2 reductases (respiratory oxidases) and NO reductases (Pereira et al. 2008; Pereira, Santana, and Teixeira 2001). The most important difference among the heme–copper oxidases is in the variation of the heme-types that occupy the low-spin and high-spin sites, such as heme *a*, heme *b*, and heme *o* (Ferguson-Miller and Babcock 1996). Accordingly, among the solved structures, it could be distinguished the aa_3-type from bovine (Yoshikawa et al. 1998; Tsukihara et al. 2003; Shinzawa-Itoh et al. 2007), from *Paracoccus denitrificans* (Iwata et al. 1995; Ostermeier et al. 1997), from *Rhodobacter sphaeroides* (Svensson-Ek et al. 2002; Qin et al. 2006), the ba_3-type oxidase from *Thermus thermophilus* (Soulimane et al. 2000; Luna et al. 2008), and the bo_3-type oxidase from *Escherichia coli* (Abramson et al. 2000). All heme–copper oxidases generally share a similar tertiary structure with high sequence similarity of subunit I and the same arrangement of the metal centers – hemes and copper complex (Cu_B) within subunit I (Noor and Soulimane 2013).

Although many experimental results, including those from the electrometric studies, are obtained for bacterial C*c*O (from *P. denitrificans* and *R. sphaeroides*), as it has been demonstrated recently, all species of the A-family (aa_3-type) show remarkable structural similarity. Moreover, the key amino acid residues' microscopic electrostatic and thermodynamic properties are almost identical, strongly suggesting a similar mechanism in all these species (Popović et al. 2010).

Structure–Function Correlation in *Cc*O

Figure 1 shows the structure of *Cc*O, the location of all four redox centers, and highlights the essential residues in proton entrance and exit channels. The calculations reported here are based on the crystal structures of mitochondrial cytochrome *c* oxidase from the bovine heart (PDB codes: 2OCC and 1V55) obtained at a resolution of 2.3 Å and 1.8 Å, solved by Yoshikawa et al. (1998) and Tsukihara et al. (2003), respectively.[1] The enzyme possesses 13 subunits and seven metal centers: heme *a*, heme a_3, Cu_A, Cu_B, Mg, Na, and Zn. Subunit I consists of 12 transmembrane helices (514 residues) arranged in quasi-threefold rotational symmetry forming the three internal channels filled with water molecules. Subunit I contains a low-spin heme *a*, a high-spin heme a_3 and Cu_B as cofactors. Histidines, $His61_I$ and $His378_I$, are axially ligated to heme *a*, while heme a_3 together with the copper atom (Cu_B) and coordinating ligands ($His376_I$ to iron and $His240_I$, $His290_I$, $His291_I$ to copper) form the binuclear center (Shapleigh et al. 1992). One of the histidines ($His240_I$) bound to Cu_B is cross-linked to a tyrosine ($Tyr244_I$) by the post-translational modification (Ostermeier et al. 1997; Yoshikawa et al. 1998; Buse et al. 1999). The O_2 reduction site known as the binuclear center is placed in subunit I and consists of a high-spin heme a_3 and a copper complex (Cu_B).

The available X-ray structures of cytochrome *c* oxidases indicate three possible pathways for proton conduction within subunit I: the D-, K-, and H-channels (Ostermeier et al. 1997; Yoshikawa et al. 1998; Svensson-Ek et al. 2002; Tsukihara et al. 2003; Dürr et al. 2008; Qin et al. 2009). The putative H-channel is most clearly defined in the structure of the bovine oxidase (Tsukihara et al. 2003) and is suggested to have the primary role in translocating the pumped protons from the *N*- to *P*-side of the membrane (Shimokata et al. 2007). It has been recently proposed as the proton pumping pathway of mammalian oxidases (Tsukihara et al. 2003; Yano et al. 2016), but mutagenesis studies on bacterial enzymes have failed to provide convincing evidence so far concerning the functional and mechanistic importance of this putative channel (Gennis 1998c; Pfitzner et al. 1998; Lee et al. 2000; Salje, Ludwig, and Richter 2005). Moreover, many computational studies could not find any significant correlation between residues in the H-channel and the proton pumping (Wikström, Verkhovsky, and Hummer 2003; Popović and Stuchebrukhov 2004a; Kaila et al. 2008; Fadda, Yu, and Pomès 2008). A recent MD simulation study on the bovine *Cc*O suggests that the H-channel is

[1] Numeration of the residues corresponds to bovine *Cc*O if not otherwise stated.

most likely not directly involved in proton uptake from *N*-side or proton pumping, but its role is more general and described as a potential dielectric well, modulating effects of buried charge changes (Sharma et al. 2017). Results of the electrostatic study by Popović et al. (2004) have shown a partial proton uptake (0.35 H^+) from the *P*-side of the membrane upon an electron transfer first from Cyt*c* to Cu_A and subsequent ET from Cu_A to heme *a* by some residues into the top of the H-channel and close surrounding. After the electron proceeds further to heme a_3, these sites release about one-third of a proton back to the membrane *P*-side (intermembrane space). We suggested that the group of residues has a role in maintaining a local electroneutrality and charge balance during the ET processes in the protein matrix within the membrane domain. However, these changes are not related to proton pumping since the protons involved in the pumping and chemical reduction of oxygen should come from the *N*-side of the membrane (the matrix space).

In contrast, studies utilizing site-directed mutants provide strong evidence that the D-channel and K-channel are functionally important (Zaslavsky and Gennis 2000; Konstantinov et al. 1997; Wikström et al. 2000). Mutants in the K-channel are defective in one or more steps associated with the reductive halve of the catalytic cycle, i.e., the reduction of the Fe_{a3}–Cu_B binuclear center (Hosler et al. 1996; Ädelroth, Gennis, and Brzezinski 1998; Brändén et al. 2001; Brändén et al. 2002). Mutants in the D-channel are defective in steps following the interaction of dioxygen with the reduced Fe_{a3}–Cu_B center, i.e., during the oxidative halve (Svensson-Ek et al. 2002; Fetter et al. 1996; Ädelroth et al. 1997; Mills et al. 2000). Based on the results of mutagenic studies, six protons are presumably delivered along the D-channel, whereas the D-channel provides all pumped protons. The K-channel provides up to two chemical protons in the reductive phase of the catalytic cycle (Ruitenberg et al. 2000; Ruitenberg et al. 2002; Kirchberg, Michel, and Alexiev 2013).

Subunit II possesses 227 residues and a bimetallic Cu_A center, and it consists of a globular domain mostly situated in the intermembrane space (*P*-side) and the two transmembrane helices, which interact with helices of subunit I. The Cu_A center consists of two copper atoms bridged with two cysteines ($Cys196_{II}$, $Cys200_{II}$) and one histidine ligand ($His161_{II}$ or $His204_{II}$). In addition, they are connected to $Met207_{II}$ (side-chain Sδ atom) or $Glu198_{II}$ (backbone oxygen), such that ligands of both copper atoms form distorted tetrahedrons.

Two other non-redox metal centers, Mg and Na, are placed in subunit I. Mg^{2+} cation, ligated by $His368_I$, $Asp369_I$, $Glu198_{II}$, and three water molecules, is placed at the interface between subunits I and II, close to the intermembrane (periplasmic) space. The whole complex is neutral since two negatively charged acidic groups compensate for the charge of the Mg^{2+} cation. Similarly, Na^{+} is neutralized by ligating $Glu40_I$ (–1), $Gly45_I$ (backbone O), $Ser441_I$, and one water molecule. This binding site is located in the proximity of propionate *A* of heme *a* (PRA*a*).

The subunits I and II from different species exhibit remarkable similarity (Iwata et al. 1995; Tsukihara et al. 1996; Ostermeier et al. 1997; Popović et al. 2010), and the two-subunit enzyme is active in proton pumping (Haltia, Saraste, and Wiksrom 1989; Hendler et al. 1991). Additional subunits may have a supporting role (Iwata et al. 1995; Tsukihara et al. 1996), maintaining the structural integrity (Bratton, Pressler, and Hosler 1999) and making an additional shield of protection around the catalytic core-subunit. Therefore, our calculations were performed only on chains I and II, including a membrane model, as an infinite low dielectric slab of 45 Å thickness (Gennis 1998a) that shields the membrane-exposed protein surface from the solvent (Figure 1). By these simplifications, the essence of the system is not lost, yet its size is significantly reduced.

To summarize the discussion of this section, the enzyme function can be described as follows. In the respiratory ETC, electrons are carried from complex III (Cytbc_1) to complex IV (C*c*O) by the water-soluble and membrane peripheral protein cytochrome *c*. Electrons are shuttled and supplied on the *P*-side by the reduced Cyt*c* to the Cu_A center and sequentially transferred via heme *a* to the catalytic binuclear center consisting of heme a_3 and Cu_B. In the sophisticated and complex catalytic cycle, electrons are transferred successively through the metal centers to the active site (BNC), where oxygen reduction occurs (Shapleigh et al. 1992; Hosler, Ferguson-Miller, and Mills 2006). A molecule of O_2 bound to the reduced form of BNC requires four electrons to be split. Two electrons are provided by Fe^{2+}, one electron by Cu^{1+} and the fourth electron and a proton are supplied by Tyr244, which turns into $Tyr^{\bullet}$ neutral radical state (Babcock 1999; Fabian et al. 1999; Proshlyakov et al. 2000; Blomberg et al. 2000; Wikström, Krab, and Sharma 2018). Once this is attained, only the different forms of oxygen intermediates remain present in the BNC (*e.g.*, =O, HO^-, H_2O). The transfer of each electron delivered by reduced Cyt*c* and used for the reduction of oxygen intermediates is accompanied by the translocation of two protons, of which one is consumed internally in the binuclear center (Fe_{a3}–Cu_B) for the chemical reduction of substrate species and the other is pumped across the membrane (Babcock and Wikström 1992; Ferguson-Miller and Babcock 1996). Both chemical and pumped protons originate from the membrane negative (*N*-) side. The chemistry catalyzed by the enzyme is deceptively simple, and the catalytic activity in the oxygen-binding site and the catalytic cycle of C*c*O itself are intricately complicated.

The fundamental question concerns the molecular mechanism of proton pumping, i.e., how the free energy of oxygen reduction in C*c*O is transformed into the driving force pushing pumped protons against the membrane electrochemical gradient.

The Proton Pumping Mechanism

The proposed proton pumping mechanism (Popović and Stuchebrukhov 2004b) is based on the following key structural features and observations made in previous studies, see Figure 2A. There are two chains of water molecules connecting Glu242 to propionate *D* of heme a_3 (PRDa_3) and the enzyme catalytic center (i.e., a hydroxyl group bound to either Fe or Cu metal centers in BNC) (Zheng et al. 2003; Wikström, Verkhovsky, and Hummer 2003). PRDa_3 forms a strong salt bridge with a nearby Arg438. No proton transfer to PRDa_3 from Glu242 is possible unless Arg438 is deprotonated or the salt bridge gets significantly weakened due to thermal fluctuations of the structure. From the mechanistic point of view, the salt bridge Arg438$^+$/PRDa_3^- and a crystallographically found water molecule Wat1, located between the propionates of heme a_3 (Figure 2A), are most likely necessary for the regulation of transfer of the pumped protons to PLS (Popović and Stuchebrukhov 2004a; Lee et al. 2009). The water molecules are formed as the catalytic reaction products and leave the enzyme's active site through the nonpolar cavity. The salt bridge facilitates this process and works as a gate for water exit from the hydrophobic catalytic cavity (Wikström et al. 2005; Sugitani and Stuchebrukhov 2009). Also, the strong salt bridge imposes the thermodynamic obstacle (Popović and Stuchebrukhov 2005) and kinetic barrier (Siegbahn and Blomberg 2007; Blomberg and Siegbahn 2010; Kaila et al. 2009), which hinder a reverse flow of protons back to the pump (to PLS or Glu242). There is a water molecule (Wat1), both predicted in the reference (Zheng et al. 2003) and visible in the crystal structures (Svensson-Ek et al. 2002; Tsukihara et al. 2003; Qin et al. 2009; Yano et al. 2016), located between two propionates of heme a_3 and hydrogen-bonded to formally charged nitrogen of His291, the "top" ligand of Cu_B center. This H_2O molecule facilitates attaining proton connectivity between Glu242 and Nδ1 of His291 via the Glu242–PRDa_3– Arg438–Wat1–His291 chain. According to electrostatic calculations (Popović and Stuchebrukhov 2004a), the Nδ1 position of His291 is protonated only when both Fe and Cu_B centers are formally in reduced states (i.e., when the total charge on the metal ion plus the ligands is 0 for Fe_{a3}, and 1+ for Cu_B). When one of the metal centers of BNC is oxidized or, equivalently, when one of the ligands in BNC accepts a proton, His291 becomes deprotonated.

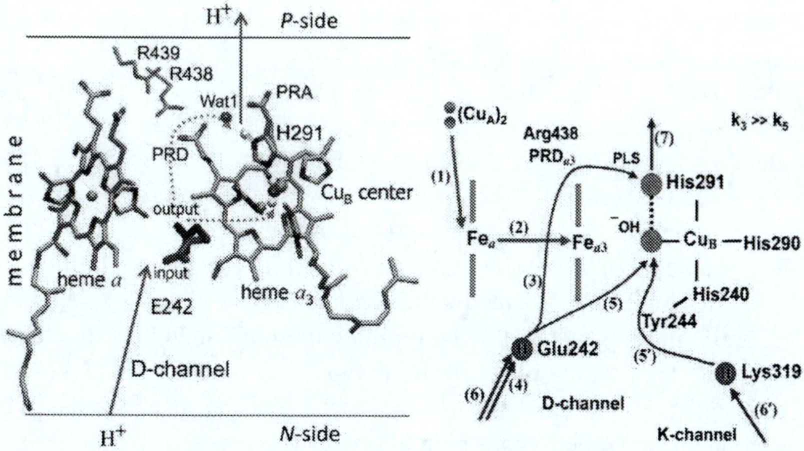

(A) (B)

Figure 2. (A) The key structural features of the proton pumping mechanism in C*c*O, including the conformational changes of the Glu242 side-chain induced by thermal fluctuations (Popović and Stuchebrukhov 2006). (B) The sequence of steps in the proton pumping mechanism of C*c*O. Red and blue arrows display the ET and PT steps, respectively. Glu242 is connected to His291 (PLS) and OH^- ligand in the active site (BNC) by two separate proton-conducting water chains with different proton-conducting rates ($k_3 >> k_5$). The faster chain delivers the pumped protons to PLS, whereas the slower chain delivers the chemical protons to BNC. The difference in rates ensures that a proton is preloaded into the pump site (PLS) before the protonation of the reduced oxygen intermediate by the chemical proton occurs (Popović and Stuchebrukhov 2004b; Belevich et al. 2007; Siletsky and Konstantinov 2012). For the role of Lys319, see the main text.

A schematic of the proposed proton pumping mechanism based on the Coulomb pump model with kinetic gating (Popović and Stuchebrukhov 2004b) is shown in Figure 2B. The enzyme steady-state during the catalytic cycle is established after a chemical proton is accepted by one of the hydroxy ligands of the BNC. In the steady-state of the catalytic center, one of the metal centers is formally oxidized (Fe^{3+}–H_2O or Cu^{2+}–H_2O), while Glu242 is protonated and His291 (Nδ1 site) is deprotonated. In order to proceed further, the enzyme needs an additional electron to be supplied to the system by reduced Cyt*c* to the Cu_A center and transferred further to heme *a* (step 1) and then to heme a_3–Cu_B binuclear center (step 2). In response to the increase in negative charge of the BNC, the proton from Glu242 now has a driving force to move closer to the catalytic center. Two proton-conducting pathways lead from Glu242 to two possible sites: one leads to the OH^- group in the BNC, and the second leads to the deprotonated His291 site (PLS). Both groups show high proton affinity, which is revealed by the high values of their pK_as (Popović and Stuchebrukhov 2004a; Quenneville, Popović, and Stuchebrukhov 2006). Since the proton translocation rate to His291 is considerably larger than the one to the OH^- group (Belevich et al. 2007; Siletsky and Konstantinov 2012), in the next step, the protonation of His291 occurs, i.e., the transfer of a pump proton to the PLS. Thus, kinetics (not thermodynamics) controls this proton transfer (PT) reaction resulting in the formation of a meta-stable state of the enzyme (Popović and Stuchebrukhov 2004b, 2005) in step 3. Step 4 is the reprotonation of Glu242 residue through the D-channel by proton uptake from the *N*-side of the membrane. Now, the second chemical proton can be transferred to the BNC protonating OH^- to H_2O by using a separate water chain path (step 5). The presence of an additional positive charge in the BNC consequently decreases the pK_a value of His291 (Popović, Quenneville, and Stuchebrukhov 2005; Quenneville, Popović, and Stuchebrukhov 2006). Clearly, the entrance of a substrate proton into the active site is an essential component of the pumping mechanism in C*c*O since the free energy is generated at the catalytic center. In step 6, Glu242, presumably the primary proton donor of the proton translocation process, once again gets reprotonated through the D-channel. It additionally increases electrostatic repulsion between a proton at His291 and H_2O ligand in the BNC, which finally, in step 7, leads to the expulsion of a preloaded proton from the PLS to the *P*-side of the membrane (Popović and Stuchebrukhov 2006). Therefore, reprotonation of the proton donor site may control a proton release from the PLS to the *P*-side of the membrane, facilitating the proton pumping event (Brzezinski and Gennis 2008; Popović and Stuchebrukhov 2006). Moreover, reprotonation of Glu242, along with a release of the pumped proton, controls the entry of the next electron into the enzyme since the reduction of heme *a* is only feasible if Glu242 is in the steady protonated state (Popović and Stuchebrukhov 2012). Such control of the flow of electrons assures that electrons are taken up and consumed, one at a time, in the enzyme active site.

Steps 5' and 6' designate that Lys319 replaces Glu242 as a donor of chemical protons in the reductive part of the catalytic cycle during the O → E → R transitions. Presumably, the negative charge in the BNC is shifted more toward the Cu_B complex, functionally opening the K-channel to provide the substrate protons.

OH^- group can be bound to Cu_B or Fe_{a3} metal ion depending on the transition state in the catalytic cycle. In both cases, the entrance of the chemical proton would form an H_2O molecule in the BNC. H_2O ligand in the BNC is causing the repulsive interactions with the proton on the PLS in the BNC oxidized state. As a result, a release of the pumped proton from the PLS, i.e., a pumping event, occurs.

Each time a proton is pumped in a single turnover of the pump, a steady-state is formed until the next electron is injected into the system. The scheme shown in Figure 2B repeats with each new electron entering the system. Since four electrons are required for the complete reduction of O_2 to $2H_2O$, the displayed sequence of steps will be repeated four times to complete an entire catalytic cycle of the enzyme. The proposed sequence of steps also ensures the stoichiometry of one pumped H^+ per one e^- passed through the system. For details on the catalytic cycle of C*c*O, we refer a reader, *e.g.*, to references (Gennis 1998b; Wikström 2004; Popović and Stuchebrukhov 2004b; Gennis 2004; Belevich, Verkhovsky, and Wikström 2006; Siletsky and Konstantinov 2012) devoted to the topic of the transient states (A, P_m, P_r, F, H, O, E, R-state) and the corresponding redox/protonation changes in the BNC during the catalytic cycle.

The ET to BNC is accompanied by the *fast* PT to PLS (k_3), while the following *slow* PT of a chemical proton into BNC (k_5) is coupled with the expulsion of the preloaded proton from PLS. The difference in the proton conducting rates, $k_3 >> k_5$, was initially a pre-request of the proposed pumping model in order to work and pump protons (Popović and Stuchebrukhov 2004a, b). Otherwise, the supplied protons would go to the BNC for chemical reduction of oxygen species and forming H_2O molecules without pumping. That indeed happens with specific mutations, which significantly disturb the proton-conducting water chains and, therefore, change the ratio of the corresponding k_3 and k_5 rates (Dürr et al. 2008). Later on, the proposed kinetic gating model received experimental support from the time-resolved electrometric measurements of the membrane potential generated by C*c*O inserted in vesicules (Belevich et al. 2007) and by studies of deuterium kinetic isotope effects (Salomonsson, Brändén, and Brzezinski 2008; Johansson et al. 2011).

The proposed model is based on the simulation of water molecules inside the protein (Zheng et al. 2003; Tashiro and Stuchebrukhov 2005) and the combined DFT and electrostatic calculations of energies of different protonation states of the enzyme cycle (Popović and Stuchebrukhov 2004a; Popović, Quenneville, and Stuchebrukhov 2005; Popović and Stuchebrukhov 2006; Quenneville, Popović, and Stuchebrukhov 2006; Popović and Stuchebrukhov 2012). Below we review these calculations and demonstrate that the results indeed provide support for the critical assumptions of the model. We focus our discussion on calculating the redox-dependent pK_a value of His291 and Glu242 – the essential elements in the proposed pump model. Other results related to the above model can be found in the original publications from this group (Popović and Stuchebrukhov 2004a; Quenneville, Popović, and Stuchebrukhov 2004; Popović and Stuchebrukhov 2005; Popović et al. 2010; Popović and Djordjević 2020).

The assumptions of the model are of two types: energetic, such as that His291 changes its protonation state depending on the redox state of the metals of the catalytic center, and kinetic, such as upon electron reduction of the catalytic center, the proton from Glu242 is transferred to His291 and *not* to OH^- group in the catalytic center. To verify the latter, the rates of the proton transfer reactions need to be calculated, the subject that is outside the scope of this chapter. We will focus here only on the thermodynamic questions.

Materials and Methods

Computational Methods for pK_a Calculations

The combination of the DFT and electrostatic calculations has been used successfully in the past to compute the pK_a values and redox potentials in various enzymes and proteins (Mouesca et al. 1994; Li et al. 1996; Li et al. 1998; Li et al. 1999; Ullmann, Noodleman, and Case 2002). This approach takes advantage of both methods by combining the accuracy of quantum chemical calculations with a detailed description of the electrostatic interactions of the whole protein–membrane–solvent system. Density functional theory is applied to a relatively small quantum-mechanical (QM) system of interest to optimize its geometry, evaluate electronic energies, and obtain the ESP atomic partial charges for different redox and protonation states of the complex. On the other hand, electrostatic calculations are used to evaluate the solvation energy of the complex (QM part) in the protein environment and the Coulomb interactions of the complex with protein charges. Both polarization of the protein medium and protein charges affect the electron distribution of the complex. Therefore, the electrostatic and DFT calculations are carried out in a self-consistent way. Generally, this method produces pK_a values in proteins in good agreement with experimental data (Chen et al. 1994; Li et al. 1996; Kallies and Mitzner 1997; Richardson et al. 1997; Ullmann, Noodleman, and Case 2002).

Overview DFT/Continuum Electrostatic Method

Here we briefly summarize the DFT/electrostatic method of pK_a calculations and introduce different energy terms discussed later in the text. In the calculation, we divide the whole system into an active site complex (QM system) and the surrounding medium – protein, membrane, and the external aqueous phase. The main idea is to treat the QM system with an accurate DFT method – in this case, we are interested in the proton affinity of a complex – and include the dielectric effects of the medium with classical electrostatic methods. The protein medium also includes the partial atomic charges, the charges of redox centers, and charges of the titratable groups, which directly affect the electronic energies of the complex.

The dielectric medium involves the protein, membrane, and surrounding solvent. The protein is an inhomogeneous dielectric medium because it has many internal cavities which contain water molecules. These cavities and channels should have a higher dielectric constant than the "dry" protein region. For the region of a dry protein, a specific value of dielectric constant ε_{prot} is assumed, and for the cavities, a higher value ε_{cav} is used. Neither of these values is well defined, although most researchers agree that ε_{prot} should be close to 4, while the cavities can have dielectric constant ε_{cav} anywhere from 4 to 80. The upper value would correspond to a dielectric response equivalent to one of the bulk water. The most reasonable values for ε_{cav} are between 4 – 20, depending on the position and shape of the cavity, i.e., whether the cavity is buried in the hydrophobic environment or is open to the external bulk solvent, see the literature (Sharp and Honig 1990; Antosiewicz, McCammon, and Gilson 1994; Simonson and Perahia 1995; Simonson and Brooks 1996; Antosiewicz et al. 1996; Warshel and Papazyan 1998; Olsson, Sharma, and Warshel 2005; Popović, Quenneville, and Stuchebrukhov 2005) and the corresponding references therein.

We begin with a review of pK_a calculations in water. The pK_a value is related to the free energy of deprotonation in an aqueous solution (ΔG_{aq}^{deprot}), which is a sum of two contributions: the free energy of deprotonation in a vacuum (ΔG_{vac}^{deprot}) and the solvation energy difference between the deprotonated and protonated forms of the protonatable group (ΔG_{solv}^{deprot}), see also (Ullmann, Noodleman, and Case 2002; Quenneville, Popović, and Stuchebrukhov 2004),

$$pK_a = \frac{1}{kT\ln 10}(\Delta G_{aq}^{deprot}) = \frac{1}{kT\ln 10}(\Delta G_{vac}^{deprot} + \Delta\Delta G_{solv}^{deprot}) \tag{2}$$

The solvation energy difference ($\Delta\Delta G_{solv}^{deprot}$) is obtained from Eq. 3:

$$\Delta\Delta G_{solv}^{deprot} = \Delta G_{solv}^{A^-} + \Delta G_{solv}^{H^+} - \Delta G_{solv}^{HA} \tag{3}$$

The solvation energy of deprotonated ($\Delta G^{A-}{}_{solv}$) or protonated ($\Delta G^{HA}{}_{solv}$) species is calculated by solving the Poisson-Boltzmann equation, as described in more detail below. A proton solvation energy ($\Delta G^{H+}{}_{solv}$) is set to –260.5 kcal/mol estimated from the experimentally measured potential of the standard hydrogen electrode (Reiss and Heller 1985).

The first term in Eq. 2, the free energy of deprotonation in a vacuum, can be formally expressed as:

$$\Delta G_{vac}^{deprot} = \Delta E_{elec} + \Delta E_{vib} + \Delta E_{trans}^{H^+} + \Delta(pV) - T(S^{H^+}) \tag{4}$$

From left to right, the terms in Eq. 4 denote the electronic and the zero-point vibrational energy differences between the deprotonated and protonated species, the energy of proton ejection, the energy due to the gas phase volume change, and finally, the entropy contribution.

As an example of the above calculations, the absolute pK_a*s* of MeIm was obtained. The vibrational energy term (ΔE_{vib}), whose main contribution is the zero-point energy of the N-H bond lost upon deprotonation, was estimated from a normal mode analysis of MeIm to be –9.0 kcal/mol. The main assumption here is that the frequency change of the N-H bond upon coordination to copper ion is very small. Therefore the zero-point energy is transferable and needs to be determined only once for the proper model compound, i.e., MeIm, in this case. The translational energy of a proton

($\Delta E^{H^+}{}_{trans}$) is estimated to be $3/2k_BT$, while the energy due to the gas phase volume change is estimated from ideal gas approximations to be k_BT. Finally, the entropic term of the gas-phase free energy of a proton was set to 7.8 kcal/mol as derived from the Sackur-Tetrode equation (Hill 1860). We obtained 7.7 and 7.8 for the δ– and ε–pK_a values, respectively (Quenneville, Popović, and Stuchebrukhov 2004), compared with established experimental values 6.6 and 7.0 for histidine protonable sites (Tanokura 1983).

The accuracy of the electronic energy calculations was probed against the experimentally measured enthalpy of the proton affinity. The proton affinity is defined as $-\Delta H_r$ for the reaction $B + H^+ \rightarrow BH^+$ and is therefore written (Bene et al. 1983),

$$PA = \Delta H_{vac}^{deprot} = \Delta E_{elec} + \Delta E_{vib} + \tfrac{5}{2} kT \tag{5}$$

where the change in vibrational energy during heating from 0 to 298 K has been neglected. The proton affinity of the nitrogen at the 3-position in 4-methylimidazole that we calculate (227.0 kcal/mol) compares well to the most recent experimental value (226.2 kcal/mol) (Catalan et al. 1990).

The *absolute* pK_a values in a protein can be obtained using the thermodynamic cycle and the following equations and energy terms shown in Figure 3. For details of the *absolute* pK_a calculations, a reader is referred to references, *e.g.*, (Li et al. 1999; Ullmann, Noodleman, and Case 2002; Quenneville, Popović, and Stuchebrukhov 2004; Popović, Quenneville, and Stuchebrukhov 2005).

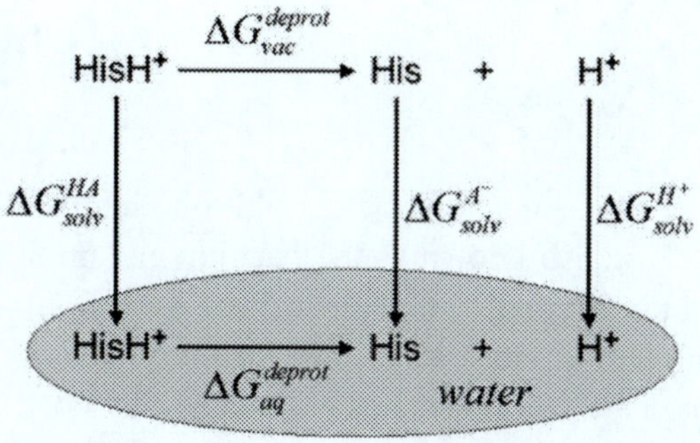

Figure 3. The thermodynamic cycle is used to calculate the "absolute pK_a value" of $HisH^+$ by computing the deprotonation energy of the titratable group in solution and protein from the gas-phase quantities (quantum-chemical approach; Jaguar 5.5) and solvation energies (the Poisson-Boltzmann solver) by applying the equations 2, 3, and 4.

Alternatively, one can use the relative approach to compute the pK_a *shift* of the group of interest relative to a suitable model compound by using the experimentally determined pK_a value of the model compound to evaluate the pK_a of the group as

$$pK_a^{site} = pK_a^{model} + \Delta pK_a \ . \tag{6}$$

The relative pK_a shift of the protonatable group relative to the model compound can be expressed as

$$\Delta pK_a = \frac{1}{kT\ln 10}\Delta\Delta G_{shift} = \frac{1}{kT\ln 10}(\Delta\Delta E_{elec} + \Delta\Delta G_{solv})\,, \tag{7}$$

where $\Delta E_{elec} = \Delta E_{elec}^{depr} - \Delta E_{elec}^{prot}$
electronic energy difference in the gas phase between the deprotonated and protonated forms, and $\Delta\Delta E_{elec}$ is its shift relative to the model compound; $\Delta G_{solv} = \Delta G_{solv}^{depr} - \Delta G_{solv}^{prot}$ is the difference in solvation energy between the deprotonated and protonated forms, and $\Delta\Delta G_{solv}$ is the shift of the solvation energy relative to the model compound. In Eq. (6), when the *difference* of pK_as is computed between two compounds, the free energy of a solvated proton and some entropy contributions cancel out; thus, only electronic energy, i.e., the enthalpic part, of the vacuum term remains.

The total solvation energy (x = *deprotonated or protonated form*) is divided into several components:

$$G_{solv}^{x} = G_{Born}^{x} + G_{strain}^{x} + G_{q}^{x} \ . \tag{8}$$

The Born solvation energy (G^x_{Born}), and the energy of interaction with protein charges (G^x_q), also known as the reaction and protein field terms, are two main contributions to the free energy of solvation of the active site complex in the protein environment. The third term (G^x_{strain}) is the so-called "strain" energy. This term reflects the energy cost associated with reorganizing the electronic density distribution of the active site complex induced by the reaction field of the polarized medium. The origin of this term is as follows. The charges of the QM system polarize the surrounding medium; in turn, the induced polarization of the medium slightly changes the charge distribution of the QM system compared with the initial charge distribution in the vacuum.

The pK_a of the active site complex can finally be calculated as

$$pK_a^{site} = pK_a^{model} + \frac{1}{kT\ln 10}(\Delta\Delta E_{elec} + \Delta\Delta G_{strain} + \Delta\Delta G_{Born} + \Delta\Delta G_q). \quad (9)$$

We have confirmed that both methods, the absolute and relative pK_a calculations, give practically identical results. We will use the above expression and the relative method in the following applications. For calculation of pK_a of His291, the model compound 4-methylimidazole will be used. For pK_a of Glu242, the model compound is propionic acid.

Density Functional Calculations

The starting structure of the active site was taken from the x-ray crystal structure of bovine heart cytochrome *c* oxidase obtained by Yoshikawa et al. (1998) at 2.3 Å resolution (PBD code: 2OCC). The simple QM model (Figure 4A) used to calculate the pK_a of a Cu_B ligand consists of the Cu atom and methylimidazole molecules representing coordinated histidines 240, 290, and 291, a methyl group representing Tyr244 (which is cross-linked to His240 (Proshlyakov et al. 2000; Hemp and Robinson 2006)), and an H_2O or HO^- Cu_B-ligand. The methyl groups attached to the imidazoles represent the β-methylene groups of histidine. Compared with our previous calculations (Popović, Quenneville, and Stuchebrukhov 2005), the new larger quantum-chemical system also includes a ferryloxy porphyrin part and an imidazole ring that represents His376 (the sixth ligand to Fe_{a3} metal ion). Thus, the binuclear complex shown in Figure 4B represents a full QM model.

Energies of each coupled protonation/redox state of the complex were calculated using density functional theory (DFT) (Hohenberg and Kohn 1964; Kohn and Sham 1965) and the Jaguar 5.5 quantum chemistry package. All calculations used the hybrid density functional B3LYP (Becke 1993), and open-shell electronic configurations (for oxidized Cu_B) were calculated with restricted open-shell DFT. Reported energies used the LACV3P+* basis set, while geometry optimizations used LACVP+*, as abbreviated in the Jaguar program (Hay and Wadt 1985). These basis sets include non-relativistic electron core potential for the Cu atom. The basis sets for the metal valence electrons (and all non-metal electrons) are of split valence quality (6-31+G*) (Ditchfield, Hehre, and Pople 1971; Hehre and Pople 1972; Hehre, Ditchfield, and Pople 1972; Binkley and Pople 1977; Hariharan and Pople 1973; Francl et al. 1982) for LACVP+* and triple split valence quality (6-311+G*) (Clark et al. 1983; Frisch, Pople, and Binkley 1984; Krishnan et al. 1980; McLean and Chandler 1980) for LACV3P+*. They include polarization and diffuse functions for all heavy atoms.

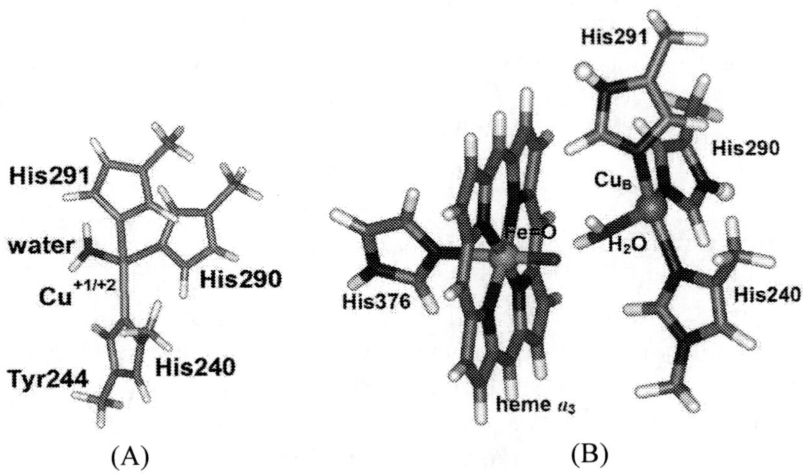

Figure 4. The two QM-model systems used to calculate the pK_as of His291 and HO^-/H_2O ligands of the Cu_B metal center in different redox states of *Cc*O: (A) the Cu_B center and (B) a full binuclear complex with Fe_{a3} porphyrin in the ferryl-oxo state.

Geometry optimizations were conducted in Cartesian space in a vacuum. In the procedure, the Cu atom, the ligating ε–nitrogen, and the methyl group of methylimidazoles were kept fixed. This scheme allows the remaining atoms to relax while preserving the crystal structure's integrity concerning the unique Cu-ligand geometry and the attachment location of the histidines to the protein backbone. Four optimizations were performed: for the protonated and deprotonated forms of His291, each for oxidized and reduced Cu_B. The geometry of the Cu-ligated water molecule was set using molecular mechanics. This water molecule is known to be very weakly bound and quite mobile (Blomberg et al. 2000; Siegbahn, Blomberg, and Blomberg 2003).

To quantify the effect of the electronic reorganization of the solute and to obtain the corresponding set of the ESP fitted charges for the QM considered system, the complex was surrounded by a continuum dielectric

($\varepsilon = 4$ or 80), and the self-consistent reaction field (SCRF) method was applied as implemented in Jaguar 5.5 (Tannor et al. 1994; Marten et al. 1996). The probe radius of the surrounding dielectric was set to 1.4 Å, and the standard Jaguar set of the van der Waals radii for atoms was used. The ESP atomic charges were generated using a modified version of the CHELPG procedure of Breneman and Wiberg (Breneman and Wiberg 1990).

To correctly describe the effects of the protein charges on the pK_a of His291, the equilibrium proton distribution in C*c*O for the corresponding redox state of metal centers is required. To this end, we performed the standard continuum electrostatic calculations (see next section) to determine the equilibrium protonation state of the titratable residues in C*c*O embedded in a membrane for different redox states of the enzyme. In different redox states, the equilibrium proton distribution is slightly different due to the proton uptake from the solution (aqueous phase) and much more different from the standard protonation state.

Continuum Electrostatic Calculations

The standard electrostatic calculations have been recently applied in studies of bacterial cytochrome *c* oxidase (Kannt, Lancaster, and Michel 1998) and several other membrane proteins that take part in energy transduction in cells, such as bacterial photosynthetic reaction center (Gunner, Nicholls, and Honig 1996; Lancaster et al. 1996; Rabenstein, Ullmann, and Knapp 1998) bacteriorhodopsin (Sampogna and Honig 1994; Spassov et al. 2001), and cytochrome bc_1 complex (Popović et al. 2001).

The protonation states of the titratable residues in C*c*O for different redox states of the enzyme were determined as described in reference (Popović and Stuchebrukhov 2004a) by numerically solving the linear Poisson-Boltzmann equation on a grid (Bashford and Karplus 1990; Nicholls and Honig 1991; Bashford and Karplus 1991; Beroza et al. 1991; Gunner and Honig 1991; Yang et al. 1993; Beroza and Case 1996; Baptista, Martel, and Soares 1999; Popović et al. 2001; Popović et al. 2002) (with a three-step focusing procedure) for the inhomogeneous dielectric of the protein–membrane–water system with the program MEAD (Bashford and Gerwert 1992; Bashford 1997). The protonation states sampling was subsequently done using a Monte-Carlo titration technique implemented in the program KARLSBERG (Rabenstein 1999). The electrostatic calculations were performed on subunits I and II, which belong to monomer I from the crystal structure of mitochondrial C*c*O from the bovine heart (PDB code: 2OCC) (Yoshikawa et al. 1998). We considered 146 sites here as protonatable, including all residues such as Asp, Glu, Arg, Lys, Tyr, Cys, heme propionates, N- and C-termini, δ– and ε–His sites, except for the residues coordinated to the metal centers. In addition, Tyr244 and His291, belonging to the Cu_B center, are considered titratable. Of three His ligands at the Cu_B center (240, 290, 291), only His 291 can possibly be a titratable site because His240 is cross-linked to Tyr244, whereas His290, according to the crystal structure, is hydrogen-bonded to nearby Thr309, which in turn is hydrogen-bonded to the carbonyl oxygen of Phe305, in all redox states (Yoshikawa et al. 1998) and therefore is always protonated at the Nδ1 position (Popović and Djordjević 2020). Several explicit water molecules included in the calculation were assigned the TIP3P charges (Jorgenson, Chandrasekhar, and Madura 1983).

In the calculations, the membrane is modeled as a low dielectric slab of 45Å covering the central part of the enzyme. Taking the thermodynamic average over different protonation states of the titratable groups, one can evaluate the equilibrium charge distribution in C*c*O for the given temperature, pH, and redox state of metal centers. The obtained charge distribution of the protein is used in electrostatic calculations of the solvation energy term $G^x{}_q$, Eq. 8. We found that using the standard protonation state of the protein leads to entirely erroneous results.

Solvation Calculations

Once the charges of the active site complex have been calculated, its geometry has been optimized, and an equilibrium charge distribution of the protein has been obtained, the DFT optimized active site cluster needs to be inserted back into the protein to set up the electrostatic solvation calculations. The MEAD program (Bashford 1997) has been used to calculate the solvation energy of the quantum-chemically treated cluster in an aqueous phase, a continuum low dielectric medium, or in the protein (see below all used models). Solvation of the active site cluster in the protein consists of two main contributions – the Born-solvation energy and the energy of the interaction with the protein charges (reaction and protein field energies).

The Poisson equation (or PBE) is solved numerically for the system divided into three dielectric regions represented by dielectric constants of 1, 4, and 80 for the quantum-chemically considered active site, a protein, and solvent region, respectively. Such choice of the values for the dielectric regions is physically reasonable and very well supported by numerous studies (Sharp and Honig 1990; Simonson and Perahia 1995; Simonson and Brooks 1996; Warshel and Papazyan 1998). Using $\varepsilon_{prot} = 4$ approximately compensates for the effects of electronic and nuclear polarization, allowing some mobility of the protein dipoles and reorientational relaxation of the protein (Li et al. 1998), which otherwise are not explicitly taken into account. However, electronic and nuclear degrees of freedom are treated explicitly in the QM system. No additional shielding is required; therefore, $\varepsilon_{QM\text{-}system} = 1$. Traditionally, $\varepsilon_{water} = 80$ is used in continuum solvent electrostatic calculations to describe the screening properties of the aqueous phase. In addition to these typical values for the dielectric regions, we examined the dependence of the results on the various values of the dielectric constant associated with water-filled cavities inside the protein.

In these calculations, the ESP fitted charges are used for the QM-system, while partial charges of the protein atoms are taken from the CHARMM22 parameter set (MacKerell et al. 1998) and modified to reflect the equilibrium protonation state of the enzyme (Popović and Stuchebrukhov 2004a). The backbone atoms of the Cu_B ligands are also included in electrostatic calculations, and they affect the energetics through the influence of the background charges (so they are not part of the QM system). The protein charges (Popović et al. 2001) and radii (MacKerell et al. 1998) that we use in this application have been used successfully before to calculate the electrostatic energies, protonation probabilities of titratable groups, and redox potentials of cofactors in different proteins (Popović et al. 2001; Popović et al. 2002; Popović and Stuchebrukhov 2004a; Vagedes et al. 2000; Rabenstein and Knapp 2001).

The electrostatic potential for the system was solved by the Poisson equation (PE) with a three-step grid focusing procedure. The lattices with the finest resolution were centered at the active site cluster, and the probe radius of the surrounding dielectric was set to 1.4 Å. Ionic strength was set to 0 since we found earlier that the mobile ions do not significantly contribute to the calculated potentials in the interior of the C*c*O (Popović and Stuchebrukhov 2004a), where the active site complex is located. Similar weak dependence on the ionic strength has been reported for other trans-membrane protein systems (Klapper et al. 1986; Gunner, Nicholls, and Honig 1996; Lancaster et al. 1996; Kannt, Lancaster, and Michel 1998; Popović et al. 2001; Spassov et al. 2001; Onufriev, Smondyrev, and Bashford 2003).

The pK_as of His291 and Glu242 reported here were calculated using the equilibrium protonation states of the enzyme, which were determined as in our previous work (Popović and Stuchebrukhov 2004a) for each redox state involved. It should be noticed that the pK_as of all groups in the enzyme should be calculated in a self-consistent manner, and in general, the protonation state of the rest of the protein should depend on the protonation state of the selected group. Here, self-consistency is easily achieved since the obtained protonation state of His291 and Glu242 is essentially 1 or 0, depending on the redox state of the enzyme.

To examine the strength of the coupling of the protonation state of His291 and other protonatable protein groups, two sets of charges for the protein were obtained – for protonated and deprotonated states of His291. The pK_a values of His291 determined for these two sets of charges differed by less than 0.3 pK_a units.

Result and Discussion

pK_a Calculations on Different Dielectric Models

Computational Models

In order to get a sense of how different factors affect the apparent pK_a values of His291, we considered several computational models shown in Figure 5.

Model (a) represents the quantum-chemically (QM) treated system (Figure 4A) solvated in a continuum dielectric of $\varepsilon = 80$. Using this model, we obtain the pK_a values of the groups in water.

Model (b) is the same as (a), only for dielectric $\varepsilon = 4$. The pK_as obtained from this model should illustrate the magnitude of the effect of a low dielectric environment on the protein.

Model (c) adds protein charges to the uniform dielectric of $\varepsilon = 4$. This is still a very rough description of the protein system because the complexity of the protein-solvent surface, inner water-filled cavities, and solvent itself is missing.

Model (d) introduces the inhomogeneity of the dielectric medium represented by the inner water cavities with $\varepsilon = 80$, the shape of the protein, membrane, and the outside aqueous phase. No protein charges are present here.

Finally, the all-inclusive model (e) comprises all components of the protein system – inhomogeneity of the protein dielectric environment, membrane, and surrounding solvent, as in (d), and protein charges, as in (c).

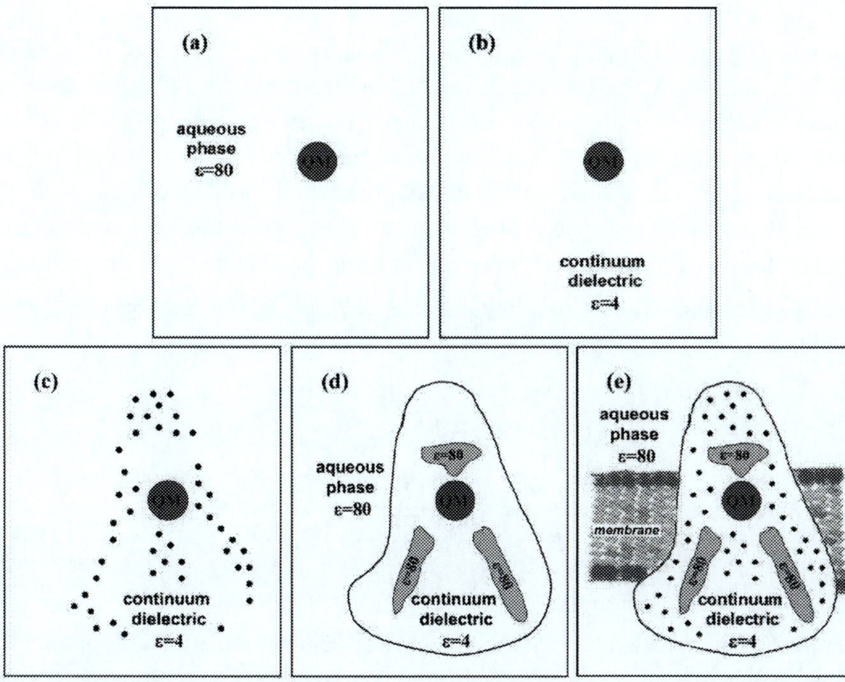

Figure 5. Schematics of different models used in the study. In model (a), the QM-system that consists of a Cu_B center and coordinated ligands (with dielectric constant $\varepsilon = 1$ inside) is solvated in the aqueous phase. (b) QM-system is solvated in the low dielectric of $\varepsilon = 4$. (c) QM-system is embedded in continuum dielectric of $\varepsilon = 4$ and surrounded with protein charges. (d) QM-system is embedded in an inhomogeneous dielectric medium represented by $\varepsilon = 4$ within the borders of the protein and $\varepsilon = 80$ inside the inner water-filled cavities and outside of the protein, while protein charges are neglected. A model (e) comprises a full protein-membrane-solvent description, including dielectric inhomogeneities and protein charges, whereas the QM-system is treated by the DFT method. Crystallographically, more than 150 water molecules have been found in the inner protein cavities (Svensson-Ek et al. 2002; Ostermeier et al. 1997; Tsukihara et al. 2003). Only the essential water-filled cavities are schematically shown in the Figure. They represent the proton input *D*- and *K*-channels leading to the active site and the large water-filled cavity on the proton output side of the enzyme. However, the calculations use the actual x-ray structure, including all cavities and structural details of C*c*O.

The Calculation of pK_a in Water

As explained in the methods section, in order to calculate pK_as in a protein, we first calculated the pK_a values of the corresponding groups in water. For a calculation of the pK_a values of δ-nitrogen of His291 ligand of Cu_B complex, we use MeIm as a reference compound, for which the experimental pK_a is known. The experimental value of pK_a of δ-nitrogen of MeIm is 6.6 (Tanokura 1983). Using this value and the difference in electronic and solvation energies between MeIm and Cu$_B$ model complex, we can evaluate the pK_a value of the latter in the aqueous phase. The results are presented in Tables 1 and 2. Here and later in the text, by protonated and deprotonated forms of MeIm, we mean protonated and deprotonated δ-nitrogen of the compound, i.e., cationic and neutral forms of MeIm, see Figure 6. Similarly, protonated and deprotonated forms of Cu$_B$ center mean protonated and deprotonated δ-nitrogen of His291 ligand.

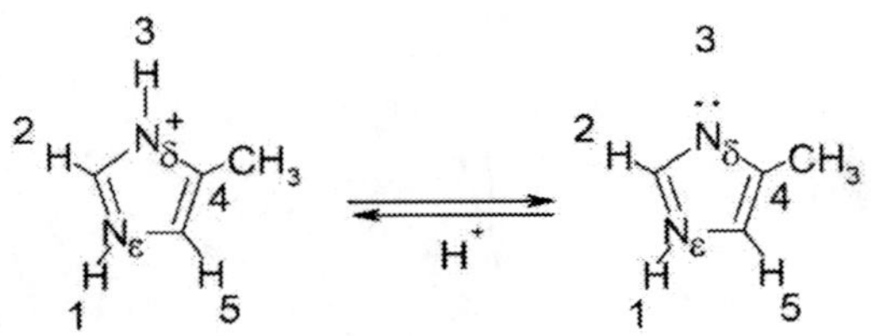

Figure 6. Deprotonation of the 4-methylimidazole (MeIm) molecule used as the model compound.

Table 1. Solvation energy difference (ΔG_{solv} kcal/mol) in an aqueous solution between the deprotonated and protonated forms of 4-methylimidazole and the Cu$_B$ complex (oxidized or reduced)

Compound [a]	$E_{reorganization}$		"Strain" energy [b]	Born energy [b]	ΔG_{solv} in water (ε = 80)
	Deprotonated form	Protonated form			
methyl-δ- imidazole	+4.0	+1.5	+2.5	+52.2	+54.7
$Cu_B^{2+}(H_2O)$ oxidized	+10.1	+3.4	+6.7	+86.7	+93.4
$Cu_B^{1+}(H_2O)$ reduced	+9.5	+3.3	+6.2	+8.5	+14.7

[a] The protonated and deprotonated forms are the corresponding states of δ-nitrogen of MeIm and His291 ligand of Cu$_B$ compound (e.g., a protonated form of MeIm is a cation, while the deprotonated MeIm is a neutral form where the δ-proton gets deprotonated).

[b] The "strain" and the Born contributions are given as a difference in energy between the deprotonated and protonated forms.

Table 1 lists the calculated values of the solvation energy changes of MeIm and Cu$_B$ complex upon protonation of these compounds. There are two contributions – the electronic reorganization energy (contribution due to solvent back reaction), called the "strain" energy, and the Born solvation energy. The total solvation energy is the sum of these two contributions.

Table 2 lists the calculated pK_a values of the His291 of Cu_B complex, individual electronic and solvation energy contributions, and the corresponding data for the reference compound MeIm. Here, the electronic energy difference in the gas phase between the deprotonated and protonated forms (ΔE_{elec}), a shift of the electronic gas-phase energy relative to the MeIm model compound ($\Delta\Delta E_{elec}$), the difference in aqueous phase solvation energy between the two forms (ΔG_{solv}), a shift of the solvation energy relative to the model compound ($\Delta\Delta G_{solv}$), and a total shift in kcal/mol and pK_a units are given separately to show their individual contributions to pK_a values.

Table 2. The calculated pK_a values of the His291 site in the oxidized and reduced Cu_B complex in the aqueous phase relative to the pK_a of the model compound (methylimidazole)

Compound	ΔE_{elec}	$\Delta\Delta E_{elec}$	ΔG_{solv}	$\Delta\Delta G_{solv}$	$\Delta\Delta G_{shift}$	ΔpK_a	pK_a
methyl-δ-imidazole	+234.6	0	+54.7	0	0	0.0	6.6 [b]
$Cu_B^{2+}(H_2O)$ oxidized	+198.7	-35.9	+93.4	+38.7	+2.8	+2.0	8.6 [c]
$Cu_B^{1+}(H_2O)$ reduced	+283.7	+49.1	+14.7	-40.0	+9.1	+6.6	13.2 [c]

[a] The corresponding electronic and solvation energy contributions to the pK_as, and their differences relative to the reference compound, are shown. Energies are reported in kcal/mol.

[b] Experimental pK_a value.

[c] Calculated pK_a value.

The side-chain of an isolated histidine has two macroscopic pK_a values of 7.0 and 14.0 for the first and the second deprotonation. Compared with the methylimidazole, which microscopic pK_a in water has been measured to be 6.6 (for the Nδ-site proton) (Tanokura 1983), the equivalent histidine ligated to Cu_B metal center possesses pK_as of 8.6 in oxidized and 13.2 in reduced form, according to our present calculations.

The pK_a of His291 and Glu242 in CcO

We summarize the results of the electrostatic solvation calculations for different dielectric models ((a), (b), (d)) applied in our study. The solvation energy of the Cu_B complex is calculated in a continuum of a low dielectric medium of $\varepsilon = 4$, the inhomogeneous dielectric of protein with the dielectric boundaries, which reflects the natural shape of the protein, and in a continuum solvent of $\varepsilon = 80$ (aqueous phase). Here we consider only the effects of the dielectric medium on solvation energies, while the effects of the surrounding protein charges are not included yet. The solvation in an inhomogeneous dielectric protein region is to some extent larger than in a uniform dielectric of $\varepsilon = 4$ but much smaller than in the aqueous phase. Also, the "strain" energy in the aqueous phase is larger than in a low dielectric medium because the high dielectric medium polarizes the charges of the QM system much stronger than the low dielectrics can. Since the reorganization of the charge densities due to the medium-induced polarization produces an energetic cost on the electronic gas-phase energies, the strain contributions always have a positive sign.

The results of electrostatic calculations for models (c) and (e) can be compared. The solvation energies are considered in a uniform low dielectric of $\varepsilon = 4$ and in the "real" protein. The surrounding protein charges are also included here. It is interesting to examine the differences in the reaction and protein field energies between the two models. Comparing the Born-energies, all four species are better solvated, as expected, in an inhomogeneous protein dielectric, including high-epsilon water cavities, than in a continuum dielectric of low epsilon. Also, the protonated forms of each redox pair are better solvated than deprotonated ones. Primarily this is true for inhomogeneous protein environments, where for instance, comparing their Born-energies, the oxidized-protonated Cu_B complex (total charge +2) is much better solvated in the dielectric environment of C*c*O than the oxidized-deprotonated form (total charge +1). One may also expect that the Born-energy of the oxidized-protonated Cu_B complex (total charge +2) should be the largest, the one of the reduced-deprotonated form (total charge 0) should be the smallest, while the oxidized-deprotonated and reduced-protonated forms (with total charges of +1) should have similar values. All discrepancies (from this) are due to the specific charge distributions in different species.

The interaction effect with the protein charges is much smaller in the "real" protein than in a uniform ε of 4. A continuum low dielectric medium significantly destabilizes the deprotonated form of His291 in both redox states. In the "real" protein, there is a large bulb filled with water molecules around the position of the His291 site, which significantly screens the influence of the surrounding charges, and therefore the deprotonated form is less destabilized by the interaction with protein charges (~ +35 vs. +9 kcal/mol). This destabilization is due mainly to the adjacent negatively charged propionates of heme a_3, and since in both redox states, OORO and OORR, they are deprotonated, the effect of the protein charges within the same model gets very similar (+34.1 vs. +35.8 or +8.0 vs. +9.6).

Table 3 shows the difference in electronic gas-phase and solvation energies reported in kcal/mol between protonated/deprotonated forms and calculated pK_a values for oxidized and reduced Cu_B complex obtained in different computational models applied in our study. The pK_a of His291 is calculated relative to the pK_a of the model compound MeIm in the aqueous phase. The results illustrate how different dielectric mediums and interactions with protein charges affect the calculated pK_a of the His291 site in various dielectric models and different redox states of the enzyme.

Table 3. The pK_a values of His291 residue as a part of oxidized or reduced Cu_B center in C*c*O

Redox state	ΔE_{elec}	$\Delta\Delta E_{elec}$	ΔG_{solv}	$\Delta\Delta G_{solv}$	$\Delta\Delta G_{shift}$	ΔpK_a	pK_a
in aqueous phase:							
methyl-δ-imidazole	+234.6	0	+54.7	0	0	0.0	6.6
$Cu_B^{2+}(H_2O)$ oxidized	+198.7	-35.9	+93.4	+38.7	+2.8	+2.0	8.6
$Cu_B^{1+}(H_2O)$ reduced	+283.7	+49.1	+14.7	-40.0	+9.1	+6.6	13.2
in continuum dielectric of $\varepsilon = 4$ (no protein charges):							
$Cu_B^{2+}(H_2O)$ oxidized	+198.7	-35.9	+68.0	+13.3	-22.6	-16.5	-9.9
$Cu_B^{1+}(H_2O)$ reduced	+283.7	+49.1	+10.0	-44.7	+4.4	+3.2	9.8
in inhomogeneous dielectric of protein, $\varepsilon = 4$ and 80 (no protein charges):							
$Cu_B^{2+}(H_2O)$ oxidized	+198.7	-35.9	+76.4	+21.7	-14.2	-10.3	-3.7
$Cu_B^{1+}(H_2O)$ reduced	+283.7	+49.1	+10.9	-43.8	+5.3	+3.8	10.4
in continuum $\varepsilon = 4$ (including protein charges):							
OORO	+198.7	-35.9	+102.1	+47.4	+11.5	+8.4	15.0
OORR	+283.7	+49.1	+45.8	-8.9	+40.2	+29.3	35.9
in protein (including dielectric inhomogeneity and protein charges):							
OORO	+198.7	-35.9	+84.4	+29.7	-6.2	-4.5	2.1
RORO	+198.7	-35.9	+85.7	+31.0	-4.9	-3.5	3.1
ORRO	+198.7	-35.9	+86.8	+32.1	-3.8	-2.7	3.9
RRRO	+198.7	-35.9	+87.0	+32.3	-3.6	-2.6	4.0
OORR	+283.7	+49.1	+20.5	-34.2	+14.9	+10.8	17.4
RORR	+283.7	+49.1	+21.8	-32.9	+16.2	+11.8	18.4
ORRR	+283.7	+49.1	+22.7	-32.0	+17.1	+12.5	19.1
RRRR	+283.7	+49.1	+23.6	-31.1	+18.0	+13.1	19.7

Similarly, Table 4 is referred to the pK_as of Glu242 residues. The pK_a of Glu242 is calculated relative to the pK_a of the model compound – propionic acid in the aqueous phase. The table illustrates how different dielectric mediums and interactions with protein charges affect the calculated pK_a of the Glu242 site. Energies are reported in kcal/mol.

Table 4. The pK_a values of Glu242 residue

Redox state	ΔE_{elec}	$\Delta\Delta E_{elec}$	ΔG_{solv}	$\Delta\Delta G_{solv}$	$\Delta\Delta G_{shift}$	ΔpK_a	pK_a
in aqueous phase:							
propionic acid	+342.7 [a]	0	-60.0	0	0	0.0	4.4
in continuum dielectric of ε = 4 (no protein charges):							
Glu242	+342.7	0	-46.1	+13.9	+13.9	+10.1	14.5
in inhomogeneous dielectric of protein, ε = 4 and 80 (no protein charges):							
Glu242	+342.7	0	-49.5	+10.5	+10.5	+7.6	12.0
in continuum ε = 4 (including protein charges):							
Glu242 in OORO	+342.7	0	-33.0	+27.0	+27.0	+19.6	24.0
Glu242 in OORR	+342.7	0	-32.9	+27.1	+27.1	+19.7	24.1
in protein (including dielectric inhomogeneity and protein charges):							
Glu242 in ORRO	+342.7	0	-49.9	+10.1	+10.1	+7.4	11.8
Glu242 in OORR	+342.7	0	-53.0	+7.0	+7.0	+5.1	9.5
Glu242 in OORO	+342.7	0	-53.1	+6.9	+6.9	+5.0	9.4
Glu242 in RRRO	+342.7	0	-50.1	+9.9	+9.9	+7.2	11.6
Glu242 in ORRR	+342.7	0	-49.9	+10.1	+10.1	+7.4	11.8
Glu242 in RRRR	+342.7	0	-49.5	+10.5	+10.5	+7.6	12.0
Glu242 in RORR	+342.7	0	-52.4	+7.6	+7.6	+5.5	9.9
Glu242 in RORO	+342.7	0	-52.4	+7.6	+7.6	+5.5	10.0
Glu242 in OOOR	+342.7	0	-53.0	+7.0	+7.0	+5.1	9.5

[a] The electronic energy difference was obtained with the B3LYP functional and the 6-311G** basis set, which improves agreement with experimentally measured proton affinities for carboxylic acids.

The calculated aqueous phase pK_a values of the His291 ligand of the Cu_B center are 8.6 and 13.2; therefore, this complex would be protonated in the aqueous phase at pH 7 in both redox states. In a low continuum dielectric without protein charges, the pK_a value of the His site already becomes redox-dependent. The His-ligand would be distinctly deprotonated with pK_a = –9.9 in oxidized and distinctly protonated in the reduced state (pK_a = +9.8). Introducing the dielectric inhomogeneities but without the protein charges, the pK_a value of oxidized Cu_B complex arises by about 6 pK_a units; however, the δ-His position is remarkably still deprotonated with the pK_a value of –3.7. There is not much change here in the pK_a of the reduced complex, such that the pK_a redox dependence remains.

The equilibrium protein charge distribution for the given redox state of all four metal centers is also required for a complete description of the protein-solvent system. Assuming the reduced state of heme a_3, the pK_a of His291 site in the oxidized Cu_B center of bovine C*c*O is between 2.1 to 4.0, depending on the redox state of other metal centers. When the binuclear center is in the two-electron reduced state, the His291 site has pK_a above 17. In the fully reduced enzyme (FR = R_4), His291 possesses the highest pK_a of 19.7, while in the mixed-valence enzyme (MV = R_2), the pK_a of His291 is just slightly lower, i.e., 17.4. Thus, adding the protein charges to dielectric inhomogeneities further increases the apparent pK_a values of the His291-ligand by an additional 5.8 to 7.7 pK_a units for the oxidized Cu_B center and about 7 to 9.3 pK_a units in the corresponding reduced states of the enzyme.

It should be noticed that by using model (c), which is the Cu_B complex embedded in a uniform low dielectric region that also includes the appropriate protein charges, the pK_a of the His291 site becomes much larger, 15 and 36 for the OORO and OORR redox state, respectively. Thus, adding the protein charges to the continuum low dielectric environment will significantly increase the apparent pK_a values by about 25 pK_a units. Compared to the most sophisticated and realistic model (e), one can conclude that the difference is made by having the inhomogeneous system with well-defined dielectric boundaries, which accounts for the shape of the protein molecule, including external solvent, and introducing the water-filled cavities inside the protein where they are in reality located. It came out that the cavities inside the cytochrome *c* oxidase appear to play a critical and crucial role. Although the Born solvation energies are not so significantly affected by the cavities, the crucial role of the water-filled cavities appears to be in the screening effects on the protein charges, which can considerably change the energetics of the protonation reactions in the protein.

The Dielectric Constant of Protein Cavities

In described calculations, the cavities in the protein's interior are assumed to be filled with water molecules, and a dielectric value of 80 is assigned, as has been done, e.g., in the references (Richardson et al. 1997; Li et al. 1998; Li et al. 1999; Ullmann, Noodleman, and Case 2002). Indeed lots of internal water are observed in the crystal structure of C*c*O (Ostermeier et al. 1997; Svensson-Ek et al. 2002; Tsukihara et al. 2003). However, there is a significant ambiguity in assigning the suitable dielectric constant to the water-filled cavities. A discussion of the appropriate value of the dielectric of the cavities is beyond the scope of this chapter; therefore, we explore here only the dependence of the main results and draw qualitative conclusions on the possible values of this important parameter of the model.

In order to clarify the importance of water-filled channels and cavities, we examine how the pK_as of Glu242 and His291 depend on the dielectric constant associated with the protein cavities. Figure 7 shows a dependence of the His291 pK_a on the dielectric constant of cavities. Here, the dependence is shown only for the OORO redox state of the enzyme, which is the most critical for the whole model of proton pumping via the His291 as a proton-loading site. The more detailed information on this dependence is presented in Table 5, where results are shown for the various ε values of the cavities.

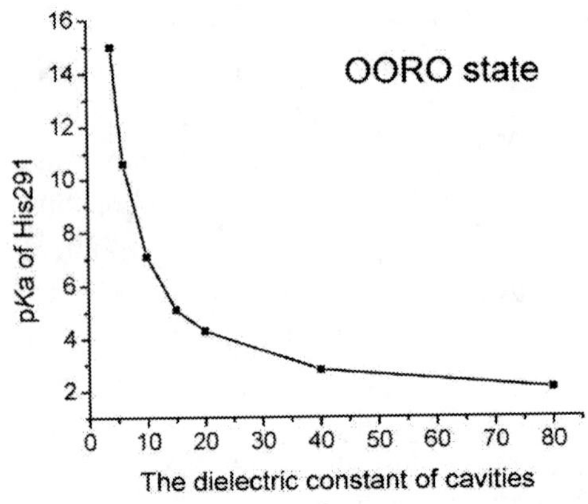

Figure 7. The pK_a dependence of His291 residue, in the OORO redox state, on the dielectric constant of water-filled cavities. The obtained dependence is affected by changes in the ΔG_{Born} and ΔG_q energy terms. In addition to the expected decrease of the ΔG_q term, the Born-energies of deprotonated and protonated species change with an increase of ε_{cavity} causing more significant stabilization of the protonated form (+2) compared with the deprotonated form (+1).

Table 5. The pK_as of Glu242 and His291 sites in different redox states of C*c*O compared for different dielectric properties of water-filled cavities

Redox state	Glu242			His291		
ε-cavity [a]	15	20	80	15	20	80
OORR [b]	12.9	11.9	9.5	22.6	21.1	17.4
OORO	13.0	12.0	9.4	5.1	4.3	2.1
RORO	13.8	12.6	10.0	6.5	5.4	3.1
ORRO	15.9	14.7	11.8	7.4	6.3	3.9
RRRO	15.8	14.6	11.6	7.7	6.5	4.0
RORR	13.7	12.6	9.9	23.8	22.3	18.4
ORRR	15.9	14.7	11.8	24.6	23.1	19.1
RRRR	16.3	15.1	12.0	25.5	23.9	19.7

[a] In all calculations, the dielectric constants of the QM system and protein part are set to 1 and 4, while the effect of different dielectric constants of water-filled cavities (ε = 15, 20, or 80) are explored.
[b] R stands for the reduced and O for the oxidized states of redox cofactors of C*c*O.

There is experimental evidence that water molecules inside C*c*O are quite mobile (Tsukihara et al. 1996; Garcia and Hummer 2000; Backgren et al. 2000). Firstly, the enzyme produces the water molecules in the binuclear center, which leave the active site. In the EPR measurements using isotopic $^{17}O_2$, their appearance is observed near the position of the Mg-center on a time scale of 10 ms (Schmidt, McCracken, and Ferguson-Miller 2003). Thus, the Cu_B center is located at the beginning of the so-called water exit channel(s). Secondly, it is believed that water molecules actively participate and facilitate a proton transfer through the enzyme (Zheng et al. 2003; Olkhova et al. 2004; Wikström, Verkhovsky, and Hummer 2003), which is associated with the significant reorganization and reorientation of water molecule dipoles. Such considerable mobility of the inner water molecules is most likely related to a higher dielectric constant value.

Our general conclusion is that the pumping scheme of C*c*O proposed in this and the previous work is valid in a surprisingly wide range of possible dielectric values for the cavities. Roughly, for values greater than 10, the general pumping scheme emerging from this work remains the same. That is an encouraging result because the exact value of this parameter would be difficult to control by the enzyme and even harder to predict by the calculations.

The Energy Diagram Comparing Different Models

We have calculated the pK_a values of His291 – the critical element of the proposed pumping model of C*c*O. In a similar way, the pK_a of another essential group Glu242 was estimated (Popović, Quenneville, and Stuchebrukhov 2005).

The active site Cu_B complex was treated by density functional theory, while the effects of the protein charges and solvation were incorporated using the continuum electrostatic calculations. Our model does not account for any protein dynamic effects and deals only with a fixed crystal structure. The dielectric constants are chosen in a usual in such calculations empirical way, and therefore the obtained results should be considered keeping the model limitations in mind. The choice of the dielectric constants for a protein and the protein cavities that may contain internal water has been a topic of active debate for a very long time; see the references (Warshel and Russel 1984; Gilson and Honig 1986; Harvey 1989; Warshel and Aqvist 1991; Gilson 1995; Warshel and Papazyan 1998), a discussion of a new procedure with optimized dielectric constants (Wojciechowski et al. 2003), and references cited therein. We choose the dielectric constants in the same way as in the recent similar

DFT/electrostatic studies (Richardson et al. 1997; Li et al. 1998; Li et al. 1999; Ullmann, Noodleman, and Case 2002). The good agreement between calculated and measured pK_a of Glu242 residue, for instance, gives us reason to believe that, although not perfect, our model still manages to account for most of the electronic and electrostatic energies of this complex system in a reasonably accurate way.

The pK_a values were calculated for several redox states of the enzyme, and the influence of different factors on the pK_a values was analyzed in detail. We were particularly interested in determining factors that would make it possible for this complex system to work as a proton pump. We considered several computational models to accomplish this goal, and the results of these calculations are briefly summarized here.

The dielectric inhomogeneities and water-filled cavities, model (d), increase the pK_as of His291 in both redox states (–3.7 and +10.4) and slightly decrease the pK_a of Glu242 to 12.0. However, qualitative conclusions here are the same – there is a redox-dependent acidity of the His291 site, but Glu possesses 2 pK-units higher acidity than His, and a hypothetical transfer of a pump proton from Glu242 to His291 would be energetically unfavorable.

Including the protein charges, which are found in a self-consistent manner, model (e) finely tunes the pK_a values to become suitable for proton pumping. Namely, the pK_as of His291 increase further by 6 to 9 pK_a units, and at the same time, the pK_a of Glu242 slightly decreases or remains the same depending on the redox state of the heme *a* (Figure 8). Table 7 compares the pK_as of these two residues for the different redox states of the enzyme. We find that additional electrons on two other redox centers (Cu_A and heme *a*) only slightly modify the Glu242 and His291 pK_a values for a given redox state of the binuclear complex.

Based on the calculated pK_a values, the proton transfer from Glu242 to His291 can only occur after electron transfer to the binuclear center of the enzyme. For example, the heme *a* reduction increases the pK_a of Glu242 by 2 units, making it more challenging to transfer a proton to His291 before an electron is transferred further to the catalytic center of the enzyme.

Comparing the effects of the addition of the protein charges to continuum and inhomogeneous dielectric environments, we concluded that the internal protein cavities, presumably filled with water, appear to play a very important role and have a significant influence on the pK_a values of the buried residues, with which they are in (direct) contact. Evidently, the reason is the high dielectric constant associated with the cavities and their screening effects.

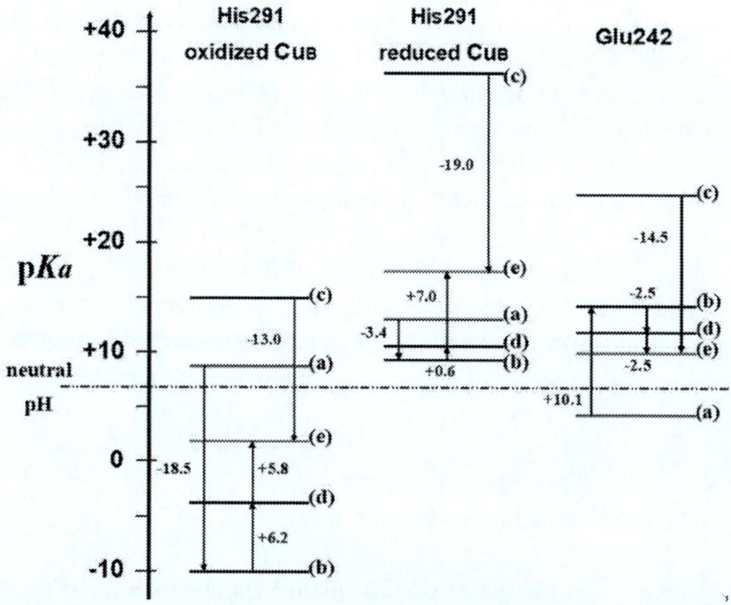

Figure 8. A schematics of the His291 pK_a values in the oxidized and reduced Cu_B complex and the Glu242 pK_as for different models ((a) – (e)) used in our study. Blue bars are related to the pK_a values in the aqueous phase, and green bars represent the calculated pK_as in the protein (cytochrome oxidase). The arrows with the numbers represent the magnitude of the pK_a shifts caused by the different structural details. See the text for the explanation.

The effects of different factors on the pK_as values are summarized in Figure 8. As shown in the figure, going from the aqueous phase to a low dielectric continuum, the pK_as change by –18.8, –3.4, and +10.1 for His-oxidized-Cu_B, His-reduced-Cu_B, and Glu-site, respectively. Introducing dielectric inhomogeneities changes the Born solvation energies by shifting the pK_a by +6.2, +0.6, and –2.5 units, respectively. Adding the protein charges causes the shifts of +5.8, +7.0, and –2.5 pK_a units, respectively. Finally, the screening effects of the protein inner water cavities and the external high dielectric solvent are estimated at –13.0, –19.0, and –14.5 units, respectively.

It has been shown that the pK_a of His 291 depends very strongly on the redox state of the Cu_B center, while a much weaker dependence is observed for Glu242, and its pK_a values remain relatively high for all redox states of the enzyme (see Table 7). It is not surprising, given that His291 is much closer to the catalytic redox center than Glu242. Since in the reduced state of Cu_B, His291 is protonated, while in the oxidized state, His291 is deprotonated, the total charge of Cu_B *together* with its ligand His291 does not change, and hence the redox state of the Cu_B center has much less effect on the pK_a of a distant Glu242 residue. Instead, we find that the pK_a of Glu242 is coupled to the redox state of the adjacent heme *a*. This dependence is less than 3 pK_a units, yet it can be substantial. Each time Glu242 donates a proton, it needs to get reprotonated; hence the obtained high pK_a values of Glu242 absolutely make sense.

In summary, the obtained results of this study support our proposal (Popović and Stuchebrukhov 2004b) that the His291 ligand of the Cu_B center works as a proton pump element in C*c*O. This prediction is currently the subject of several experimental investigations which aim to probe the role of His291 in C*c*O.

Effects of Glu242 Conformational Changes on pK_as

Next, we considered the effect of two distinct conformations of the Glu242 side-chain, as shown in Figure 2A. The two conformations are a proton loading downward (*input*) conformation and a proton releasing upward (*output*) conformation. The downward conformation is taken from the crystal structure of bovine C*c*O (Tsukihara et al. 2003), while the upward conformation is obtained by a quantum-chemical optimization of the system consisting of the Glu242 side-chain, three intermediate water molecules, and PRD of heme a_3 represented by propionic acid.

By rotational isomerization of its side-chain, Glu242 can adopt two distinct conformations – downward and upward, apparently the proton *input* and *output* conformations (Popović and Stuchebrukhov 2006; Brzezinski and Ädelroth 2006). Due to the thermal fluctuations, Glu242 can flip out from the stable downward conformation to the upward conformation (3–6 kcal/mol less stable depending on its protonation state (Popović and Stuchebrukhov 2012; Kaila et al. 2009)), facilitating in this way a proton transfer to the PLS or OH^- ligand of the BNC. In contrast, Glu can get reprotonated through the D-channel by flipping back to the downward conformation.

If Glu is in its “down” (*input*) conformation, it is in protonic contact via water molecules and protonatable groups in the D-channel with the *N*-side of the membrane. When it is in its “up” (*output*) conformation, Glu can transport a proton to the PLS or BNC by establishing the H-bond connectivity via internal water molecules inside the hydrophobic cavity. Thus, in “up” conformation, Glu is in contact with the *P*-side of the membrane, see Figure 2A. Rotational isomerization of the Glu side-chain may be considered an essential component of the mechanism that prevents simultaneous contact with the pump site, the site of oxygen reduction, and the proton-conducting D-channel. Glu can form the H-bonded proton-conducting pathway only to one site at a time, leaving the other two pathways shut down in the meantime. The idea of gating through conformational changes of the Glu side-chain (Popović and Stuchebrukhov 2006) is virtually the same as the "E242 valve model" proposed by the Wikström group (Kaila et al. 2008). Thus, by the *conformational gating*, Glu242 functions as a valve that enables the vectorial transport of protons in one direction from *N*- to *P*-side and prevents a proton leak in the opposite direction. However, a recent computational study (Yang and Cui 2011) has cast some doubts on the "E242 valve model" and the water-wire gating via water reorientation.

The Equilibrium pK_as of Glu242

The dielectric properties of proteins are poorly understood. In order to evaluate the dependence of the results on protein dielectric models, several computational schemes have been employed, similar to those in the references (Popović, Quenneville, and Stuchebrukhov 2005; Quenneville, Popović, and Stuchebrukhov 2006). The reported here pK_a values of Glu242 and His291 were calculated in several models, in which different dielectric constants ε were assigned to protein (ε between 4 and 20), water cavities inside the protein (ε in the range of 10 to 80), the membrane ($\varepsilon = 4$), and the surrounding water ($\varepsilon = 80$). The results are summarized in Tables 6 and 7 and Figure 9.

Table 6 shows the equilibrium Glu242 pK_a values for two conformations of its side-chain in all relevant redox states of the enzyme. The two conformations are referred to as a proton loading conformation pointing downwards and a releasing conformation pointing upwards. In its loading conformation, the Glu242 side-chain presumably interacts with water molecules of the D-channel, providing a proton for reprotonation. In the upper (releasing) conformation, Glu242 can form a stable hydrogen bond with the chain of three water molecules leading to propionate D of heme a_3, see Figure 2B. The latter, in turn, is hydrogen-bonded to His291 via an additional water molecule (Ostermeier et al. 1997; Svensson-Ek et al. 2002), thus providing the proton transfer pathway between Glu242 and His291, according to our model (Popović and Stuchebrukhov 2004a, b). In the calculations reported in Table 6, the protonation state of all other titratable residues, including the His291 site, is an equilibrium protonation state for a given redox state of metal centers.

For His291, this means that this residue is deprotonated in the RO redox state of the binuclear complex and protonated in the binuclear center fully reduced (RR) state.

Table 6. The pK_as of Glu242 residue in *down* loading conformation and *upper* releasing conformation for the different redox states of bovine C*c*O [a]

Redox state [b]	pK_a [c]	
	Glu242 down	Glu242 up
OORO	9.0	9.4
RORO	9.2	10.0
ORRO	10.6	11.8
RRRO	10.5	11.6
OORR	9.2	9.5
RORR	9.4	9.9
ORRR	10.7	11.8
RRRR	10.9	12.0

[a] Protonation state of all other residues, including His291, is the equilibrium protonation state for a given redox state of metal centers.

[b] The four-letter code refers to the redox state (R = reduced, O = oxidized) of four redox-active metal centers: Cu_A, heme *a*, heme a_3 and Cu_B complex, respectively.

[c] The calculations are done for $\varepsilon_{prot} = 4$ and $\varepsilon_{cavity} = 80$.

From Table 6, one can see that there is no significant difference in the acidity of Glu242 for different conformations of its side-chain. The pK_as in all redox states are generally somewhat smaller in the downward conformation. For both conformations, the pK_as are higher for the states where heme *a* is reduced. Naturally, an additional electron on heme *a* and the proton on Glu242 attract each other, increasing the pK_a value of Glu.

In Table 6, we assume that the conformational changes of Glu242 do not produce protonation changes of other titratable groups (mainly of His291), and the protonation state of the enzyme is equilibrium one for a given redox state of metal centers. The non-equilibrium states will be discussed in the following subsection.

Comparing the equilibrium states OO*RO* and OO*RR*, in the latter, there is an additional electron on Cu_B, but also there is an additional proton on His291; hence, it is not surprising that pK_as of Glu242 in these two states are very similar, showing that His291 and Cu_B center are almost equidistant from Glu242 in both conformations.

The pK_as in OORO, ORRO, and OORR Redox State

In this subsection, we analyze different redox states, and within each redox state, we calculate pK_as of His291 depending on the conformation and protonation state of Glu242; we also evaluate pK_as of Glu242 depending on the conformation of the Glu242 side-chain and protonation state of His291. Table 7 and Figure 9 summarize the results for OORO, ORRO, and OORR states. Different dielectric models of the protein medium were considered. We varied two parameters ε – dielectric constant of the protein matrix (ε_{prot}) and that of the inner water cavities (ε_{cavity}). We first kept constant $\varepsilon_{prot} = 4$ and gradually increased ε_{cavity} from 4, 15, 20, to 80. Then, we probed the effect of increasing ε of the protein (4, 10, 20) while keeping constant $\varepsilon_{cavity} = 20$. The last two entries, $\varepsilon_{prot}/\varepsilon_{cavity} = 10/20$ and 20/20, have also been probed because some authors suggest (Schutz and Warshel 2001; Wojciechowski et al. 2003; Olsson, Sharma, and Warshel 2005) that proteins should have higher dielectric constant than 4, a value assumed and used in this kind of calculation for a very long time (Sharp and Honig 1990; Bashford et al. 1993; Simonson and Perahia 1995; Simonson and Brooks 1996; Li et al. 1998; Warshel and Papazyan 1998; Onufriev, Smondyrev, and Bashford 2003). Figure 9 summarizes the results for a model of $\varepsilon_{prot} = 4$, $\varepsilon_{cavity} = 20$. For all other dielectric conditions, refer to the corresponding Table 7.

Generally, the results for the homogeneous model with a dielectric of ε = 4 show exceedingly high pK_as of both, Glu242 and His291 in all states, see Table 7, which means they stay protonated in all redox states, and therefore this dielectric model does not make much sense. The different models are presented here to demonstrate how much the results can vary depending on the dielectric conditions.

In this study, we have applied DFT/continuum electrostatic calculations to explore the coupling between Glu242 and His291 residues in several redox states of bovine cytochrome oxidase (Yoshikawa et al. 1998) and for two conformations of the Glu242 side-chain. We have calculated the pK_as of His291 and Glu242 for different redox states of the metal centers, protonation states of the two residues, and different conformations of the Glu242 side-chain. In addition to the equilibrium states of the catalytic cycle, we have covered all non-equilibrium states concerning the protonation/ conformation changes of Glu242 and His291 residues. Some of these states are just transient; others are actual intermediate states, while several states are only virtual states that likely never occur.

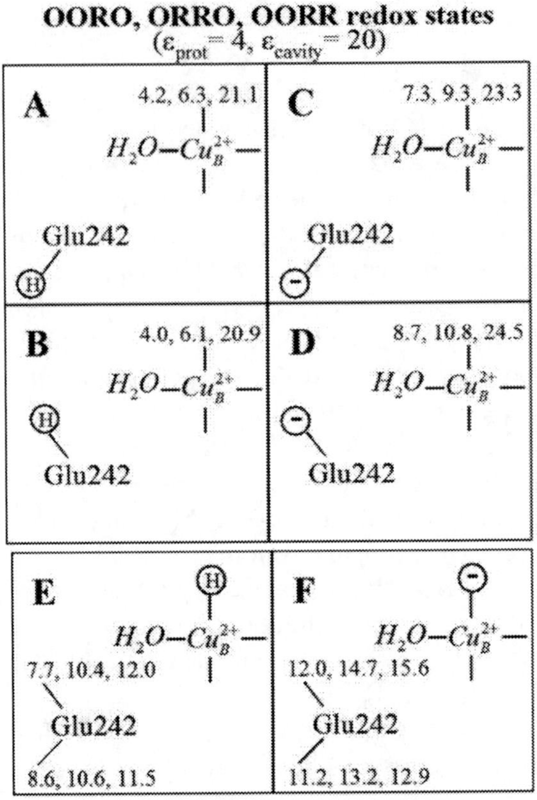

Figure 9. The mutual dependence of pK_a values of His291 and Glu242 on their protonation states and conformation of the Glu242 side-chain for the dielectric model of $\varepsilon_{prot} = 4$ and $\varepsilon_{cavity} = 20$. See Table 7 for all other dielectric models. The sequences of numbers correspond to OORO, ORRO, and OORR redox states, respectively. The pK_as of His291 depend on the protonation and conformation state of Glu242: A) Glu protonated and downwards, B) Glu protonated and upwards, C) Glu deprotonated and downwards, D) Glu deprotonated and upwards. The Glu242 pK_as depend on the orientation of its side-chain (down/up) and protonation state of His291: E) His protonated, F) His deprotonated.

Any calculation that includes a heterogeneous system has to deal with the problem of choosing the appropriate dielectric constants. Considering that the actual values of the dielectric constant of the protein and the inner (presumably) water-filled cavities are unknown, we examined several different dielectric models. The results were compared with experimental data, and the viability of different dielectric models in the light of the proposed proton pumping mechanism of *CcO* was evaluated.

Table 7. The mutual dependence of the pK_a values of Glu242 and His291 on their protonation states and the conformation state of Glu242 in OORO, ORRO, and OORR redox states of the enzyme [a]

Dielectric medium	redox state	p*Ka* of His291				p*Ka* of Glu242			
		A	B	C	D	E		F	
$\varepsilon_{prot}/\varepsilon_{cavity}$ [b]		Glu prot Glu down	Glu prot Glu up	Glu depr Glu down	Glu depr Glu up	His prot Glu down	His prot Glu up	His depr Glu down	His depr Glu up
continuum 4	OORO	15.0	14.8	20.4	21.8	18.3	17.7	22.9	24.1
	ORRO	18.2	17.9	23.2	24.7	21.4	21.8	26.0	28.2
	OORR	35.9	35.6	39.0	40.3	23.0	24.1	25.5	29.2
4/15	OORO	5.2	5.0	8.4	9.9	9.4	8.5	12.2	13.1
	ORRO	7.5	7.3	10.6	12.1	11.5	11.4	14.3	15.9
	OORR	22.6	22.3	24.8	26.0	12.4	12.9	13.9	16.8
4/20	OORO	4.2	4.0	7.3	8.7	8.6	7.7	11.2	12.0
	ORRO	6.3	6.1	9.3	10.8	10.6	10.4	13.2	14.7
	OORR	21.1	20.9	23.3	24.5	11.5	12.0	12.9	15.6
4/80	OORO	2.1	1.9	4.5	6.1	6.7	5.7	8.9	9.4
	ORRO	3.9	3.7	6.2	7.6	8.4	8.1	10.6	11.8
	OORR	17.5	17.3	19.3	20.4	9.2	9.5	11.0	12.7
10/20	OORO	7.1	7.0	8.9	9.4	4.8	5.9	6.1	7.9
	ORRO	8.2	8.1	9.7	10.4	5.9	7.2	7.2	9.2
	OORR	17.4	17.3	18.4	19.0	6.3	7.8	7.0	9.7
20/20	OORO	8.1	8.1	9.2	9.5	3.5	5.3	4.5	6.7
	ORRO	8.9	8.7	9.8	10.1	4.3	6.2	5.1	7.4
	OORR	15.9	15.8	16.5	16.7	4.5	6.7	5.0	7.6

[a] See Figure 9 for the case of $\varepsilon_{prot} = 4$, $\varepsilon_{cavity} = 20$. The p$K_a$s of His291 and Glu242 for all other dielectric conditions can be obtained from this table. Notation of the states A-F from the table corresponds to the notation in Figure 9A-F.

[b] Calculations are done for the different dielectric conditions: in continuum dielectric of $\varepsilon = 4$ (including protein charges); and for the several cases where dielectric of protein (ε_{prot}) or the solvent filled cavities (ε_{cavity}) was varied.

The work objectives are to answer the following questions: How does a presence/absence of a proton on the Glu242 and/or His291 site affect the pK_as of other sites? How do the conformational changes of the Glu242 side-chain affect the pK_as of the two residues in different redox states of the enzyme? How strong is the coupling between the redox states of heme *a* or Cu_B center and the protonation states of Glu242 and His291 residues? How do the redox state and redox potential of heme *a* depend on the protonation state of Glu242 and vice versa?

Here it is particularly interesting to compare the obtained pK_as of Glu242 in different redox states of the enzyme with the results of recent work (Nyquist et al. 2003), which suggests that Glu242 residue might be deprotonated in some of the states in the catalytic cycle. Several other studies, based on the pH-dependent proton transfer (PT) rates and FTIR measurements, suggest that Glu242 is protonated at pH 7 with the apparent $pK_a \geq 9$ (Lübben and Gerwert 1996; Hellwig et al. 1996; Puustinen et al. 1997; Hellwig et al. 1998; Lübben et al. 1999; Namslauer et al. 2003). An interesting model of reprotonation of Glu242 was published recently by Xu and Voth (Xu and Voth 2005).

The calculations show that in a broad range of $\varepsilon_{prot} = 4 - 8$ and $\varepsilon_{cavity} = 4 - 80$, the Glu242 pK_as are above 7 in all redox states and, therefore, Glu should always be protonated at equilibrium. The best agreement with the available experimental data is obtained for $\varepsilon_{prot} = 4$ and $\varepsilon_{cavity} = 80$; however, dielectric models ($\varepsilon_{prot} = 4$, $\varepsilon_{cavity} = 15$ or 20) also give reasonable agreement.

Within these dielectric models, the only case when Glu242 residue was found to be deprotonated is for the OORO redox state and $\varepsilon_{prot}/\varepsilon_{cavity} = 4/80$. Glu242 can remain deprotonated here after it delivers a chemical proton to the binuclear center to form a water molecule and before the His291 site gets rid of its Hδ1 proton, provided that Glu242 is somehow trapped in its upward conformation. For the case ($\varepsilon_{prot}/\varepsilon_{cavity} = 4/80$), Glu has a pK_a of 5.7 in the upper conformation prior to the ejection of a proton from the His site. Even in the downward conformation, its pK_a is low (6.7) if His291 still keeps the proton. According to the present calculation, unless the release of a proton from His291 is somehow blocked and Glu stays in the upper configuration, this group remains protonated at equilibrium.

Interestingly, in agreement with present calculations is our previous finding that OH^- ligand to Cu_B has a computed pK_a of 8.5 to 9.6 in the OORO redox state of the enzyme prior to the pumping event, i.e., when His291 is still protonated (Quenneville, Popović, and Stuchebrukhov 2006). Thus, the OH^- ligand possesses higher pK_a than Glu242 residue (5.7 to 6.7); therefore, it can pull out H^+ from the Glu site, as our model assumes (Popović and Stuchebrukhov 2004b). The ejection of the proton from the His291 site can only further increase the pK_a of the Cu_B–OH^- ligand, which, therefore, will remain protonated, i.e., like H_2O.

Other dielectric models, $\varepsilon_{prot} = 10$ and $\varepsilon_{cavity} = 10$ or 20, give too low pK_as of Glu242 (below 7) in all redox states, which does not agree with experiments. At the same time, the His291 site possesses pK_a values higher than 7, making it always protonated. As already discussed, in these two models, gating through the conformational changes of the Glu side-chain is impossible because, in the OORO redox state, even an additional proton on Glu is not enough to push out a proton from His site. Also, the pK_as of Glu in the downward conformation are so low that Glu cannot get reprotonated via the D-channel. Therefore, the results of these two dielectric models do not fit either the proton pumping mechanism that we proposed,(Popović and Stuchebrukhov 2004a, b) or the results of the experimental measurements. (Lübben and Gerwert 1996; Hellwig et al. 1996; Puustinen et al. 1997; Hellwig et al. 1998; Lübben et al. 1999; Namslauer et al. 2003)

The main findings of this work can be summarized as follows. The results of this work support our earlier proposed proton pumping mechanism of C*c*O (Popović and Stuchebrukhov 2004b). The pK_a of His291 depends on the redox state of the enzyme, whereby, His291 tends to be protonated in the OORR redox state but deprotonated in the OORO state. In the OORR state, Glu242 also possesses the $pK_a > 7$, but lower than the pK_a of His291, and therefore, Glu can transfer a proton to His-site, but afterward, it should quickly get reprotonated, unless Glu242 is somehow trapped in the upper conformation.

In the OORO redox state, we find that after donating a proton to OH^- ligand in the BNC, Glu242 needs to be reprotonated before His291 can release a proton, and the pumping occurs. Thus, the reprotonation of Glu can be said to displace an H^+ from the PLS. This result agrees with the recent work of Brzezinski and coworkers (Faxen et al. 2005). Moreover, our calculations suggest that the enzyme might employ a subtle gating mechanism through conformational changes of the Glu side-chain. Namely, in the OORO state with deprotonated Glu in the upper conformation, the proton-pump site His291 still remains protonated and has pK_a higher than that of Glu242. In its upward conformation, Glu also has $pK_a > 7$, but since it cannot get a proton from any residue from the upper side, the only way to get protonated is to swing down. This will simultaneously decrease the proton affinity of His site and further increase the pK_a of the Glu site. In the downward conformation, the Glu residue will re-establish the hydrogen-bond connectivity with the D-channel and quickly get reprotonated. The reprotonation of Glu ultimately decreases the pK_a of the His291 site, releasing the pumped proton into the intermembrane space (*P*-side).

As we examined, Glu most likely is never deprotonated in the stable equilibrium states of the catalytic cycle. Of course, since Glu242 residue is the proton donor of both chemical and pump protons, in some of the *transient* states, Glu gets temporarily deprotonated after transferring a proton to an acceptor group with a higher pK_a (His291 or OH^-). However, in all these intermediate (transient) states, the obtained pK_as of Glu are high enough to pull up a proton from the D-channel and get reprotonated once again.

Full QM Model of the Binuclear Complex

Data are presented for the DFT-B3LYP and DFT-PBE0 functionals. Details of such calculations are reported in the reference (Quenneville, Popović, and Stuchebrukhov 2006). Below, we summarize the main results relevant to the pumping mechanism of the enzyme.

pK_a of His

In B3LYP, the calculated pK_a values of His291 ligand of the Fe_{a3}–Cu_B complex in C*c*O are found to be 5.4 and 21.9 for oxidized [Fe=O (H_2O)Cu^{2+}(HisH)] and reduced [Fe=O (H_2O)Cu^{+}(HisH)] complexes, respectively. In PBE0, the corresponding values are 5.0 and 26.1. These values compare reasonably well with 2.1 (for the OORO redox state of the protein) and 17.4 (OORR) obtained previously for the Cu_B center alone (Popović, Quenneville, and Stuchebrukhov 2005). The agreement is even better with the results of the new procedure that fits the protein and reaction field potentials in the region of the QM system with artificial point charges placed on a surface of a sphere surrounding the QM system, which allows for performing DFT calculations in an inhomogeneous dielectric environment in a self-consistent way (Makhov, Popović, and Stuchebrukhov 2006). The procedure improves the accuracy of calculations in comparison with previous work. The improved results show, however, that although the absolute energies change considerably, the relative energies of the protonated and deprotonated states of His291 remain close to the previously reported ones and therefore do not change the pK_a values significantly (6.1 – oxidized and 22.4 – reduced BNC). Thus, we confirm that the protonation state of His291 is redox dependent when the full protein model, with charges and dielectric inhomogeneities, is used.

The effects of different dielectric and protein charge models on the pK_a of His291 are given in reference (Quenneville, Popović, and Stuchebrukhov 2006). We find that for our model to work, i.e., for His291 to have a redox-dependent protonation state, one needs to assume that the dielectric constant of water cavities in the protein is higher than approximately 15. The actual value of this important parameter is unknown; however, a value higher than 15 does not seem unreasonable. (Work is underway in this group to determine the value of the dielectric constant of water cavities in the direct simulation of water molecules in the enzyme cavities.)

The main conclusion of this part of the calculations is that the pK_a of His291 depends on the redox state of the Fe_{a3}–Cu_B center so that in the OORR state, it is indisputably protonated, while in the OORO state, it is most likely deprotonated. It means that when a chemical proton arrives at the Fe_{a3}–Cu_B center, the state, according to our notation, changes from OORR to OORO, and the repulsion between the chemical proton and the proton on His291 is strong enough (Popović and Djordjević 2020) so that the latter becomes deprotonated. This expulsion of the proton from His291 by the chemical proton is crucial for the proposed pumping mechanism.

The following principal question concerns the chemical proton itself. The question of whether it would go to the Fe_{a3}–Cu_B center at the expense of expelling the proton from His291 is discussed next.

pK_a of H_2O/OH^- Ligand to Cu_B

Considering two structures, [Fe=O $(HO^-)Cu^{2+}$(HisH)] and [Fe=O $(H_2O)Cu^{2+}$(HisH)], we calculated the pK_a of the H_2O ligand of the Cu_B center. Methylimidazole (MeIm) was used as a reference. Within the protein, the computed pK_a of H_2O/OH^- ligand is found to be 8.5–9.6 in the state of the enzyme prior to pumping, i.e., when His291 is protonated. It means that a small driving force allows a chemical proton to enter the active site and form a water molecule in the presence of Cu^{2+}, even when His291 is protonated. Given the uncertainties of calculations (see discussion of possible error bars below), the exact value of the driving force is difficult to compute accurately, and the conclusion should be taken cautiously.

In fact, for the pump to work, the driving force for the chemical proton when His291 is protonated is not necessary. What is necessary, however, is to have a driving force for the exchange of the chemical proton and proton on His291. In other words, the question is will the chemical proton go to the HO^- group in the Fe_{a3}–Cu_B center at the expense of expelling the proton from His291? This is a question about the energy of the pumping step. Not only should this energy be available, but it also should be large enough to push the proton through the electrochemical proton gradient of the membrane.

The Free Energy of the Pumping Step

As proposed earlier, based on pure continuum electrostatic calculations (Popović and Stuchebrukhov 2004a, b), the entrance of a chemical proton into the binuclear complex of C*c*O that converts the HO^- ligand to H_2O is a trigger for a proton-pumping event which causes the deprotonation of the His291 site and a release of a proton on the intermembrane (positive) side of the membrane. The reaction free energy (ΔG_R) of a proton-pumping step

$$Fe{=}O\ (HO^-)Cu^{2+}\ (HisH) \rightarrow Fe{=}O\ (H_2O)Cu^{2+}\ (His^-) \tag{10}$$

initial state (1) final state (2)

was calculated as follows:

$$\Delta G_R = G_{final(2)} - G_{initial(1)} = \Delta E_{elec} + \Delta G_{vib} + \Delta G_{Born} + \Delta G_q + \Delta G_{strain} \quad (11)$$

where ΔG_{vib} is the difference in the zero-point vibrational energy of two states, and other terms have the same meaning as explained earlier.

The values of various energy terms (in kcal/mol) have been evaluated and used to calculate the reaction free energy of the proton-pumping step for the DFT-B3LYP and DFT-PBE0 density functionals. The gas-phase DFT-B3LYP electronic energy (–2.26 kcal/mol), the difference in vibrational energy (–0.5 kcal/mol), and the Born-solvation energy due to the polarization (–3.23 kcal/mol) favor the exergonicity of the reaction, while the effect of the protein charges slightly disfavors the proton pumping step by +1.58 kcal/mol. The effect of the strain energy term is almost negligible. The total B3LYP reaction energy is –4.20 kcal/mol for the inhomogeneous dielectric of the protein–membrane–solvent system where $\varepsilon_{\text{prot}} = \varepsilon_{\text{memb}} = 4$ and $\varepsilon_{\text{sol}} = \varepsilon_{\text{cavity}} = 80$. The calculation based on PBE0 density functional gives similar results with the corresponding terms: $\Delta E_{elec} = -4.68$, $\Delta G_{vib} = -0.10$, $\Delta G_{Born} = -3.54$, $\Delta G_q = +1.61$, $\Delta G_{strain} = +0.35$, and a total reaction free energy of –6.36 kcal/mol.

We also examine the dependence of the reaction energy on the various possible values for the dielectric constant of solvent within the water-filled cavities and epsilon of a protein region itself. A discussion of the appropriate values of these important parameters of the model system is beyond the scope of this chapter, and we are here only interested in exploring the dependence of the main results and making qualitative conclusions. The obtained results show that a proton-pumping step, as described in our proposed pumping mechanism (Popović and Stuchebrukhov 2004b), is energetically possible (and favorable) for several dielectric models examined in this study. The DFT-PBE0 gives the reaction energies of a pumping step slightly higher than those of B3LYP, mainly due to the difference in electronic energy.

The calculations, therefore, show that the examined pumping step is energetically feasible (Figure 10). The energy generated in this step is relatively small yet enough to push a proton against the electrochemical gradient of the membrane. Indeed, the difference in the electrochemical potential across the mitochondrial membrane is about 200 meV (4.6 kcal/mol). Assuming the high efficiency of C*c*O, one may expect the actual energy of the pumping step to be around 5 kcal/mol, which perfectly agrees with our results.

Given the uncertainties of the calculations, however, this is likely a fortuitous agreement with the expectation. The most important and more reliable conclusion perhaps is that the free energy of the proposed pumping step is clearly not of a large and positive value, a result that would make the proposed scheme very unlikely. As it stands now, however, we conclude that the present calculations indeed support the proposal that the chemical proton expels the proton from His291 and triggers the pumping event.

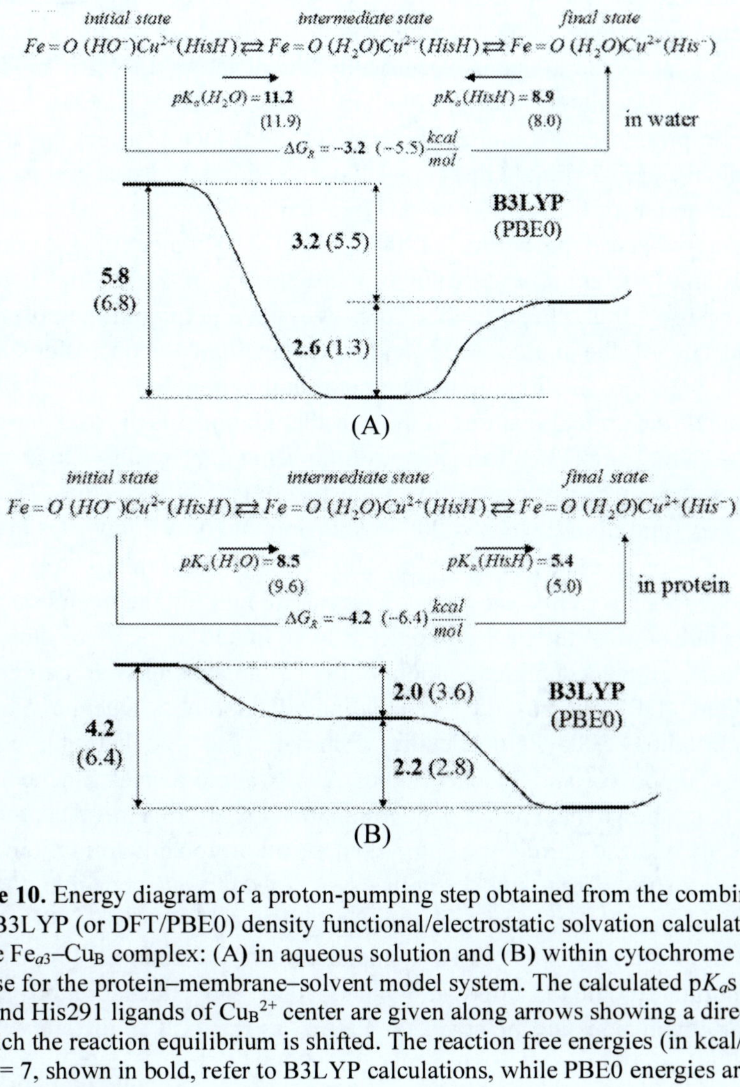

Figure 10. Energy diagram of a proton-pumping step obtained from the combined DFT/B3LYP (or DFT/PBE0) density functional/electrostatic solvation calculations for the Fe_{a3}–Cu_B complex: (A) in aqueous solution and (B) within cytochrome *c* oxidase for the protein–membrane–solvent model system. The calculated pK_as of H_2O and His291 ligands of Cu_B^{2+} center are given along arrows showing a direction in which the reaction equilibrium is shifted. The reaction free energies (in kcal/mol) at pH = 7, shown in bold, refer to B3LYP calculations, while PBE0 energies are reported in parentheses.

Conclusion

The combined density functional theory (DFT) and continuum electrostatic calculations (QM/CE method) are employed to explore the coupling of the conformational changes of Glu242 residue, the primary proton donor of both chemical and pump protons, to its pK_a, and the pK_a of His291, a putative proton-loading site (PLS) of the pumping model. The pK_a values of His291 and Glu242, the two essential residues of the model, are calculated for different redox states of the enzyme, and the influence of different factors on the pK_as is analyzed in detail. Specifically, the different dielectric models are explored to elucidate the effect of the inner water-filled cavities and their dielectric properties. To understand how different factors affect the apparent pK_a values of His291 and Glu242, we have considered several computational QM/CE models of the membrane–enzyme–cavities–solvent system and tested different dielectric properties of the protein and water-filled cavities. A discussion of the appropriate value of the dielectric of the cavities is beyond the scope of this chapter; therefore, we have explored here only the dependence of the main results and draw qualitative conclusions on the possible values of this important parameter of the model.

One of the critical features of the considered model is the existence of the proton-loading site (PLS) located above the heme porphyrins. The identity of PLS has been recently examined in the theoretical kinetic study (Sugitani, Medvedev, and Stuchebrukhov 2008), showing that only a few sites may play that role – propionates of heme a_3, His291, Trp126, Trp236, Arg438, and Arg439, all at roughly the same dielectric depth in the membrane. The protonation state of the PLS group needs to be linked to the redox state of the binuclear Fe_{a3}–Cu_B complex, although it is not an absolute requirement. The combined DFT/electrostatic calculations (Popović, Quenneville, and Stuchebrukhov 2005; Quenneville, Popović, and Stuchebrukhov 2004; Makhov, Popović, and Stuchebrukhov 2006; Popović and Stuchebrukhov 2012) have shown that His291 is deprotonated if one of the metal centers, Fe_{a3} or Cu_B, is oxidized or after the entry of chemical proton and formation of H_2O molecule in the active site. In both cases, an additional positive charge is present in the BNC, significantly decreasing the His291 pK_a. In contrast, His291 is protonated if both BNC metal centers are reduced (Popović and Stuchebrukhov 2004a; Popović, Quenneville, and Stuchebrukhov 2005). Similar conclusions are obtained for PRAa_3 as the PLS in different models (Blomberg and Siegbahn 2010; Kaila et al. 2009).

In our recent calculations (Quenneville, Popović, and Stuchebrukhov 2006; Popović and Stuchebrukhov 2012), using the similar QM/CE computational method, we have also calculated the pK_a of OH^- group in the binuclear center of the enzyme and have shown that energetically it is a favorable process to protonate OH^- and form water, at the expense of deprotonation of His291 and ejection of a proton into intermembrane space. This is the crucial energy-generating step of the proton pump. Thus, the calculations have provided the basis for and indeed helped to discover the proposed pumping model of cytochrome oxidase.

The new set of calculations shows that including an accurate self-consistent reaction field at the *ab initio* stage of the computation gives rise to an increase in the absolute values of the solvation energy (Makhov, Popović, and Stuchebrukhov 2006). However, this increase is approximately equal for protonated and deprotonated states of His291. As a result, new pK_a values of His291 are very close to those previously reported by our group and consistent with the proposed model for the C*c*O proton pump.

This chapter introduces a new QM/CE method of bioenergetic calculations developed by Popović et al. (2005). The method is applied here to examine the energetics of biological electron and proton transfer reactions in cytochrome *c* oxidase. The other biologically relevant and highly valuable feature of the study refers to the treatment of internal solvent-filled cavities and the effects they produce in proteins, an often underestimated problem in most bio-computational studies. We have recognized the importance of these effects on the energetics of electron and proton transfer reactions in C*c*O and generally on most chemical reactions in the active site of enzymes. Combining quantum mechanics with molecular mechanics or electrostatic calculations also offers better accuracy of the results, as we have recently demonstrated and illustrated here.

Acknowledgments

This work has been financially supported by the Ministry of Education, Science and Technological Development of the Republic of Serbia (MPNTR): Grant No. 451-03-68/2022-14/200026 and project OI/127035. This article/publication is based upon work from COST Action CA21101, supported by COST (European Cooperation in Science and Technology).

References

Abramson, J., S. Riistama, G. Larsson, A. Jasaitis, M. Svensson-Ek, L. Laakkonen, A. Puustinen, S. Iwata, and M. Wikström. 2000. "The structure of the ubiquinol oxidase from Escherichia coli and its ubiquinone binding site." *Nature Struct. Biol.* 7:910–917.

Ädelroth, P., R. B. Gennis, and P. Brzezinski. 1998. "Role of the pathway through K(I-362) in proton transfer in cytochrome c oxidase from R. Sphaeroides." *Biochemistry* 37:2470–2476.

Ädelroth, P., M. Svensson-Ek, D. M. Mitchell, R. B. Gennis, and P. Brzezinski. 1997. "Glutamate 286 in cytochrome aa3 from *Rhodobacter sphaeroides* is involved in proton uptake during the reaction of the fully-reduced enzyme with dioxygen." *Biochemistry* 36:13824–13829.

Antonini, E., M. Brunori, C. Greenwood, and B. G. Malmström. 1970. "Catalytic mechanism of cytochrome oxidase." *Nature* 228:936–937.

Antosiewicz, J., J. M. Briggs, A. H. Elcock, M. K. Gilson, and J. A. McCammon. 1996. "Computing ionization states of proteins with a detailed charge model." *J. Comput. Chem.* 17:1633–1644.

Antosiewicz, J., J. A. McCammon, and M. K. Gilson. 1994. "Prediction of pH-dependent properties of proteins." *J. Mol. Biol.* 238:415–436.

Babcock, G. T. 1999. "How oxygen is activated and reduced in respiration." *Proc. Natl. Acad. Sci. USA* 96:12971–12973.

Babcock, G. T., and M. Wikström. 1992. "Oxygen activation and the conservation of energy in cell respiration." *Nature* 356:301–309.

Backgren, C., G. Hummer, M. Wikström, and A. Puustinen. 2000. *Biochemistry* 39:7863–7867.

Baptista, A. M., P. J. Martel, and C. M. Soares. 1999. "Simulation of electron-proton coupling with a Monte Carlo method: Application to cytochrome c3 using continuum electrostatics." *Biophys. J.* 76:2978–2998.

Bashford, D. 1997. "An object-oriented programming suite for electrostatic field effects in biological molecules." In *Scientific Computing in Object-Oriented Parallel Environments*, edited by Y. Ishikawa, R. R. Oldehoeft, J. V. W. Reynders and M. Tholburn, 233–240. Berlin: Springer.

Bashford, D., D. A. Case, C. Dalvit, L. Tennant, and P. E. Wright. 1993. "Electrostatic calculations of side-chain pKa values in myoglobin and comparison with NMR data for histidines." *Biochemistry* 32:8045–8056.

Bashford, D., and K. Gerwert. 1992. "Electrostatic calculations of the pKa values of ionizable groups in bacteriorhodopsin." *J. Mol. Biol.* 224:473–486.

Bashford, D., and M. Karplus. 1990. "pKas of ionizable groups in proteins: Atomic detail from a continuum electrostatic model." *Biochemistry* 29:10219–10225.

Bashford, D., and M. Karplus. 1991. "Multiple-site titration curves of proteins: An analysis of exact and approximate methods for their calculations." *J. Phys. Chem.* 95:9557–9561.

Becke, A. D. 1993. "Density-functional thermochemistry. III. The role of exact exchange." *J. Chem. Phys.* 98:5648–5652.

Belevich, I., D. A. Bloch, N. Belevich, M. Wikström, and M. I. Verkhovsky. 2007. "Exploring the proton pump mechanism of cytochrome c oxidase in real time." *Proc. Natl. Acad. Sci. USA* 104:2685–2690.

Belevich, I., M. I. Verkhovsky, and M. Wikström. 2006. "Proton-coupled electron transfer drives the proton pump of cytochrome c oxidase." *Nature* 440:829–832.

Bene, J. E. Del, H. D. Metter, M. J. Frisch, B. J. Luke, and J. A. Pople. 1983. *J. Phys. Chem.* 87:3279.

Beroza, P., and D. A. Case. 1996. "Including side chain flexibility in continuum calculations of protein titration." *J. Phys. Chem.* 100:20156–20163.

Beroza, P., D. R. Fredkin, M. Y. Okamura, and G. Feher. 1991. "Protonation of interesting residues in a protein by a Monte Carlo method: Application to lysozyme and the photosynthetic reaction center." *Proc. Natl. Acad. Sci. USA* 88:5804–5808.

Binkley, J. S., and J. A. Pople. 1977. "Self-consistent molecular orbital methods. XIX. Split-valence Gaussian-type basis sets for beryllium." *J. Chem. Phys.* 66 (2):879–880.

Bloch, D., I. Belevich, A. Jasaitis, C. Ribacka, A. Puustinen, M. I. Verkhovsky, and M. Wikström. 2004. "The catalytic cycle of cytochrome c oxidase is not the sum of its two halves." *Proc. Natl. Acad. Sci. USA* 101 (2):529–533.

Blomberg, M. R. A. 2021. "The redox-active tyrosine is essential for proton pumping in cytochrome c oxidase." *Front. Chem.* 9: 640155, pp. 1-11.

Blomberg, M. R. A., and P. E. M. Siegbahn. 2012. "The mechanism for proton pumping in cytochrome c oxidase from an electrostatic and quantum chemical perspective." *Biochim. Biophys. Acta* 1817:495-505.

Blomberg, M. R. A., P. E. M. Siegbahn, G. T. Babcock, and M. Wikström. 2000. "Modeling cytochrome oxidase: A quantum chemical study of the O-O bond cleavage mechanism." *J. Am. Chem. Soc.* 122:12848–12858.

Blomberg, M. R., and P. E. M. Siegbahn. 2010. "A quantum chemical study of the mechanism for proton-coupled electron transfer leading to proton pumping in cytochrome c oxidase." *Mol. Phys.* 108:2733–2743.

Brändén, M., H. Sigurdson, A. Namslauer, R. B. Gennis, P. Ädelroth, and P. Brzezinski. 2001. "On the role of the K-proton transfer pathway in cytochrome coxidase." *Proc. Natl. Acad. Sci. USA* 98:5013–5018.

Brändén, M., F. Tomson, R. B. Gennis, and P. Brzezinski. 2002. "The entry point of the K-proton transfer pathway in cytochrome c oxidase." *Biochemistry* 41:10794–10798.

Bratton, M. R., M. A. Pressler, and J. P. Hosler. 1999. "Suicide inactivation of cytochrome c oxidase: Catalytic turnover in the absence of subunit III alters the active site." *Biochemistry* 38:16236–16245.

Breneman, C. M., and K. B. Wiberg. 1990. "Determining atom-centered monopoles from molecular electrostatic potentials. The need for high sampling density in formamide conformational analysis." *J. Comput. Chem.* 11 (3):361–373.

Brzezinski, P. 2004. "Redox-driven membrane-bound proton pumps." *Trends Biochem. Sci.* 29:380–387.

Brzezinski, P., and P. Ädelroth. 2006. "Design principles of proton-pumping haem-copper oxidases." *Curr. Opin. Struct. Biol.* 16 (4):465–472.

Brzezinski, P., and R. B. Gennis. 2008. "Cytochrome c oxidase: exciting progress and remaining mysteries." *J. Bioenerg. Biomembr.* 40:521–531.

Buse, G., T. Soulimane, M. Dewor, H. E. Meyer, and M. Bluggel. 1999. *Protein Sci.* 8:985–990.

Catalan, J., J. L. G. d. Paz, M. Yanez, R. M. Claramunt, C. Lopez, J. Elguero, F. Anvia, J. H. Quian, M. Taagepera, and R. W. Taft. 1990. *J. Am. Chem. Soc.* 112:1303.

Chen, J. L., L. Noodleman, D. A. Case, and D. Bashford. 1994. "Incorporating solvation effects into density functional electronic structure calculations." *J. Phys. Chem.* 98:11059–11068.

Clark, T., J. Chandrasekhar, G. W. Spitznagel, and P. V. R. Schleyer. 1983. "Efficient diffuse function-augmented basis sets for anion calculations. III. The 3-21+G basis set for first-row elements, Li-F." *J. Comput. Chem.* 4 (3):294–301.

Das, T. K., C. M. Gomes, M. Teixeira, and D. L. Rousseau. 1999. "Redox-linked transient deprotonation at the binuclear site in the aa3-type quinol oxidase from Acidianus ambivalens: Implications for proton translocation." *Proc. Natl. Acad. Sci. USA* 96:9591–9596.

Ditchfield, R., W. J. Hehre, and J. A. Pople. 1971. "Self-consistent molecular-orbital methods. IX. An extended Gaussian- type basis for molecular-orbital studies of organic molecules." *J. Chem. Phys.* 54 (2):724–728.

Du, W.-G. H., A. W. Götz, and L. Noodleman. 2018. "A water dimer shift activates a proton pumping pathway in the Pr → F transition of ba3 cytochrome c oxidase." *Inorg. Chem.* 57 (3):1048–1059.

Dürr, K. L., J. Koepke, P. Hellwig, H. Müller, H. Angerer, G. Peng, E. Olkhova, O. M. H. Richter, B. Ludwig, and H. Michel. 2008. "A D-pathway mutation decouples the *Paracoccus denitrificans* cytochrome c oxidase by altering the side-chain orientation of a distant conserved glutamate." *J. Mol. Biol.* 384:865–877.

Fabian, M., W. W. Wong, R. B. Gennis, and G. Palmer. 1999. "Mass spectrometric determination of dioxygen bond splitting in the ''peroxy'' intermediate of cytochrome c oxidase." *Proc. Natl. Acad. Sci. USA* 96:13114–13117.

Fadda, E., C.-H. Yu, and R. Pomès. 2008. "Electrostatic control of proton pumping in cytochrome c oxidase." *Biochim. Biophys. Acta* 1777:277–284.

Faxen, K., G. Gilderson, P. Ädelroth, and P. Brzezinski. 2005. "A mechanistic principle for proton pumping by cytochrome c oxidase." *Nature* 437 (8):286–289.

Fee, J. A., D. A. Case, and L. Noodleman. 2008. "Toward a chemical mechanism of proton pumping by the B-type cytochrome c oxidases: application of density functional theory to cytochrome ba(3) of Thermus thermophilus." *J. Am. Chem. Soc.* 130:15002–15021.

Ferguson-Miller, S., and G. T. Babcock. 1996. "Heme-copper oxidase." *Chem. Rev.* 7:2889–2907.

Fetter, J. R., M. Sharpe, J. Qian, D. Mills, S. Ferguson-Miller, and P. Nicholls. 1996. "Fatty acids stimulate activity and restore respiratory control in a proton channel mutant of cytochrome c oxidase." *FEBS Lett.* 393:155–160.

Francl, M. M., W. J. Pietro, W. J. Hehre, J. S. Binkley, M. S. Gordon, D. J. DeFrees, and J. A. Pople. 1982. "Self-consistent molecular orbital methods. XXIII. A polarization-type basis set for second-row elements." *J. Chem. Phys.* 77 (7):3654–3665.

Frisch, M. J., J. A. Pople, and J. S. Binkley. 1984. "Self-consistent molecular orbital methods 25. Supplementary functions for Gaussian basis sets." *J. Chem. Phys.* 80 (7):3265–3269.

Garcia, A. E., and G. Hummer. 2000. *Proteins* 38:261–272.

Gennis, R. B. 1998a. "Cytochrome c oxidase: one enzyme, two mechanisms." *Science* 280:1712–1714.

Gennis, R. B. 1998b. "How does cytochrome oxidase pump protons?" *Proc. Natl. Acad. Sci. USA* 95:12747–12749.

Gennis, R. B. 1998c. "Multiple proton-conducting pathways in cytochrome oxidase and a proposed role for the active-site tyrosine." *Biochim. Biophys. Acta* 1365:241–248.

Gennis, R. B. 2004. "Coupled proton and electron transfer reactions in cytochrome oxidase." *Front. Biosci.* 9:581–591.

Gilson, M. K. 1995. "Theory of electrostatic interactions in macromolecules." *Curr. Opin. Struct. Biol.* 5:216–223.

Gilson, M. K., and B. Honig. 1986. "The dielectric constant of a folded proteins." *Biopolymers* 25:2097–2119.

Gunner, M. R., and B. Honig. 1991. "Electrostatic control of midpoint potentials in the cytochrome subunit of the *Rhodopseudomonas viridis* reaction center." *Proc. Natl. Acad. Sci. USA* 88:9151–9155.

Gunner, M. R., A. Nicholls, and B. Honig. 1996. "Electrostatic potentials in *Rhodopseudomonas viridis* reaction centers: Implications for the driving force and directionality of electron transfer." *J. Phys. Chem.* 100:4277–4291.

Haltia, T., M. Saraste, and M. Wiksrom. 1989. "Subunit III of cytochrome c oxidase is not involved in proton translocation: a site-directed mutagenesis study." *EMBO J.* 10:2015–2021.

Hammes-Schiffer, S., and A. A. Stuchebrukhov. 2010. "Theory of coupled electron and proton transfer reactions." *Chem. Rev.* 110:6939–6960.

Han, S., S. Takahashi, and D. L. Rousseau. 2000. "Time dependence of the catalytic intermediates in cytochrome c oxidase." *J. Biol. Chem.* 275 (3):1910–1919.

Hariharan, P. C., and J. A. Pople. 1973. "The influence of polarization functions on molecular orbital hydrogenation energies." *Theor. Chim. Acta* 28 (3):213–222.

Harvey, S. C. 1989. "Treatment of electrostatic effects in macromolecular modeling." *Proteins* 5:78–92.

Hay, P. J., and W. R. Wadt. 1985. "Ab initio effective core potentials for molecular calculations. Potentials for K to Au including the outermost core orbitals." *J. Chem. Phys.* 82 (1):299–310.

Hehre, W. J., R. Ditchfield, and J. A. Pople. 1972. "Self-consistent molecular orbital methods. XII. Further extensions of Gaussian—type basis sets for use in molecular orbital studies of organic molecules." *J. Chem. Phys.* 56 (5):2257–2261.

Hehre, W. J., and J. A. Pople. 1972. "Self-consistent molecular orbital methods. XIII. An extended Gaussian-type basis for boron." *J. Chem. Phys.* 56 (8):4233–4234.

Hellwig, P., J. Behr, C. Ostermeier, O.-M. H. Richter, U. Pfitzner, A. Odenwald, B. Ludwig, H. Michel, and W. Mantele. 1998. "Involvement of glutamic acid 278 in the redox reaction of the cytochrome c oxidase from *Paracoccus denitrificans* investigated by FT-IR spectroscopy." *Biochemistry* 37:7390–7399.

Hellwig, P., B. Rost, U. Kaiser, C. Ostermeier, H. Michel, and W. Mantele. 1996. "Carboxyl group protonation upon reduction of the *Paracoccus denitrificans* cytochrome c oxidase: direct evidence by FTIR spectroscopy." *FEBS Lett.* 385:53–57.

Hemp, J., and D. E. Robinson. 2006. "Evolutionary migration of a posttranslationally modified active-site residue in the proton-pumping heme–copper oxygen reductases." *Biochemistry* 45 (51):15405–15410.

Hendler, R. W., K. Pardhasaradhi, B. Reynafarje, and B. Ludwig. 1991. "Comparison of energy-transducing capabilities of the two- and three-subunit cytochrome aa3, from *Paracoccus denitrificans* and the 13-subunit beef heart enzyme." *Biophys. J.* 60:415–423.

Hill, T. L. 1860. *An introduction to statistical thermodynamics.* New York: Dover.

Hohenberg, P., and W. Kohn. 1964. "Inhomogeneous Electron Gas." *Physical Review* 136:B864–B871.

Hosler, J. P., S. Ferguson-Miller, and D. A. Mills. 2006. "Energy transduction: proton transfer through the respiratory complexes." *Annu. Rev. Biochem.* 75:165–187.

Hosler, J. P., J. P. Shapleigh, D. M. Mitchell, Y. Kim, M. A. Pressler, C. Georgiou, G. T. Babcock, J. O. Alben, S. Ferguson-Miller, and R. B. Gennis. 1996. "Polar residues in helix VIII of subunit I of cytochrome c oxidase influence the activity and the structure of the active site." *Biochemistry* 35:10776–10783.

Iwata, S., C. Ostermeier, B. Ludwig, and H. Michel. 1995. "Structure at 2.8 Å resolution of cytochrome c oxidase from *Paracoccus denitrificans*." *Nature* 376:660–669.

Jaguar 5.5, S., L.L.C., Portland, OR, 1991-2003.

Jasaitis, A., M. I. Verkhovsky, J. E. Morgan, M. L. Verkhovskaya, and M. Wikström. 1999. "Assignment and charge translocation stoichiometries of the electrogenic phases in the reaction of cytochrome c with dioxygen." *Biochemistry* 38:2697–2706.

Johansson, A.-L., J. Carlsson, M. Högbom, J. P. Hosler, R. B. Gennis, and P. Brzezinski. 2013. "Proton Uptake and pKa Changes in the Uncoupled Asn139Cys Variant of Cytochrome c Oxidase." *Biochemistry* 52:827–836.

Johansson, A.-L., S. Chakrabarty, C. L. Berthold, M. Högbom, A. Warshel, and P. Brzezinski. 2011. "Proton-transport mechanisms in cytochrome c oxidase revealed by studies of kinetic isotope effects." *Biochim. Biophys. Acta* 1807 (9):1083-1094.

Jorgenson, W. L., J. Chandrasekhar, and J. D. Madura. 1983. "Comparison of simple potential functions for simulating liquid water." *J. Chem. Phys.* 79:926–935.

Kaila, V. R. I., M. I. Verkhovsky, G. Hummer, and M. Wikström. 2008. "Glutamic acid 242 is a valve in the proton pump of cytochrome c oxidase." *Proc. Natl. Acad. Sci. USA* 105:6255–6259.

Kaila, V. R. I., M. I. Verkhovsky, G. Hummer, and M. Wikström. 2009. "Mechanism and energetics by which glutamic acid 242 prevents leaks in cytochrome c oxidase." *Biochim. Biophys. Acta* 1787:1205–1214.

Kallies, B., and R. Mitzner. 1997. "pKa values of amines in water from quantum mechanical calculations using a polarized dielectric continuum representation of the solvent." *J. Phys. Chem. B* 101:2959–2967.

Kannt, A., R. D. Lancaster, and H. Michel. 1998. "The coupling of electron transfer and proton translocation: electrostatic calculations on *Parracocus denitrificants* cytochrome c oxidase." *Biophys. J.* 74:708–721.

Karlsberg online manual. http://lie.chemie.fu-berlin/karlsberg/.

Kirchberg, K., H. Michel, and U. Alexiev. 2013. "Exploring the entrance of proton pathways in cytochrome c oxidase from *Paracoccus denitrificans*: Surface charge, buffer capacity and redox-dependent polarity changes at the internal surface." *Biochim. Biophys. Acta* 1827:276–284.

Klapper, I., R. Fine, K. A. Sharp, and B. H. Honig. 1986. "Focusing of electric fields in the active site of Cu-Zn superoxide dismutase: Effects of ionic strength and amino-acid modification." *Prot. Struct. Funct. Genet.* 1:47–59.

Kohn, K., and L. J. Sham. 1965. "Self-consistent equations including exchange and correlation effects." *Physical Review* 140:A1133–1136.

Konstantinov, A. A., S. Siletsky, D. Mitchell, A. Kaulen, and R. B. Gennis. 1997. "The role of the two proton input channels in cytochrome c oxidase from *Rhodobacter sphaeroides* probed by the effects of site-directed mutations on time-resolved electrogenic intraprotein proton transfer." *Proc. Natl. Acad. Sci. USA* 94:9085–9090.

Krishnan, R., J. S. Binkley, R. Seeger, and J. A. Pople. 1980. "Self-consistent molecular orbital methods. XX. A basis set for correlated wave functions." *J. Chem. Phys.* 72:650–654.

Lancaster, C. R., H. Michel, B. Honig, and M. R. Gunner. 1996. "Calculated coupling of electron and proton transfer in the photosynthetic reaction center of *Rhodopseudomonas viridis*." *Biophys. J.* 70:2469–2492.

Lee, H.-M., D. L. Das, D. A. Rousseau, S. Mills, and R. B. Ferguson-Miller. 2000. "Mutations in the putative H-channel in the cytochrome c oxidase from *Rhodobacter sphaeroides* show that this channel is not important for proton conduction but reveal modulation of the properties of heme a." *Biochemistry* 39:2989–2996.

Lee, H. J., L. Ojemyr, A. Vakkasoglu, P. Brzezinski, and R. B. Gennis. 2009. "Properties of Arg481 mutants of the aa(3)-type cytochrome c oxidase from *Rhodobacter sphaeroides* suggest that neither R481 nor the Nearby D-propionate of heme a(3) is likely to be the proton loading site of the proton pump." *Biochemistry* 48:7123–7131.

Li, J., C. L. Fischer, J. L. Chen, D. Bashford, and L. Noodleman. 1996. "Calculation of redox potentials and pKa values of hydrated transition metal cations by a combined density functional and continuum dielectric theory." *J. Phys. Chem.* 96:2855–2866.

Li, J., C. L. Fisher, R. Konecny, D. Bashford, and L. Noodleman. 1999. "Density functional and electrostatic calculations of manganese superoxide dismutase active site complexes in protein environments." *Inorg. Chem.* 38:929–939.

Li, J., M. R. Nelson, C. Y. Peng, D. Bashford, and L. Noodleman. 1998. "Incorporating protein environments in density functional theory: A self-consistent reaction field calculation of redox potentials of [2Fe2S] clusters in ferredoxin and phthalate dioxygenase reductase." *J. Phys. Chem. A* 102:6311–6324.

Lübben, M., and K. Gerwert. 1996. "Redox difference spectroscopy using caged electrons reveals contributions of carboxyl groups to the catalytic mechanism of heam-copper oxidases." *FEBS Lett.* 397:303–307.

Lübben, M., A. Prutsch, B. Mamat, and K. Gerwert. 1999. "Electron transfer induces side-chain conformational changes of glutamate-286 from cytochrome bo3." *Biochemistry* 38:2048–2056.

Luna, V. M., Y. Chen, J. A. Fee, and C. D. Stout. 2008. "Crystallographic studies of Xe and Kr binding within the large internal cavity of cytochrome ba3 from Thermus thermophilus: structural analysis and role of oxygen transport channels in the heme–Cu oxidases." *Biochemistry* 47 (16):4657–4665.

MacKerell, Jr., A. D., D. Bashford, M. Bellot, Jr. Dunbrack, R. L., J. D. Evanseck, M. J. Field, S. Fischer, J. Gao, H. Guo, S. Ha, D. Joseph-McCarthy, L. Kuchnir, K. Kuczera, F. T. K. Lau, C. Mattos, S. Michnick, T. Ngo, D. T. Nguyen, B. Prodholm, III Reiher, W. E., B. Roux, M. Schlenkrich, J. C. Smith, R. Stote, J. Straub, M. Watanabe, J. Wiórkiewicz-Kuczera, D. Yin, and M. Karplus. 1998. "All-atom empirical potential for molecular modeling and dynamics studies of proteins." *J. Phys. Chem.* 102:3586–3616.

Makhov, D. V., D. M. Popović, and A. A. Stuchebrukhov. 2006. "Improved density functional theory/electrostatic calculation of the His291 protonation state in cytochrome c oxidase: self-consistent charges for solvation energy calculation." *J. Phys. Chem. B* 110:12162–12166.

Marten, B., K. Kim, C. Cortis, R. A. Friesner, R. B. Murphy, M. N. Ringnalda, D. Sitkoff, and B. Honig. 1996. "New Model for Calculation of Solvation Free Energies: Correction of Self-Consistent Reaction Field Continuum Dielectric Theory for Short-Range Hydrogen-Bonding Effects." *J. Phys. Chem.* 100:11775–11788.

McLean, A. D., and S. G. Chandler. 1980. "Contracted Gaussian basis sets for molecular calculations. I. Second row atoms, Z=11-18." *J. Chem. Phys.* 72 (10):5639–5648.

Michel, H. 1998. "The mechanism of proton pumping by cytochrome c oxidase." *Proc. Natl. Acad. Sci. USA* 95:12819–12824.

Michel, H., J. Behr, A. Harrenga, and A. Kannt. 1998. "Cytochrome c oxidase: Structure and spectroscopy." *Annu. Rev. Biophys. Biomol. Struct.* 27:329–356.

Mills, D. A., L. Florens, C. Hiser, J. Qian, and S. Ferguson-Miller. 2000. "Where is "outside" in cytochrome c oxidase and how and when do protons get there?" *Biochim. Biophys. Acta* 1458:180–187.

Mills, D. A., Z. Tan, S. Ferguson-Miller, and J. Hosler. 2003. "A role for subunit III in proton uptake into the D pathway and a possible proton exit pathway in *Rhodobacter sphaeroides* cytochrome c oxidase." *Biochemistry* 42 (24):7410–7417.

Mitchell, P. 1961. "Coupling of phosphorylation to electron transfer by a chemiosmotic type of mechanism." *Nature* 191:144–148.

Mitchell, P. 1976. "Possible molecular mechanisms of the protonmotive function of cytochrome systems." *J. Theor. Biol.* 62:327–367.

Mouesca, J. M., J. L. Chen, L. Noodleman, D. Bashford, and D. A. Case. 1994. "Density functional/Poisson-Boltzmann calculations of redox potentials for iron-sulfur clusters." *J. Am. Chem. Soc.* 116:11898–11914.

Namslauer, A., A. Aagaard, A. Katsonouri, and P. Brzezinski. 2003. "Redox-coupled proton translocation in biological systems: Proton shuttling in cytochrome c oxidase." *Biochemistry* 42:1488–1498.

Nicholls, A., and B. Honig. 1991. "A rapid finite difference algorithm, utilizing successive overrelaxation to solve the Poisson-Boltzmann equation." *J. Comp. Chem.* 12:435–445.

Noor, M. R., and T. Soulimane. 2013. "Structure of caa3 cytochrome c oxidase – a nature-made enzyme-substrate complex." *Biol. Chem.* 394 (5):579–591.

Nyquist, R. M., D. Heitbrink, C. Bolwien, R. B. Gennis, and J. Heberle. 2003. "Direct observation of protonation reactions during the catalytic cycle of cytochrome c oxidase." *Proc. Natl. Acad. Sci. USA* 100 (15):8715–8720.

Olkhova, E., M. C. Hutter, M. A. Lill, V. Helms, and H. Michel. 2004. "Dynamic water networks in cytochrome c oxidase from *Paracoccus denitrificans* investigated by molecular dynamics simulations." *Biophys. J.* 86 (4):1873–1889.

Olsson, M. H. M., P. K. Sharma, and A. Warshel. 2005. "Simulating redox coupled proton transfer in cytochrome c oxidase: Looking for the proton bottleneck." *FEBS Lett.* 579:2026–2034.

Olsson, M. H. M., P. E. M. Siegbahn, M. R. A. Blomberg, and A. Warshel. 2007. "Exploring pathways and barriers for coupled ET/PT in cytochrome c oxidase: A general framework for examining energetics and mechanistic alternatives." *Biochim. Biophys. Acta* 1767:244–260.

Olsson, M. H. M., and A. Warshel. 2006. "Monte Carlo simulations of proton pumps: On the working principles of the biological valve that controls proton pumping in cytochrome c oxidase." *Proceedings of the National Academy of Sciences of the United States of America* 103 (17):6500–6505.

Onufriev, A., A. Smondyrev, and D. Bashford. 2003. "Affinity changes driving unidirectional proton transport in the bacteriorhodopsin photocycle." *J. Mol. Biol.* 332 (5):1183–1193.

Ostermeier, C., A. Harrenga, U. Ermler, and H. Michel. 1997. "Structure at 2.7 Å resolution of the *Paracoccus denitrificans* two-subunit cytochrome c oxidase complexed with an antibody Fv fragment." *Proc. Natl. Acad. Sci. USA* 94:10547–10553.

Pawate, A. S., J. Morgan, A. Namslauer, D. Mills, P. Brzezinski, S. Ferguson-Miller, and R. B. Gennis. 2002. "A mutation in subunit I of cytochrome oxidase from *Rhodobacter sphaeroides* results in an increase in steady-state activity but completely eliminates proton pumping." *Biochemistry* 41 (45):13417–13423.

Pereira, M. M., M. Santana, and M. Teixeira. 2001. "A novel scenario for the evaluation of haem-copper oxygen reductases." See also reviews - Plasticity of proton pathways in haem-copper oxygen reductases. *Biochim. Biophys. Acta* 1505:185–208.

Pereira, M. M., F. L. Sousa, A. F. Veríssimo, and M. Teixeira. 2008. "Looking for the minimum common denominator in haem-copper oxygen reductases: towards a unified catalytic mechanism." *Biochim. Biophys. Acta* 1777 (7–8):929–934.

Pfitzner, U., A. Odenwald, T. Ostermann, L. Weingard, B. Ludwig, and O.-M. H. Richter. 1998. *J. Bioenerg. Biomembr.* 30:89–97.

Pisliakov, A. V., P. K. Sharma, Z. T. Chu, M. Haranczyk, and A. Warshel. 2008. "Electrostatic basis for the unidirectionality of the primary proton transfer in cytochrome c oxidase." *Proc. Natl. Acad. Sci. USA* 105:7726–7731.

Popović, D. M. 2013. "Current advances in research of cytochrome c oxidase." *Amino Acids* 45:1073–1087.

Popović, D. M., and I. S. Djordjević. 2020. "Catalytic center of cytochrome c oxidase: Effects of protein environment on pKa values of CuB histidine ligands." *Journal of Serbian Chemical Society* 85:1429–1444.

Popović, D. M., I. V. Leontyev, D. G. Beech, and A. A. Stuchebrukhov. 2010. "Similarity of cytochrome c oxidases in different organisms." *Proteins Struct. Funct. Bioinform.* 78:2691–2698.

Popović, D. M., J. Quenneville, and A. A. Stuchebrukhov. 2005. "DFT/electrostatic calculations of pKa values in cytochrome c oxidase." *J. Phys. Chem. B* 109:3616–3626.

Popović, D. M., and A. A. Stuchebrukhov. 2004a. "Electrostatic study of proton pumping mechanism of bovine heart cytochrome c oxidase." *J. Am. Chem. Soc.* 126:1858–1871.

Popović, D. M., and A. A. Stuchebrukhov. 2004b. "Proton pumping mechanism and catalytic cycle of cytochrome c oxidase: Coulomb pump model with kinetic gating." *FEBS Lett.* 566 (1-3):126–130.

Popović, D. M., and A. A. Stuchebrukhov. 2005. "Proton exit channels in bovine cytochrome c oxidase." *J. Phys. Chem. B* 109:1999–2006.

Popović, D. M., and A. A. Stuchebrukhov. 2006. "Two conformational states of Glu242 and pKas in bovine cytochrome c oxidase." *Photochem. Photobiol. Sci.* 5:611–620.

Popović, D. M., and A. A. Stuchebrukhov. 2012. "Coupled electron and proton transfer reactions during the O→E transition in bovine cytochrome c oxidase." *Biochim. Biophys. Acta* 1817 (4):506–517.

Popović, D. M., S. D. Zarić, B. Rabenstein, and E. W. Knapp. 2001. "Artificial cytochrome b: Computer modeling and evaluation of redox potentials." *J. Am. Chem. Soc.* 123:6040–6053.

Popović, D. M., A. Zmirić, S. D. Zarić, and E. W. Knapp. 2002. "Energetics of radical transfer in DNA photolyase." *J. Am. Chem. Soc.* 124:3775–3782.

Proshlyakov, D. A., M. A. Pressler, C. DeMaso, J. F. Leykam, D. L. DeWitt, and G. T. Babcock. 2000. "Oxygen activation and reduction in respiration: involvement of redox-active tyrosine 244." *Science* 290:1588–1591.

Puustinen, A., J. A. Bailey, R. B. Dyer, S. L. Mecklenberg, and M. Wikström. 1997. "Fourier transform infrared evidence for connectivity between CuB and glutamic acid 286 in cytochrome bo3 from Escherichia coli." *Biochemistry* 36:13195–13200.

Qin, L., C. Hiser, A. Mulichak, R. M. Garavito, and S. Ferguson-Miller. 2006. "Identification of conserved lipid/detergent binding sites in a high-resolution structure of the membrane protein cytochrome c oxidase." *Proc. Natl. Acad. Sci. USA* 103: 16117–16122.

Qin, L., J. Liu, D. A. Mills, D. A. Proshlyakov, C. Hiser, and S. Ferguson-Miller. 2009. "Redox-dependent conformational changes in cytochrome c oxidase suggest a gating mechanism for proton uptake." *Biochemistry* 48:5121–5130.

Quenneville, J., D. M. Popović, and A. A. Stuchebrukhov. 2004. "Redox-dependent pKa of CuB histidine ligand in cytochrome c oxidase." *J. Phys. Chem. B* 108:18383–18389.

Quenneville, J., D. M. Popović, and A. A. Stuchebrukhov. 2006. "Combined DFT and electrostatics study of the proton pumping mechanism in cytochrome c oxidase." *Biochim. Biophys. Acta* 1757:1035–1046.

Rabenstein, B., and E. W. Knapp. 2001. "Calculated pH-dependent population of CO-myoglobin conformers." *Biophys. J.* 80:1141–1150.

Rabenstein, B., G. M. Ullmann, and E. W. Knapp. 1998. "Energetics of electron transfer and protonation reactions of the quinones in the photosynthetic reaction center of *Rhodopseudomonas viridis*." *Biochemistry* 37:2488–2495.

Reiss, H., and A. Heller. 1985. *J. Phys. Chem.* 89:4207–4213.

Richardson, W. H., C. Peng, D. Bashford, L. Noodleman, and D. A. Case. 1997. "Incorporating solvation effects into density functional theory: Calculation of absolute acidities." *Int. J. Quant. Chem.* 61:207–217.

Riistama, S., G. Hummer, A. Puustinen, B. R. Dyer, W. H. Woodruff, and M. Wikström. 1997. "Bound water in proton translocation mechanism of the heme-copper oxidase." *FEBS Lett.* 414:275–280.

Ruitenberg, M., A. Kannt, E. Bamberg, K. Fendler, and H. Michel. 2002. "Reduction of cytochrome c oxidase by a second electron leads to proton translocation." *Nature* 417:99–102.

Ruitenberg, M., A. Kannt, E. Bamberg, B. Ludwig, H. Michel, and K. Fendler. 2000. "Single-electron reduction of the oxidized state is coupled to proton uptake via the K pathway in *Paracoccus denitrificans* cytochrome c oxidase." *Proc. Natl. Acad. Sci. USA* 97 (9):4632–4636.

Salje, J., B. Ludwig, and O. M. Richter. 2005. "Is a third proton-conducting pathway operative in bacterial cytochrome c oxidase?" *Biochem. Soc. Trans.* 33:829–831.

Salomonsson, L., G. Brändén, and P. Brzezinski. 2008. "Deuterium isotope effect of proton pumping in cytochrome c oxidase." *Biochim. Biophys. Acta* 1777 (4):343–350.

Sampogna, R. V., and B. Honig. 1994. "Environmental effects on the protonation states of active site residues in bacteriorhodopsin." *Biophys. J.* 66:1341–1352.

Sarti, P., E. Forte, D. Mastronicola, A. Giuffrè, and M. Arese. 2012. "Cytochrome c oxidase and nitric oxide in action: Molecular mechanisms and pathophysiological implications." *Biochim. Biophys. Acta* 1817:610–619.

Schmidt, B., J. McCracken, and S. Ferguson-Miller. 2003. "A discrete water exit pathway in the membrane protein cytochrome c oxidase." *Proc. Natl. Acad. Sci. USA* 100:15539–15542.

Schutz, C. N., and A. Warshel. 2001. "What are the dielectric ''constants'' of proteins and how to validate electrostatic models." *Proteins: Struct. Funct. Genet.* 44:400–417.

Shapleigh, J. P., J. P. Hosler, M. J. Tecklenburg, Y. Kim, G. T. Babcock, R. B. Gennis, and S. Ferguson-Miller. 1992. "Definition of the catalytic site of cytochrome c oxidase: specific ligands of heme a and the heme a3-CuB center." *Proc. Natl. Acad. Sci. USA* 89:4786–4790.

Sharma, V., P. G. Jambrina, M. Kaukonen, E. Rosta, and P. R. Rich. 2017. "Insights into functions of the H channel of cytochrome c oxidase from atomistic molecular dynamics simulations." *Proc. Natl. Acad. Sci. USA* 114 (48):E10339–E10348.

Sharp, K., and B. Honig. 1990. "Electrostatic interactions in macromolecules: Theory and applications." *Ann. Rev. Biophys. Biophys. Chem.* 19:301–332.

Sharpe, M. A., and S. Ferguson-Miller. 2008. "A chemically explicit model for the mechanism of proton pumping in heme–copper oxidases." *J. Bioenerg. Biomembr.* 40:541–549.

Shimokata, K., Y. Katayama, H. Murayama, M. Suematsu, T. Tsukihara, K. Muramoto, H. Aoyama, S. Yoshikawa, and H. Shimada. 2007. "The proton pumping pathway of bovine heart cytochrome c oxidase." *Proc. Natl. Acad. Sci. USA* 104:4200–4205.

Shinzawa-Itoh, K., H. Aoyama, Muramoto K, Terada H, Kurauchi T, Tadehara Y, Yamasaki A, Sugimura T, Kurono S, Tsujimoto K, Mizushima T, Yamashita E, Tsukihara T, and Yoshikawa S. 2007. "Structures and physiological roles of 13 integral lipids of bovine heart cytochrome c oxidase." *EMBO J.* 26 (6):1713–1725.

Siegbahn, P. E. M., and M. R. Blomberg. 2007. "Energy diagrams and mechanism for proton pumping in cytochrome c oxidase." *Biochim. Biophys. Acta* 1767:1143–1156.

Siegbahn, P. E. M., M. R. A. Blomberg, and M. L. Blomberg. 2003. "Theoretical Study of the Energetics of Proton Pumping and Oxygen Reduction in Cytochrome Oxidase." *J. Phys. Chem. B* 107:10946–10955.

Siletsky, S. A., and A. A. Konstantinov. 2012. "Cytochrome c oxidase: Charge translocation coupled to single-electron partial steps of the catalytic cycle." *Biochim. Biophys. Acta* 1817 (4):476–488.

Siletsky, S. A., A. S. Pawate, K. Weiss, R. B. Gennis, and A. A. Konstantinov. 2004. "Transmembrane charge separation during the ferryl-oxo to oxidized transition in a nonpumping mutant of cytochrome c oxidase." *J. Biol. Chem.* 279:52558–52565.

Simonson, T., and C. L. Brooks. 1996. "Charge separation and the dielectric constant of proteins: Insights from molecular dynamics." *J. Am. Chem. Soc.* 118:8452–8458.

Simonson, T., and D. Perahia. 1995. "Internal and interfacial dielectric properties of cytochrome c from molecular dynamics in aqueous solution." *Proc. Natl. Acad. Sci. USA* 92:1082–1086.

Soares, C. M., A. M. Baptista, M. M. Pereira, and M. Teixeira. 2004. "Investigation of protonatable residues in Rhodothermus marinus caa3 haem-copper oxygen reductase. Comparison with *Paracoccus denitrificans* aa3 haem-copper oxygen reductase." *J. Biol. Inorg. Chem.* 9:124–134.

Song, Y., E. Michonova-Alexova, and M. R. Gunner. 2006. "Calculated Proton Uptake on Anaerobic Reduction of Cytochrome c Oxidase: Is the Reaction Electroneutral?" *Biochemistry* 45:7959–7975.

Soulimane, T., G. Buse, G. P. Bourenkov, H. D. Bartunik, R. Huber, and M. E. Than. 2000. "Structure and mechanism of the aberrant ba(3)-cytochrome c oxidase from Thermus thermophilus." *EMBO J.* 19:1766–1776.

Spassov, V. Z., H. Luecke, K. Gerwert, and D. Bashford. 2001. "pKa calculations suggest storage of an excess proton in a hydrogen-bonded water network in Bacteriorhodopsin." *J. Mol. Biol.* 312:203–219.

Stuchebrukhov, A. A. 2003a. "Electron transfer reactions coupled to proton translocation. Cytochrome oxidase, proton pumps, and biological energy transduction." *J. Theor. Comp. Chem* 2:91–118.

Stuchebrukhov, A. A. 2003b. "Long-distance electron tunneling in proteins." *Theor. Chem. Acc.* 110 (5):291–306.

Sugitani, R., E. S. Medvedev, and A. A. Stuchebrukhov. 2008. "Theoretical and computational analysis of the membrane potential generated by cytochrome c oxidase upon single electron injection into the enzyme." *Biochim. Biophys. Acta* 1777 (9): 1129–1139.

Sugitani, R., and A. A. Stuchebrukhov. 2009. "Molecular dynamics simulation of water in cytochrome c oxidase reveals two water exit pathways and the mechanism of transport." *Biochim. Biophys. Acta* 1787 (9):1140–1150.

Svensson-Ek, M., J. Abramson, G. Larsson, S. Tornroth, P. Brzezinski, and S. Iwata. 2002. "The x-ray crystal structures of wild-type and EQ(I-286) mutant cytochrome c oxidases from *Rhodobacter sphaeroides*." *J.Mol.Biol.* 321:329–335.

Tannor, D. J., B. Marten, R. Murphy, R. A. Friesner, D. Sitkoff, A. Nicholls, M. Ringnalda, W. A. Goddard, and B. Honig. 1994. "Accurate first principles calculation of molecular charge distributions and solvation energies from ab initio quantum mechanics and continuum dielectric theory." *J. Am. Chem. Soc.* 116:11875–11880.

Tanokura, M. 1983. "1H-NMR study on the tautomerism of the imidazole ring of histidine residues: I. Microscopic pK values and molar ratios of tautomers in histidine-containing peptides." *Biochim. Biophys. Acta* 742:576–585.

Tashiro, M., and A. A. Stuchebrukhov. 2005. "Thermodynamic properties of internal water molecules in the hydrophobic cavity around the catalytic center of cytochrome c oxidase." *J. Phys. Chem. B* 109 (2):1015–1022.

Tsukihara, T., H. Aoyama, E. Yamashita, T. Tomizaki, H. Yamaguchi, K. Shinzawa-Itoh, R. Nakashima, R. Yaono, and S. Yoshikawa. 1996. "The whole structure of the 13-subunit oxidized cytochrome c oxidase at 2.8 Å." *Science* 272:1136–1144.

Tsukihara, T., K. Shimokata, Y. Katayama, H. Shimada, K. Muramoto, H. Aoyama, M. Mochizuki, K. Shinzawa-Itoh, E. Yamashita, M. Yao, Y. Ishimura, and S. Yoshikawa. 2003. "The low-spin heme of cytochrome c oxidase as the driving element of the proton-pumping process." *Proc. Natl. Acad. Sci. USA* 100:15304–15309.

Ullmann, G. M., L. Noodleman, and D. A. Case. 2002. "Density functional calculation of pKa values and redox potentials in the bovine Rieske iron-sulfur protein." *J. Biol. Inorg. Chem.* 7:632–639.

Vagedes, P., B. Rabenstein, J. Åqvist, J. Marelius, and E. W. Knapp. 2000. "The deacylation step of acetylcholinesterase: Computer simulation studies." *J. Am. Chem. Soc.* 122:12254–12262.

Warshel, A., and J. Aqvist. 1991. "Electrostatic energy and macromolecular function." *Ann. Rev. Biophys. Biophys. Chem.* 20:267–298.

Warshel, A., and A. Papazyan. 1998. "Electrostatic effects in macromolecules: Fundamental concepts and practical modeling." *Curr. Op. Struct. Biol.* 8:211–217.

Warshel, A., and S. T. Russel. 1984. "Calculations of electrostatic interaction in biological systems and in solution." *Quart. Rev. Biophys.* 17:283–422.

Wikström, M. 1977. "Proton pump coupled to cytochrome c oxidase in mitochondria." *Nature* 266:271–273.

Wikström, M. 1998. "Proton translocation by bacteriorhodopsin and heme-copper oxidases." *Curr. Op. Struct. Biol.* 8:480–488.

Wikström, M. 2000. "Mechanism of proton translocation by cytochrome c oxidase: a new four-stroke histidine cycle." *Biochim. Biophys. Acta* 1458:188–198.

Wikström, M. 2004. "Cytochrome c oxidase: 25 years of the elusive proton pump." *Biochim. Biophys. Acta* 1655:241–247.

Wikström, M., A. Jasaitis, C. Backgren, A. Puustinen, and M. I. Verkhovsky. 2000. "The role of the D- and K-pathways of proton transfer in the function of the heam-copper oxidases." *Biochim. Biophys. Acta* 1459:514–520.

Wikström, M., K. Krab, and V. Sharma. 2018. "Oxygen activation and energy conservation by cytochrome c oxidase." *Chem. Rev.* 118:2469–2490.

Wikström, M., C. Ribacka, M. Molin, L. Laakkonen, M. Verkhovsky, and A. Puustinen. 2005. "Gating of proton and water transfer in the respiratory enzyme cytochrome c oxidase." *Proc. Natl. Acad. Sci. USA* 102:10478–10481.

Wikström, M., and M. I. Verkhovsky. 2007. "Mechanism and energetics of proton translocation by the respiratory heme–copper oxidases." *Biochim. Biophys. Acta* 1767:1200–1214.

Wikström, M., M. I. Verkhovsky, and G. Hummer. 2003. "Water-gated mechanism of proton translocation by cytochrome c oxidase." *Biochim. Biophys. Acta* 1604 (2):61–65.

Wojciechowski, M., T. Grycuk, J. M. Antosiewicz, and B. Lesyng. 2003. "Prediction of secondary ionization of the phosphate group in phosphotyrosine peptides." *Biophys. J.* 84:750–756.

Xu, J., and G. A. Voth. 2005. "Computer simulation of explicit proton translocation in cytochrome c oxidase: The D-pathway." *Proc. Natl. Acad. Sci. USA* 102 (19):6795–6800.

Yang, A.-S., M. R. Gunner, R. Sompogna, and B. Honig. 1993. "On the calculation of pKas in proteins." *Prot. Struct. Funct. Genet.* 15:252–265.

Yang, S., and Q. Cui. 2011. "Glu-286 rotation and water wire reorientation are unlikely the gating elements for proton pumping in cytochrome c oxidase." *Biophys. J.* 101:61–69.

Yano, N., K. Muramoto, A. Shimada, S. Takemura, J. Baba, H. Fujisawa, M. Mochizuki, K. Shinzawa-Itoh, E. Yamashita, T. Tsukihara, and S. Yoshikawa. 2016. "The Mg2+-containing water cluster of mammalian cytochrome c oxidase collects four pumping proton equivalents in each catalytic cycle." *J. Biol. Chem.* 291: 23882−23894.

Yoshikawa, S., K. Shinzawa-Itoh, R. Nakashima, R. Yaono, E. Yamashita, N. Inoue, M. Yao, M. J. Fei, C. P. Libeu, T. Mizushima, H. Yamaguchi, T. Tomizaki, and T. Tsukihara. 1998. "Redox-coupled structural changes in bovine heart cytochrome c oxidase." *Science* 280:1723–1729.

Zaslavsky, D., and R. B. Gennis. 2000. "Proton pumping by cytochrome oxidase: progress, problems and postulates." *Biochim. Biophys. Acta* 1458:164–179.

Zheng, X., D. M. Medvedev, J. Swanson, and A. A. Stuchebrukhov. 2003. "Computer simulation of water in cytochrome c oxidase." *Biochim. Biophys. Acta* 1557:99–107.

Biographical Sketch

Dragan M. Popović

Affiliation: Department of Chemistry, University of Belgrade – Institute of Chemistry, Technology and Metallurgy, National Institute of the Republic of Serbia, Belgrade, Serbia

Education:

2003-2010, University of California, Davis, USA, postdoctoral scholar and Assistant research professor in the field of biochemistry, biophysics, and theoretical/computational chemistry

1999-2002, Freie Universität (Free University) Berlin, Germany, a Ph.D. degree in Chemistry (*magna cum laude*); dissertation: "Modeling of conformation and redox potentials of hemes and other cofactors in proteins"

1988-1993, Graduated with a B.Sc. in Chemistry with honors (*GPA: 9.87/10.0*) from the Faculty of Chemistry, University of Belgrade, Serbia

Business Address: ICTM – University of Belgrade, Group for Molecular Modeling and Dynamics, Studentski Trg 12, 11000 Belgrade, Serbia

Research and Professional Experience: Associate research professor at the Institute of Chemistry, Technology and Metallurgy, University of Belgrade, since 2015. He was visiting professor teaching postgraduate students "Computational methods in biochemistry" at the University of the Philippines Diliman, Manila, in 2017.

Professional Appointments: Coordinator of the Group for Molecular Modeling and Bioinformatics. Associate editor of the journal Frontiers in Chemistry. Author of several chapters in international books.

Honors: Award of Serbian Chemical Society for the best-graduated student. Scholarship of honor to academic excellence. Professional recognition: research fellowships from the German Research Foundation (Deutsche Forschungsgemeinschaft), DFG (SFB 498 A5 and GRK 268).

Publications from the Last 3 Years:

1. Protić-Rosić, I., Nešić, A., Lukić, I., Miljković, R., Popović, D. M., Atanasković-Marković, M., Stojanović, M. and Gavrović-Jankulović M. (2021). Recombinant Bet v 1-BanLec chimera modulates functional characteristics of peritoneal murine macrophages by promoting IL-10 secretion, *Molecular Immunology* 138: 58-67.
2. Popović, D. M. (2021). Photolyase – molecular mechanism for repair of UV-damaged DNA. *Hemijski Pregled* 62: 50-62.
3. Lopandić, Z., Dragačević, L., Popović, D. M., Andjelković, U., Minić, R., Gavrović-Jankulović, M. D. (2021). BanLec-eGFP chimera as a tool for evaluation of lectin binding to high-mannose glycans on microorganisms, *Biomolecules* 11: 180, pp 1-13.
4. Popović, D. M. and Djordjević, I. S. (2020). Catalytic center of cytochrome *c* oxidase: Effects of protein environment on p*Ka* values of CuB histidine ligands. *Journal of Serbian Chemical Society* 85: 1429-1444.
5. Lazić, A. M., Djordjević, I. S., Radovanović, L. D., Popović, D. M., Rogan, J. R., Janjić, G. V. and Trisović, N. P. (2020). Self-assembly and biorecognition of a spirohydantoin derived from α-tetralone: Interplay between chirality and intermolecular interactions. *ChemPlusChem* 85: 1220-1232.
6. Janjić, G. V., Jelić, S. T., Trisović, N. P., Popović, D. M., Djordjević, I. S. and Milčić, M. K. (2020). New theoretical insight into fluorination and fluorine-fluorine interactions as a driving force in crystal structures. *Crystal Growth & Design* 20: 2943-2951.

Chapter 2

Detection of Amyloid Fibrils Using Unsymmetric Cationic Cyanine Dyes

O. Zhytniakivska[1,*], PhD, A. Kurutos[2,3], PhD, U. Tarabara[1], PhD, K. Vus[1], PhD, V. Trusova[1], PhD and G. Gorbenko[1], PhD

[1]Department of Medical Physics and Biomedical Nanotechnologies, V. N. Karazin Kharkiv National University, Kharkiv, Ukraine
[2]Institute of Organic Chemistry with Centre of Phytochemistry, Bulgarian Academy of Sciences, Sofia, Bulgaria
[3]Department of Pharmaceutical and Applied Organic Chemistry, Sofia University "St. Kliment Ohridski," Sofia, Bulgaria

Abstract

Amyloid fibrils, β-sheet-rich linear protein aggregates formed by a large variety of proteins and peptides, have received much interest in the recent decades due to their involvement in a range of human disorders among which are neurodegenerative diseases, type II diabetes, systemic amyloidoses, etc. One of the simplest and inexpensive methods for the identification and characterization of amyloid fibrils *in vitro* and the visualization of intracellular and intercellular amyloid inclusions *in vivo* is based on the use of fluorescent dyes sensitive to the beta-pleated motifs. Several classes of fluorescent compounds, such as near-infrared fluorescent DANIR probes, benzanthrone dyes, curcumine-based fluorescence imaging probes, cyanine, BODIPY and styryl derivatives have been successfully employed as amyloid sensors along with classical amyloid markers thioflavin T and Congo Red. This contribution is focused on the application of cyanine dyes to differentiating between the

* Corresponding Author's Email: olga.zhytniakivska@karazin.ua.

In: Advances in Biology. Volume 2
Editor: Charles D. Grant
ISBN: 979-8-88697-422-5

fibrillar and monomeric states of protein molecules. More specifically, a series of novel unsymmetric trimethine cyanine dyes have been tested for their ability to identify and characterize amyloid fibrils of insulin and lysozyme. The spectral properties of the trimethines were examined in a free state and in the presence of non-fibrillized and fibrillar proteins. The fluorescence studies showed that the examined trimethines have negligible emission in the buffer solution as well as in the presence of the non-fibrillized insulin and lysozyme due to their H-type self-association. However, the binding of the examined cyanines to amyloid fibrils was followed by a significant increase of their fluorescence, with the magnitude of this effect being dependent on the dye chemical structure. The quantitative parameters of the trimethine association with fibrillar proteins have been derived from the results of fluorimetric titration. The spectral properties and the binding characteristics of cyanines proved to be different for the lysozyme and insulin amyloid fibrils, highlighting the sensitivity of these dyes to the distinctions in fibril morphology. The molecular docking studies revealed the possible binding sites, as well as the nature of interactions involved in the dye–protein complexation in the monomeric protein and amyloid assemblies. The possibility of differentiating between the non-fibrillized and fibrillar states of insulin using the Förster resonance energy transfer between Thioflavin T as a donor and one of a series of trimethine cyanine dyes as an acceptor has been demonstrated. By comparing the dye preference to the fibrillar protein aggregates, the most prospective amyloid probes have been selected. Overall, our findings demonstrate that the novel usymmetric trimethine cyanine dyes are prospective amyloid tracers capable of operating in both fluorescence and absorption modes.

Keywords: amyloid fibrils, trimethine cyanine dyes, amyloid tracers, insulin, lysozyme.

Introduction

The aggregation of proteins and peptides into their filamentous states, amyloid fibrils, is associated with a variety of human disorders, including Alzheimer's, Parkinson's, Huntington's diseases, type II diabetes, rheumatoid arthritis, cardiac arrhythmias and systemic amyloidosis among others [1-5]. Considering the crucial link of amyloid fibrils with these debilitating disorders, detection of amyloid aggregates, along with understanding of their structure, molecular mechanisms underlying fibril formation and cytotoxicity of amyloid species are of paramount diagnostic and therapeutic importance [6-

9]. Apart from this, a unique cross-β-sheet structure of amyloid aggregates in which β-strands run transversely to the main fiber axis makes amyloid fibrils considerably attractive as components of biosensors and hydrogels for drug delivery in nanotechnology and biomedicine [10-12]. In recent years, different microscopic and spectroscopic techniques have been used to investigate the protein fibrillation mechanism and to characterize the structure of amyloid fibrils, including electron microscopy [13-16], atomic-force microscopy [17-20], fluorescence microscopy [21-25], dynamic light scattering [26-28], circular dichroism spectroscopy [29-31], stimulated emission depletion nanoscopy [22, 32], X-ray scattering [33, 34], Fourier transform infrared spectroscopy [30, 35], fluorescent spectroscopy [36-46], positron emission tomography [47, 48], and so forth. Among them, fluorescence approaches with the use of optical probes sensitive to a beta-pleated sheet structure are particularly advantageous, owing to their simplicity, high sensitivity, low cost, and rapidity [25, 36-46]. In recent years, the classical amyloid markers Thioflavin T and Congo Red have generally been employed as amyloid markers [36, 47]. Among other prospective fluorescent dyes demonstrating excellent amyloid-sensing potential are the polarity-sensitive fluorescent dye Nile red [46], the ultrafast molecular rotor Auramine O [37], DCVJ [40], near-infrared fluorescence DANIR probe [48], dyes based on intermolecular charge transfer (benzanthrone dyes [42], styryl fluorophores [43], BODIPY dyes [38]), Michler' s hydrol blue [41], β-ketoenole derivatives [39], naphthalenes [44] oligothiophenes [45], the list is far from being exhaustive. Along with this, numerous studies demonstrated that dyes of the cyanine family are particularly effective for detection of amyloid fibrils [23, 24, 49-55]. More specifically, Yang et al. showed that a hemicyanine probe ZT-1 is a promising agent for *in vitro* detection and *in vivo* fluorescent imaging of Aβ plaques in the Alzheimer's disease (AD) brains [23]. Wang and coworkers demonstrated the potential of the cyanine-conjugated gadolinium(III) complex for one- and two-photon excited fluorescence and magnetic resonance imaging of Aβ in transgenic mouse models of AD [49]. Moreover, it was shown that this cyanine-based dyad can inhibit Aβ aggregation and suppress Aβ induced production of reactive oxygen species [49]. Likewise, the neutral merocyanine dyes have been reported to be especially promising for *in vivo* fluorescence imaging of amyloid β-plaques [50]. In addition, Mu et al. have synthesized a cyanine-derived molecular rotor for ratiometric imaging of amyloid aggregates [24]. Moreover, a novel cyanine-based dye Cy-aa proved to be more effective for monitoring the fibrillation kinetics of amyloid paired helical filaments (AcPHF6) relative to ThT [51]. The higher amyloid-sensing potential in comparison with the classical amyloid markers ThT and CR was also demonstrated for pinocyanol chromophore [52] and trimethine cyanine dyes [53]. In addition, a very strong fluorescence enhancement when bound to amyloid fibrils was observed also for the cyanine dye Thiazole Orange [54]. Moreover, numerous studies indicate that cyanine derivatives are capable of preventing and modulating protein fibrillazation [55-58]. Specifically, mono-, tri-, penta- and heptamethine cyanine platforms have been reported to inhibit the insulin fibril formation under physiological conditions *in vitro* at low dye-protein ratios [55]. Besides, the phthalocyanine dyes have been shown to modify the insulin fibril formation affecting the morphology of the formed fibrils [57, 58]. Remarkably, the potency of phthalocyanines to interfere with protein fibrillization correlates with their tendency to self-aggregation in solution [57]. The aforementioned applications of cyanine dyes for detection and inhibition of amyloid fibrils are mainly dictated by their advantages such as: i) long wavelength absorption and emission maxima [23, 24]; ii) a significant enhancement of fluorescence emission upon binding to amyloid fibrils [24, 50-53]; iii) the characteristic fluorescence turn-on mechanism in the presence of aggregated proteins [54, 59]; iv) the ability to form self-assembled supramolecular complexes [55, 57-59].

During the past decade our research efforts have been directed to: i) searching effective and highly sensitive amyloid markers and fibrillation inhibitors [42, 55, 59, 60]; and ii) the development of the novel fluorescence-based strategies for detection of amyloid fibrils [61-64]. Herein, we tested the ability of a series of novel unsymmetric trimethine cyanine dyes to identify and characterize amyloid fibrils. More specifically, we described the photophysical properties of the cyanine dyes in the buffer solution and in the presence of nonfibrillar and aggregated insulin and lysozyme. The binding parameters of the fluorophores were calculated and the possible binding sites of cyanines were identified. The obtained results showed that trimethine cyanine dyes are prospective amyloid probes capable of operating in both fluorescence and absorption modes.

Materials and Methods

Experimantal

Bovine insulin, hen egg yolk lysozyme and thioflavin T (ThT) were purchased from Sigma, USA. The trimethine cyanine dyes (Figure 1) were synthesized and purified in the University of Sofia, Bulgaria, as described previously [59]. All other starting materials and solvents were commercial products of analytical grade and were used without further purification.

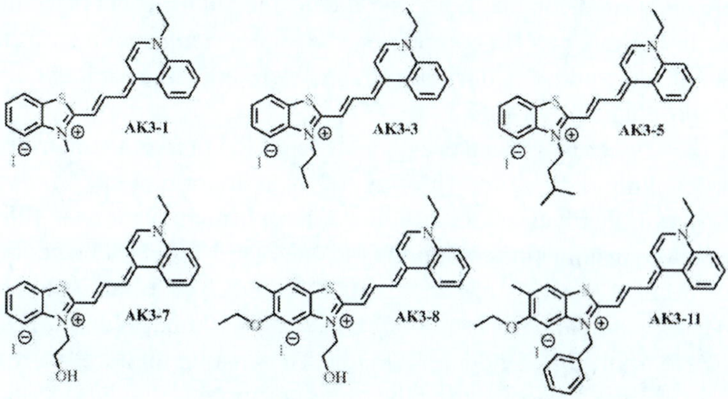

Figure 1. Chemical structures of the thrimethine cyanine dyes.

Preparation of Working Solutions

The stock solutions of the trimethine cyanines were prepared immediately before measurements by dissolving the dyes in dimethyl sulfoxide. The ThT stock solution was prepared in 5 mM sodium phosphate buffer, pH 7.4. The concentrations of the dyes were determined spectrophotometrically, using their extinction coefficients [59]. The insulin and lysozyme stock solutions (10 mg/ml) were prepared by dissolving the proteins in 10 mM glycine buffer (pH 2.0). These solutions were used as references for non-aggregated proteins. Hereafter, the non-fibrillar insulin and lysozyme forms are designated as InsN and LzN, respectively.

The insulin amyloid fibrils were obtained *in vitro* by the protein incubation in glycine buffer (pH 2) at 37ºC under continuous shaking for 6 days [65]. The amyloid fibrils of lysozyme were prepared by the protein incubation in 10 mM glycine buffer at pH 2 and 60°C for 14 days. The amyloid nature of the protein aggregates was confirmed by ThT assay and the transmission electron microscopy. Hereafter, the fibrillar insulin and lysozyme are designated as InsF and LzF, respectively. The working solutions of proteins were prepared by dissolving the stock solutions of the non-fibrillar and fibrillar insulin or lysozyme in 5 mM sodium phosphate buffer (pH 7.4). To measure the absorption spectra of cyanine dyes in the aqueous phase or in the complexes with proteins, appropriate amounts of the stock dye solutions in DMSO were added either to 5 mM sodium phosphate buffer (pH 7.4) or directly to the working solutions of the non-aggregated and fibrillar proteins, and were incubated for one hour. To record the fluorescence spectra of the cyanines in the dye-protein complexes, the appropriate amounts of the stock solutions of the control or fibrillar proteins were added to each dye in 5 mM sodium phosphate buffer (pH 7.4).

For the Förster resonance energy transfer (FRET) measurements between the classical amyloid marker Thioflavin T as a donor and one of a series of trimethine cyanine dyes as an acceptor, 1 cm plastic cuvette was filled with ThT (4.5 μM), insulin stock solution (54.5 μM) and Tris-HCl buffer (10 mM, pH 7.4, 150 mM NaCl, 1 mM EDTA, 0.01% NaN_3) to a final volume of 1.6 ml. The insulin amyloid fibrils were obtained at 50 °C under continuous orbital rotation (155 rpm) for 18 hours. Then, the ThT-insulin mixtures were titrated with the cyanine dyes. Specifically, the appropriate amounts of the stock solutions of the dyes in DMSO were added to 10 mM Tris-HCl buffer containing ThT and non-aggregated / fibrillar insulin and incubated for 5 minutes.

Spectroscopic Measurements

The absorption spectra of the dyes were acquired on Shimadzu UV-2600 Spectrophotometer (Japan) at 25°C using 10-mm path-length quartz cuvettes. The steady-state fluorescence spectra were collected on a Shimadzu RF-6000 Fluorescence Spectrometer or on a FL-6500 spectrofluorimeter (Perkin-Elmer Ltd., Beaconsfield, UK) at 20°C using 10 mm path-length quartz cells. Emission spectra were recorded in buffer and in the presence of non-aggregated or fibrillar proteins with excitation wavelength 590 nm. The excitation and emission slit widths were set at 10 nm. In the FRET measurements, the fluorescence spectra were recorded from 460 to 800 nm with the excitation wavelength 440 nm. Excitation and emission bandpasses were set at 10 nm.

Deconvolution Analysis

The deconvolution of the dye absorption spectra was performed in the Origin 9.0 (OriginLab Corporation, Northampton, USA) using the log-normal asymmetric function (LN) [66]:

$$A = A_{\max} \exp\left[-\frac{ln2}{\ln^2(\rho)} \ln^2\left(\frac{a-v}{a-v_c}\right) \right] \quad (1)$$

where A denominates the absorbance, $A_{\max}$ v is the wavenumber, v_c ρ expresses the asymmetry of the function defined as:

$$\rho = \frac{v_c - v_{\min}}{v_{\max} - v_c} \quad (2)$$

where $v_{\min}$ and $v_{\max}$
The parameter a designates the limiting wavenumber:

$$a = v_c + \frac{(v_{\max} - v_{\min}) \cdot \rho}{\rho^2 - 1} \quad (3)$$

Binding Model

The quantitative characteristics of the dye-protein complexation were determined in terms of the Langmuir adsorption model by analyzing the protein-induced changes in the probe fluorescence intensity at the wavelengths corresponding to the emission maximum of a certain dye [67]. Briefly, the approximation of the experimental dependencies ΔI
increase) on C_P
the dye-protein binding parameters – association constant (K_a
stoichiometry (n) and molar fluorescence (α), characterizing the difference between the molar fluorescences of the bound and free dye:

$$\Delta I = 0.5\alpha[Z_0 + nC_P + 1/K_a - \sqrt{(Z_0 + nC_P + 1/K_a)^2 - 4nC_P Z_0})] \quad (4)$$

where Z_0

Docking Studies

To define the most energetically favorable binding sites for the examined dyes on the insulin amyloid fibrils, the molecular docking studies were performed using the the AutoDock (version 4.2) incorporated in the PyRx software (version 0.8) [68]. To further characterize the nature of the dye-protein interactions, the protein-ligand interaction profiler PLIP was used [69]. The crystal structures of hen egg white lysozyme (PDB ID: 3A8Z) and bovine insulin (PDB ID: 2ZP6) were taken from the Protein Data Bank. The lysozyme model fibril was built from the K-peptide, GILQINSRW (residues 54–62 of the wild-type protein), using the CreateFibril tool as described previously [70]. The model of the human insulin fibril was obtained from http://people.mbi.ucla.edu/sawaya/jmol/fibrilmodels/. The dye structures were built in MarvinSketch (version 18.10.0) and optimized in Avogadro (version 1.1.0) [71, 72]. The selected docking poses were visualized with the UCSF Chimera software (version 1.14) [73].

Result and Discussion

Fluorescence Response of Trimethine Cyanine Dyes in the Buffer Solution and in the Presence of Non-Fibrillar and Fibrillar Proteins

To explore the interaction of the novel trimethine cyanine dye with proteins in non-fibrillar and amyloid states, at the first step of our study the excitation and emission spectra of the examined trimethine compounds were measured for the free dyes in buffer and after addition of the control (non-fibrillar) and fibrillar insulin and lysozyme. Insulin and lysozyme were selected as model proteins since they are well-suited to test the robustness of amyloid sensing and differ in their charge and hydrophobicity. The representative fluorescence spectra of trimethines in the unbound state and in the presence of the native and fibrillar proteins are presented in Figure 2 and spectral characteristics are summarized in Table 1. The trimethines AK3-1, AK3-3, AK3-5 and AK3-7 were found to exhibit a negligible fluorescence in buffer solution with the excitation maxima centered in the range 626-631 nm and the emission maxima located between 640 and 651 nm, depending on the dye chemical structure. The corresponding Stokes shifts between the excitation and emission maxima were found to be rather small, which is typical for cyanine dyes [19, 20, 36].

The trimethines AK3-8 and AK3-11 are characterized by a negligible fluorescence intensity in buffer and do not show clear emission and excitation maxima. Allowing for a high probability of Coulombic interactions, the cationic cyanine dyes are expected to exhibit favorable binding to the negatively charged insulin monomers at neutral pH. However, no significant enhancement in fluorescence was observed (Table 1).

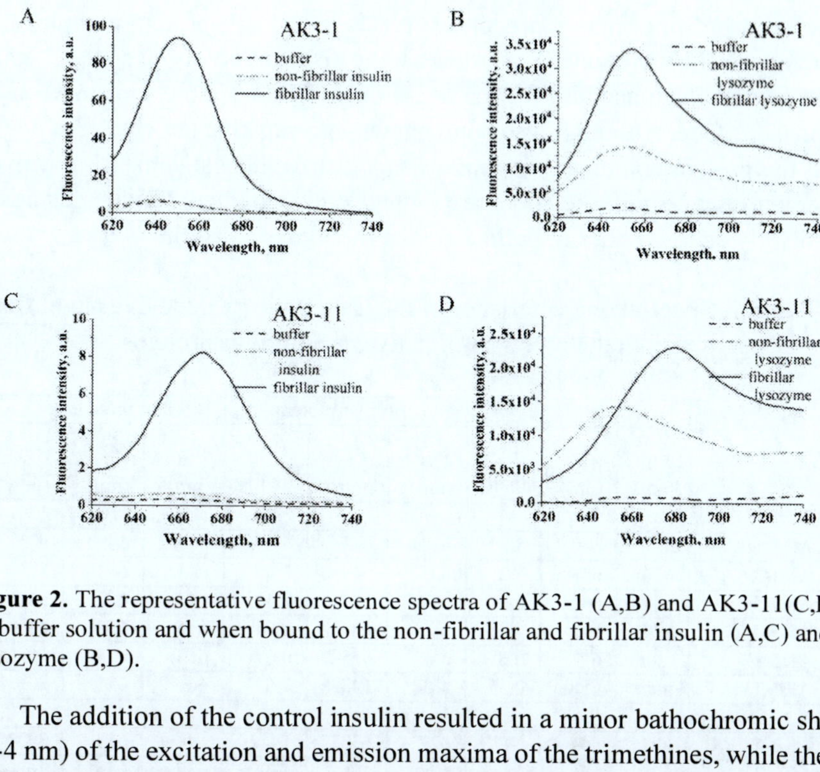

Figure 2. The representative fluorescence spectra of AK3-1 (A,B) and AK3-11(C,D) in buffer solution and when bound to the non-fibrillar and fibrillar insulin (A,C) and lysozyme (B,D).

The addition of the control insulin resulted in a minor bathochromic shift (1-4 nm) of the excitation and emission maxima of the trimethines, while their fluorescence intensity remained almost unchanged. Furthermore, the alterations in the spectral responses of the trimethines in the presence of native lysozyme appeared to be more pronounced than those observed for non-fibrillar insulin. In particular, the addition of the control lysozyme to the AK31, AK33, AK38 and AK3-11 in the buffer solution was accompanied by a significant enhancement of the dye fluorescence intensity along with a slight bathochromic shift (1-6 nm) of the excitation and emission maxima. Such behavior may be a consequence of a higher affinity of trimethines for lysozyme monomers in comparison with insulin. In turn, the association of AK3-5 and AK3-7 with LzN was followed by a slight fluorescence increase coupled with a ~ 70-nm shift of the dye excitation maxima to the shorter wavelengths. Simultaneously, a ~ 15 nm bathochromic shift of the AK3-5 and AK3-7 emission maxima was observed in the presence of non-fibrillar lysozyme.

Moreover, taking into account the positive charge of lysozyme, the negative charge of insulin at neutral pH [74, 75], and the positive charge of the studied trimethyne compounds, we can hypothesize that electrostatic interactions do not play a predominant role in the dye-protein complexation with nonfibrillar proteins. Nevertheless, the distribution of the charged amino acid residues can account for the anchoring of fluorophore molecule on the protein surface. Although lysozyme possesses a positive net charge at neutral pH, the protein contains 9 negatively charged residues against 17 positively charged ones. Similarly, there are 4 negatively charged residues against 2 positively charged ones in both chains of monomeric insulin.

Table 1. Spectral characteristics of the trimethine cyanine dyes in buffer and in the presence of native and fibrillar proteins

Dye	Free dye			Non-fibrillar protein			Fibrillar protein		
	λ_{ex} (nm)	λ_{em} (nm)	I_0, (a.u.)	λ_{ex} (nm)	λ_{em} (nm)	I_n, (a.u.)c	λ_{ex} (nm)	λ_{em} (nm)	I_f, (a.u.)
				Insulin					
AK3-1	627	640	5.8*	631	643	6.6	640	649	92
AK3-3	626	643	5.2	631	644	7.6	644	652	89.6
AK3-5	631	642	5.0	632	645	7.4	642	652	91.2
AK3-7	628	642	7.25	636	645	6.84	646	653	59.4
AK3-8	n.d.[a]	651	0.6	n.d.[a]	655	0.67	663	673	6.64
AK3-11	n.d.	658	0.37	n.d.[a]	643	0.62	662	671	8.26
				Lysozyme					
AK3-1	627	644	3470	627	653	14420	640	654	34701
AK3-3	626	646	1827	634	656	13928	637	656	16066
AK3-5	631	645	7423	559	661	8586	635	655	63481
AK3-7	628	642	2052	559	661	9364	634	654	33768
AK3-8	n.d.[a]	651	2212	625	650	19945	668	676	17232
AK3-11	n.d.	658	1070	630	656	14832	668	677	23654

λ_{ex} – maximum of the excitation spectra; λ_{em} –maximum of the emission spectra; n.d.[a] – not determined because of negligible fluorescence intensity; I_0, I_n, and I_f – fluorescence intensity of the dyes in buffer and in the presence of control and fibrillar proteins, respectively. The concentration of nonfibrillar and fibrillar proteins was 8 μM. The dye concentration was 0,5 μM. *The steady-state fluorescence spectra in the presence of non-fibrillar and fibrillar insulin were collected on a Shimadzu RF-6000 Fluorescence Spectrometer, whereas the dye spectra in the presence of lysozyme were obtained using a FL-6500 spectrofluorimeter (Perkin-Elmer Ltd., Beaconsfield, UK).

As shown in Table 1, the spectral response of the trimethine dyes (excepting AK3-8 in the presence of fibrillar lysozyme) to the fibrillar proteins lies in a strong increase of the dye fluorescence (I_f) as opposed to that in buffer (I_0) and in the presence of nonfibrillar protein (I_n
of the fluorescence intensity increase being more pronounced in the presence of the insulin amyloid fibrils. The fluorescence maxima of the trimethines are shifted by approximately 9-22 nm (depending on the dye chemical structure) towards longer wavelengths compared to those observed either in the non-aggregated proteins or in the buffer solution, as illustrated in Table 1.

It appeared that AK3-8 demonstrates a higher affinity to the nonaggregated lysozyme in comparison with the fibrillar protein, since the higher fluorescence increase was observed in the presence of LzN. It is worth noting that the emission spectra for all trimethine dyes bound to the lysozyme fibrils were red shifted by 4-8 nm (depending on the dye structure) in comparison with those in the presence of the fibrillar insulin. Notably, the sensitivity of the dye spectral properties to the changes in the protein environment was previously reported for Michler's hydrol blue, ThT derivatives and Nile red [41, 46, 76]. The above red shifts in the position of the trimethine fluorescence maxima cannot be explained by the sensitivity of the examined dyes to the polarity of their environment. Notably, the emission maxima of lysozyme incorporated ThT derivatives and Nile red were shifted to the shorter wavelengths indicating about the lower polarity of lysozyme fibrillar aggregates [46, 76]. Likewise, it was reported previously that the surface hydrophobicity of lysozyme fibrils appeared to be higher than that of insulin aggregates [76, 77]. Specifically, the fluorescent probe 1,8-ANS showed the blue shift of the emission maximum *ca.* 46 (51) nm in the presence of insulin (lysozyme) amyloid fibrils which can be attributed to the presence of nonpolar binding sites for the probes (the solvent-exposed hydrophobic patches) on the fibril surface [77].

To determine the specificity of the examined trimethines to amyloid fibrils, the fluorescence spectral data were processed to the amyloid detection factor (ADF) (Figure 3) characterizing the ability of a dye to selectively detect the fibrillar state over its native structure relative to the background fluorescence of the dye in buffer [59, 78]:

$$ADF = \frac{I_f - I_n}{I_0} \tag{5}$$

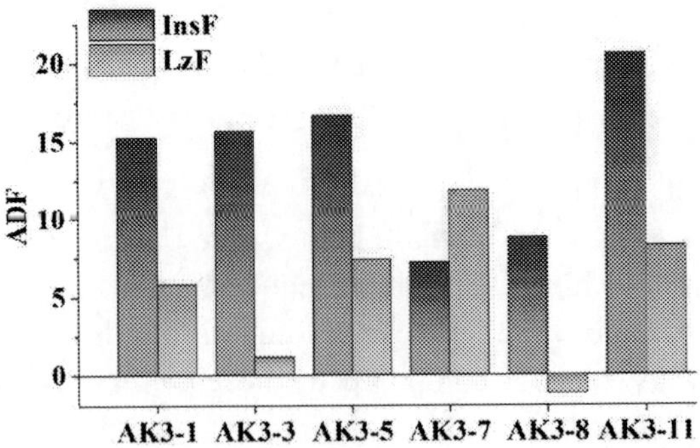

Figure 3. ADF values of the cyanine dyes for insulin and lysozyme amyloid fibrils. Dye and protein concentrations were 0.5 μM and 8 μM, respectively.

It appeared that the trimethines under study (excepting the system AK3-8 + lysozyme amyloid fibrils) are characterized by the positive ADF values in the presence of the fibrillar insulin and lysozyme, being indicative of their higher sensitivity to the fibrillar protein aggregates compared to the non-aggregated state. Although the enhanced fluorescence was observed for AK3-8 in the presence of both LzN and LzF, the magnitude of fluorescence intensity increase was higher in the presence of monomeric lysozyme, as judged from the negative amyloid detection factor for this dye.

The ADF values for the aggregated insulin were found to rise in the order AK3-11→AK3-5 → AK3-3 → AK3-1 → AK3-8→ AK3-7, indicating the decrease of insulin amyloid selectivity from AK3-11 to AK3-7. The AK3-11 dye possessing the largest ADF value of 20.6 appeared to be the most sensitive and selective sensor for the insulin amyloid fibrils among the screened compounds. The relatively high ADF values in the presence of InsF (15.3-16.7) were also observed for AK3-3, AK3-5 and AK3-1 bearing pure aliphatic alkyl substitutions on the *N*-alkyl fragment of the benzothiazolic moiety. In contrast, the presence of aliphatic OH-groups of the aliphatic substituents on the nitrogen atom of AK3-7 and AK3-8 turned out to significantly reduce the insulin amyloid sensing ability (ADF = 7.2-8.8) of the trimethines. Besides, in the presence of lysozyme fibrils the amyloid specificity was found to decrease in the row AK3-7 → AK3-11→ AK3-5 → AK3-1 → AK3-3→ AK3-8. It appeared that AK3-7, possessing the lowest sensitivity to InsF, has demonstrated the largest ADF value in the presence of the lysozyme amyloid fibrils. Similarly to the insulin amyloid fibrils, the relatively high amyloid detection factors were observed in the presence of lysozyme for AK3-11 and AK3-5. Remarkably, the ADF values of the trimethines appeared to be significantly higher in the presence of insulin amyloid fibrils compared to lysozyme, suggesting a sensitivity of the examined cyanines to the fibril morphology. Although a β-*pleated* sheet is the common motif of the fibrillar lysozyme and insulin, there are distinctions in the structure of the filaments formed by these proteins. It should be noted that the lysozyme amyloid fibrils formed under acidic condition and heating (10 mg/ml, pH 2, incubation at 60°C, no agitation) had the diameter *ca.* 7–20 nm and consisted of 2–3 filaments [79,80]. In turn, the insulin fibrils prepared at pH 2 under agitation, were *ca.* 20–100 nm in diameter and were composed of 2–6 filaments [81, 82]. Moreover, the bending ridigity (the physical characteristic of amyloid fibrils which depends on a generic inter-backbone hydrogen-bonding network) was equal to *ca.* ~ .9.1 × 10^{-26} and 4 × 10^{-26} N·m^2 for the insulin and lysozyme amyloid fibrils, respectively [83]. In view of this, it could be assumed that the lysozyme fibrils possess lower rigidity than the insulin amyloid aggregates due to a positive correlation between the bending ridigity and the number of fibril filaments [83]. Moreover, the packaging of longer lysozyme polypeptide chain into the cross-β amyloid core could lead to the additional constraints in the fibril packing [83].

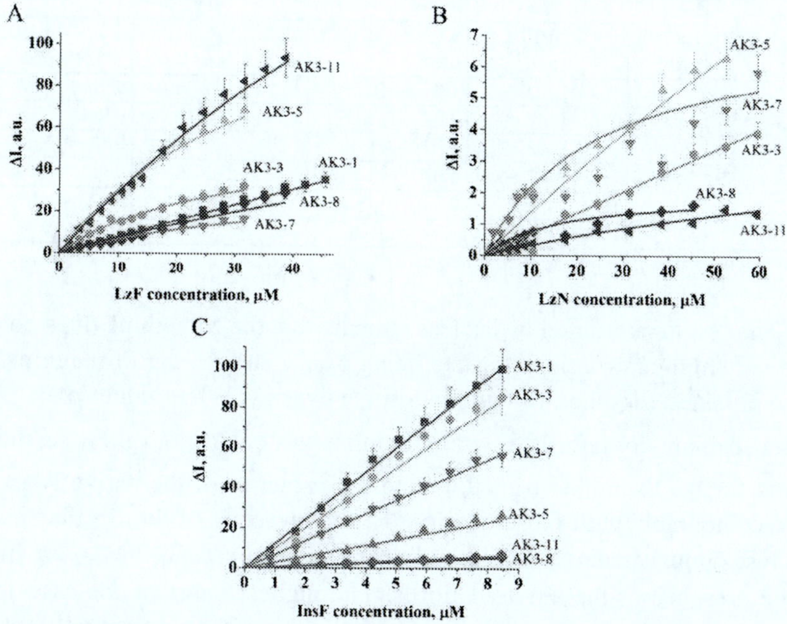

Figure 4. The isotherms of the cyanine binding to LzF (A), LzN (B) and InsF (C). The dye concentration was 0.5 μM.

Next, to derive the quantitative parameters of the dye-protein binding, the fluorescence spectra were recorded upon titration of cyanine dyes with proteins in the globular and amyloid states (Figure 4). The obtained experimental dependencies of the fluorescence intensity increase (ΔI) *vs* protein concentration (C_p) (Figure 4) were approximated by the Langmuir adsorption model (Eq. 4). Notably, we failed to obtain the binding isotherms for the cyanine dyes in the presence of non-fibrillar insulin since the fluorescence intensity in the presence of InsN remained almost unaffected.

The quantitative analysis of the dependencies $\Delta I(C_p)$ allowed us to estimate the parameters of the trimethine dye binding to the native and fibrillar proteins – the association constant (K_a), the binding stoichiometry (n) and the difference between the molar fluorescence of the bound and free dye (α) (Table 2).

Table 2. Binding characteristics of cyanine dyes in the presence of the non-aggregated and fibrillar proteins

Dye	LzN			LzF			InsF		
	$Ka \cdot 10^3$, μM^{-1}	n	$\alpha \cdot 10^{-1}$, μM^{-1}	$Ka \cdot 10^3$, μM^{-1}	n	$\alpha \cdot 10^{-3}$, μM^{-1}	$Ka \cdot 10^3$, μM^{-1}	n	$\alpha \cdot 10^{-3}$, μM^{-1}
AK3-1	n.d.a	n.d.	n.d.	1.07	0.50	15.84	1.08	0.98	9.64
AK3-3	0.43	0.99	147.69	2.98	1.06	1.18	0.72	0.96	27.57
AK3-5	2.02	0.88	103.62	6.87	1.00	1.11	2.55	0.51	138.07
AK3-7	1.58	1.47	182.26	4.65	1.04	1.83	0.20	0.98	70.18
AK3-8	1.24	1.39	4.35	0.34	1.05	0.06	0.03	0.94	44.75
AK3-11	12.73	1.0	1.72	1.34	1.02	2.95	3.55	1.02	0.42

n.d.a – not determined.

The results presented in Table 2 indicate that the trimethine dyes possess relatively high association constant being by the same order of magnitude as those obtained for other trimethine cyanine dyes [84]. The values of n in the presence of amyloid fibrils were approximately equal to 1.0. Likewise, the K_a values for the examined trimethines in the presence of the native lysozyme were comparable with those obtained in the presence of the fibrillar insulin and lysozyme, whereas the values of the molar fluorescence (α) were higher in the presence of the amyloid fibrils. The highest value of the association constant in the presence of fibrillized insulin was observed for AK3-11 and AK3-5, whereas the largest binding preference to the lysozyme fibrillar aggregates was observed for AK3-3, AK3-5 and AK3-7. Recently it was shown that the nature of *N,N′* -substituents in the trimethine structure could strongly affect the fluorescence response and the binding preference of these dyes in the presence of fibrillar protein [59, 84]. Specifically, Kuperman et al. demonstrated that the long-chain alkyl groups, hydroxyalkyl and phenylalkyl groups as *N,N′* -substituents, lead to the increase in the fluorescence sensitivity of the dye molecule to the lysozyme fibrils [84]. Interestingly, the higher ADF value and the relatively high association constant in the presence of LzF was observed for AK3-7 bearing long aliphatic alkyl substitutions on the *N*-alkyl fragment. In turn, the trimetine dye AK3-8 having the same aliphatic alkyl substitutions on the *N*-alkyl fragment demonstrated the lower binding preference to both insulin and lysozyme amyloid fibrils. A possible reason for the observed effect is a carboxyl group in the AK-38 structure, most probably introducing the constraints in the dye binding ability. Notably, the *N,N′* -substituents in the trimethine structure can exert influence on the geometry of the dyes and on the amplitude of the fluorophore rotation, thus providing better fitting of the dye geometry to the amyloid fibril architecture. Moreover, the *N,N′* -substituents in the cyanine structures can serve as the anchors for the dye binding to proteins in the native and amyloid states. In both cases, the fluorophore binding ability depends on its environment, namely the availability of the interaction-able dye groups near the binding sites. Furthermore, cyanine dyes are known to possess a very high polarizability of the π-electron system along the polymethine group in the ground state, which gives rise to strong dispersion forces between two cyanine molecules in solution leading to the formation of extended dye aggregates [85]. Importantly, the changes in the cyanine environmental conditions greatly affect the cyanine aggregation [60, 85-87]. Numerous studies indicate that proteins in the native and amyloid states could serve as a scaffold for cyanine aggregation or disaggregation [59, 60, 86]. The structure of the dye, particularly, the presence of substituents in heterocyclic residues and the N-alkyl chain length can affect the aggregation of cyanine dyes on a protein template [59, 60, 86]. The cyanine aggregation on the protein template introduces uncertainties in the interpretation of the binding data, since H-aggregates (two dipoles are in a side-to-side orientation) are non-fluorescent in nature and only binding of the dye monomers and J-aggregates can be traced by fluorescence spectra. In addition, it was demonstrated previously that ssociation of the heptamethine cyanine dyes with the fibrillar and native lysozyme involves the complexation of the dye monomers and dimers with the protein [60]. Specifically, the heptamethines showed high affinity for lysozyme fibrils and different aggregation extent in the presence of monomeric and fibrillar protein [60].

Spectroscopic Characterization of Trimethine Cyanine Dyes in Buffer Solution and in the Presence of Non-Fibrillar and Fibrillar Proteins

To interpret the observed fluorescence responses and the binding data of the cyanine dyes, in the following studies we analyzed their absorption spectra free in buffer solution as well as in the presence of control and fibrillar proteins (Figure 5).

With an exception of AK3-1, all examined trimethine dyes are characterized by one well-defined absorption maximum ranging from 637 to 658 nm in DMSO (shown in the insets of Figure 5). Besides, the absorption spectrum of AK3-1 was featured by a vibronic two-peak shape with a well-defined high-intensity absorption band corresponding to the 0-0 transition and a less intense sub-peak at the short-wavelength side of the band depicting 0-1 transition. However, in the aqueous medium the spectral behavior of the examined cyanines undergoes significant changes (Figure 5). In particular, AK3-1 and AK3-3 are characterized by a broad absorption spectrum with two well-defined peaks. The bathochromic peak (at 621 nm and 627 nm for AK3-1 and AK3-3, respectively) corresponds to the absorbance of monomeric dye species, while the hypsochromic peak (centered at 558 nm and 553 nm, respectively) is a hallmark of H-dimers. The intensity ratio of these bands was found to strongly depend on the dye chemical structure, but in all cases the monomer peak prevailed over that of the dye self-associates. Moreover, the shoulder around 510 nm in the absorption spectrum of AK3-1 can be regarded as an evidence for the presence of the higher order face to face staked aggregates. At the same time, the absorption spectra of AK3-5 and AK3-7 were featured by a one-peak structure with absorption maxima at 623 nm and 618 nm, in aqueous phase respectively, consistent with the solvatochromic behavior of many cyanine derivatives [53, 59, 88, 89]. This implies that monomers are the dominating species in the case of AK3-5 and AK3-7 in buffer. However, the absorption spectra of AK3-5 and AK3-7 were found to be broader compared to these in DMSO pointing to the presence of at least a small portion of H-dimers. In turn, AK3-8 and AK3-11 distinguished by ethoxy and methyl substituents on the benzothiazolic moiety, exhibited somewhat different behavior. The UV-Vis spectra of these chromophores were blue-shifted relative to the other dyes with the maximum centered around 523 nm and 560 nm, respectively.

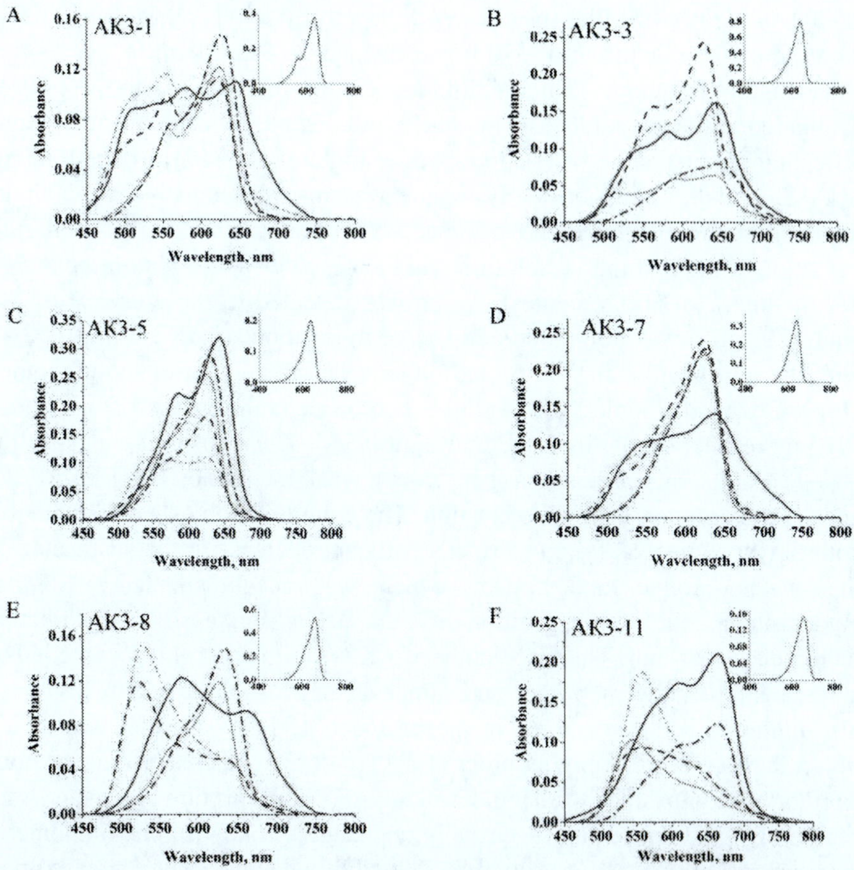

Figure 5. Absorption spectra of trimethine cyanine dyes in buffer solution (dashed blue line), in the presence of non-fibrillized insulin (dash dot dot orange line) and lysozyme (short dot dark-green line), and fibrillar insulin (solid violet line) and lysozyme (dash dot brown line). Shown in the insets are the absorption spectra of dyes in DMSO. The insulin, lysozyme and dye concentrations were 31.3 μM, 26.2 μM and 8 μM, respectively.

The peak of monomers was virtually unstructured and resembled the spectrum sub-peak, suggesting that H-aggregates dominate over other dye species in the buffer solution.

To test the efficacy of the trimethines in detecting amyloid protein aggregates, the absorption spectra of each dye were recorded after incubation with the control or fibrillar proteins. The changes produced by non-aggregated insulin and lysozyme in the absorption spectra of trimethines were found to differ from those observed in the buffer solution. Specifically, the binding of AK3-1 to the non-fibrillar insulin was followed by a hyperchromism – 20% increase of the maximum at 550 nm along with a 3-nm shift to the shorter wavelengths (Figure 3A). In addition, association of AK3-1 with the control insulin led to the appearance of the blue-shifted shoulder centered at 544 nm, reflecting the formation of H-aggregates of higher order. In turn, the binding of AK3-1 to the non-fibrillar lysozyme was manifested only in the slight hyperchromism of the dye monomer absorption band coupled with the reduction of the H-dimer absorption, but the shoulder with a maximum at 544 nm (hallmark of H-aggregates), which was detected in the presence of the control insulin, was not observed. The complexation of AK3-3 and AK3-7 with the non-fibrillar insulin brought about the attenuation of the monomer absorption band, coupled with a slight absorbance increase at ~522 nm and ~ 510 nm, respectively (Figure 3B and Figure 3D). These insignificant spectral changes can be interpreted as arising from a weak interaction between AK3-3 and AK3-7 and non-aggregated insulin. The complexation of AK3-3 with the control lysozyme was followed by a significant decrease in the absobance of the dye monomers and dimers, which was accompanied by~ 15 nm batochromic shift of the position of both peaks and a slight absorbance increase at ~ 510 nm. The association of AK3-7 with control lysozyme led to a weakening of the monomer and dimer absorption bands, comparable by magnitude with that observed in the presence of InsN. Besides, we didn't detect the formation of higher order H-aggregates in the presence of the non-fibrillar lysozyme since the shoulder in the AK-33 absorption spectrum at the region of ~ 510 nm remained virtually unchanged. Likewise, the monomeric sub-peak of AK3-5, AK3-11 and AK3-8 exhibited significant decrease upon formation of the dye-protein complexes in the presence of non-fibrillar insulin, and the contributions of the aggregated dye species dominated over the monomeric ones (Figure 3C, Figure 3E and Figure 3F). In turn, the binding of AK3-5 to the non-fibrillar lysozyme was followed by the significant absorbance increase of the dye monomer and dimer peaks coupled with a ~7 nm shift of the position of the monomer band. It should be noted that the changes in the absorption spectra of AK3-11 in the presence of the monomeric lysozyme were similar to those observed in the presence of control insulin: a significant decrease in the band of monomers, coupled with the increase in the peak of H-aggregates. In turn, the spectral responses of AK3-8 in the presence of non-fibrillar proteins were quite different. In particular, the absorbance at

626 nm corresponding to the dye monomers was increased four times and the dimer peak was expressed only by the shoulder at 561 nm. Furthermore, the contribution of the higher order H-aggregates, which were the dominating dye fraction in the buffer and in the presence of control insulin, was negligible.

While comparing the absorption spectra of the trimethine cyanine dyes in the presence of the non-aggregated and aggregated proteins, it becomes evident that the spectral changes induced by the fibrillar proteins are more drastic than those observed for the non-fibrillized counterparts. The absorption maxima of all bands in the presence of the fibrillar insulin are shifted by approximately 10-24 nm (depending on the dye chemical structure) towards longer wavelengths relative to those for the control insulin or buffer solution. The observed bathochromic shifts of the dye absorption maxima could presumably result from the decrease of polarity of the dye microenvironment in the presence of amyloid fibrils [11, 46]. This effect was accompanied by the absorbance increase of AK3-1 monomers (644 nm), dimers (576 nm) and H-aggregates (506 nm) in the presence of the fibrillar insulin, being much more pronounced than that brought about by the non-fibrillized insulin. In turn, the association of AK3-3 and AK3-7 with aggregated insulin was followed by the absorption spectrum broadening (more pronounced in the case of AK3-7), reflecting the formation of H-aggregates. In addition, the formation of AK3-7 complexes with fibrillar insulin led to the appearance of the red-shifted shoulder centered at 719 nm, which could be attributed to the dye J-aggregates formed on the fibrillar matrix. Moreover, the binding of AK3-5, AK3-8 and AK3-11 to the insulin fibrils was followed by ~2, 2.8 and 4-fold increases of the absorbance of monomers, dimers and H-aggregates, respectively. After incubation with fibrillar insulin, AK3-5 and AK3-11 monomers became a dominating dye fraction, whereas H-aggregates prevailed in the absorption spectrum of AK3-8. Similarly to InsF, the complexation of AK3-1, AK3-5, AK3-8, and AK 3-11 with the lysozyme fibrils was accompanied by the hyperchromism of the dye monomer (all dyes) and dimer (excepting AK3-1) absorption bands. In the meantime, the reduction of the dimer absorbance of AK3-1 was observed in the presence of the aggregated lysozyme. Notably, after the incubation with fibrillar lysozyme, AK3-8 and AK3-11 monomers became a dominating dye fraction, whereas H-aggregates (being a dominating fraction in the presence of non-fibrillar lysozyme and buffer) have minimal contribution to the total absorption spectrum. The complexation of AK3-3 and AK3-7 with the fibrillar lysozyme brought about the attenuation of the monomer and dimer absorption bands (more pronounced in the case of AK3-3). It is noteworthy that in the presence of the fibrillar lysozyme, neither the appearance of a long-wavelength shoulder located at 719 nm for AK3-7 in the InsF nor the broadening of the absorption spectrum of AK3-3 and AK3-7 were observed. Moreover, the absorption maxima of all bands in the presence of the fibrillar lysozyme are shifted only for 3-8 nm towards longer wavelengths relative to those for the monomeric lysozyme or buffer solution.

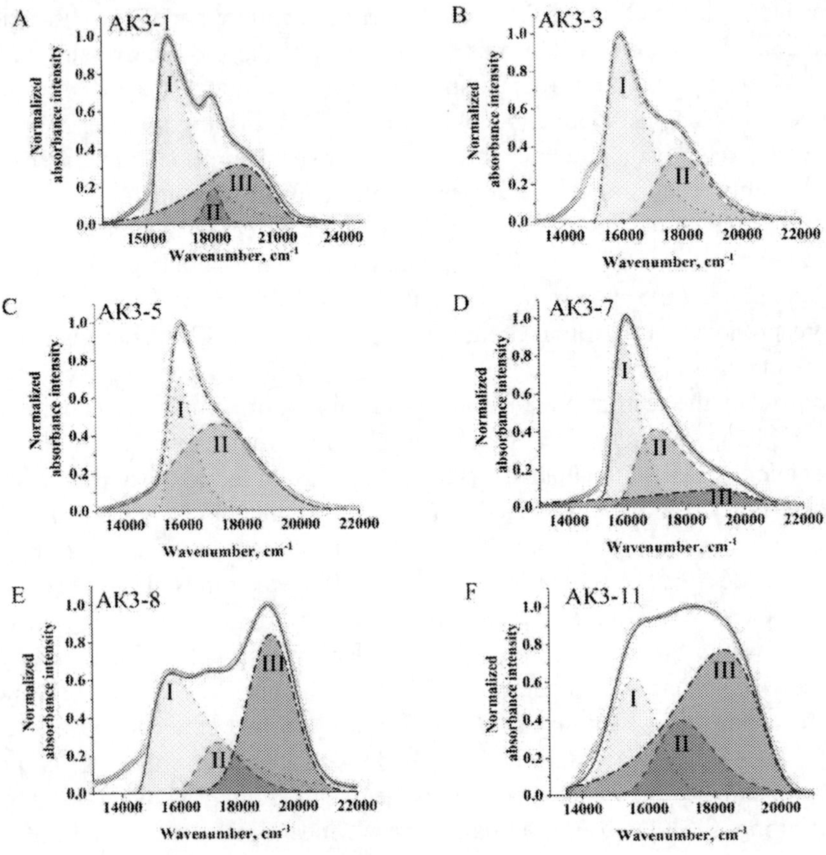

Figure 6. Deconvolution of the absorption spectra of trimethines in the buffer solution.

Next, to analyze the spectral behavior of the trimethine cyanines in the non-fibrillar or fibrillar proteins in more detail, we performed the decomposition of the dye absorption spectra using the log-normal asymmetric function (Eq. 1). The choice of LN function was dictated mainly by a highly asymmetric nature of the absorption spectra of the examined dyes. Recent studies indicate that in the case of asymmetry in a spectral shape, the decomposition procedure is more accurate with the log-normal function [90]. It appeared, that the absorption spectra of AK3-5 and AK3-3 in buffer solution can be represented as a sum of two separate bands, with a short-wavelength and long-wavelength spectral components presumably corresponding to the dimeric and monomeric dye species, respectively (Figure 6).

Table 3. Deconvolution parameters of the absorption spectra of trimethines in the buffer solution

Dye	Band	A_{max}	ν_c, cm^{-1}	FWHM, cm^{-1}	ρ, cm^{-1}
AK3-1	I	40000	16007	1767	0.42
	II	8500	18100	910	1.02
	III	14000	19300	3795	1.53
AK3-3	I	97000	15900	1422	0.54
	II	35171	17900	2000	0.81
AK3-5	I	50000	15900	1000	0.67
	II	33680	17200	3251	1.05
AK3-7	I	84168	15900	1000	0.67
	II	39000	17000	2159	0.59
	III	8000	19000	4803	2.66
AK3-8	I	20739	15700	2347	0.45
	II	8816	17300	1750	0.75
	III	27500	19100	1832	1.00
AK3-11	I	15353	15550	1582	0.91
	II	9711	17000	2027	0.76
	III	18934	18300	2935	1.54

The corresponding absorption spectra of AK3-1, AK3-7, AK3-8 and AK3-11 in aqueous phase were deconvoluted into three bands representing the dye monomers, dimers and higher order aggregates (Figure 6). In all cases R^2 was found to be 0.98-0.99. The deconvolution of the absorption spectra allowed us to calculate a set of parameters, viz.: i) the amplitude (A_{max}

peak position ν_c

at half-maximum of the band (FWHM); and iii) the peak asymmetry parameter (ρ). The calculated parameters are presented in Table 3.

Table 4 represents the changes in the deconvolution parameters in the dye-protein systems containing the non-fibrillar insulin and lysozyme. Specifically, in the presence of the control insulin the following changes of the amplitude were observed: i) a marked drop of the band I corresponding to the monomeric dye form (all dyes); ii) a rise in the relative contribution of the band II (AK3-5, AK3-11); iii) a 1.4-fold (AK3-1, AK3-8 and AK3-11), 1.1-fold (AK3-3) and 1.6-fold absorbance increase of the band III (Table 4). These effects were accompanied by the increase of the full width at half-maximum of the band II (AK3-3, AK3-5, AK3-11) and band III (AK3-8), coupled with the rise of the asymmetry parameter of band II (AK3-5, AK3-7 and AK3-8). The presence of the non-fibrillar insulin did not affect the position of the peaks, with the exception of ν_c

AK3-8 where a small short-wavelength shift was observed. Taken together, these results mirror the ability of non-fibrillized insulin to increase the H-aggregation propensity of the trimethine cyanine dyes. The similar results were previously obtained for pentamethine dyes in the presence of bovine serum albumin [91]. The formation of the self-associated assemblies of the thiacarbocyanine dyes was observed in the presence of ribonuclease A, lysozyme, trypsin and gelatin [91, 92]. The peculiar features of the thiacarbocyanine-protein binding are that the dye aggregation was cooperative in nature and the complexation was dependent on both the dye and protein structures [91]. Moreover, the Coloumbic interactions were assumed to play a predominant role in the aggregation of these dyes upon the formation of the dye-protein complexes [91, 92]. Probably, by analogy to the pentamethine and tricarbocyanine dyes, the trimethine self-assembly is facilitated by electrostatic interactions between the negatively charged protein residues and cationic dye molecules, as well as by the hydrophobic dye-protein interactions in the presence of the non-fibrillized insulin.

The incubation of AK3-1, AK3-5 та AK3-8 (Figure 7 A) with the control lysozyme resulted in the following changes of the decomposition parameters compared to those obtained in the buffer solution: i) the rise of the amplitude of the bands I (a 1.1-fold, 1.4-fold and 1.3-fold increase) and II (a 1.4-fold, 1.6-fold та 2.8-fold increase); ii) the increase of the asymmetry parameter and the full width at half-maximum of the band II (AK3-1 and AK3-7); iii) the drop of the amplitude of the band III for AK3-1 (1.8 times) та AK3-8 (2.9 times) and the band constriction. In the meantime, the presence of the non-fibrillar lysozyme led to a 3.3-fold, 1.7-fold and 1.1-fold decrease in the amplitude of the band I for AK3-3, AK3-7 and AK3-11, respectively. This effect was accompanied by the decrease of the full width at half-maximum of the band I coupled with the rise of the amplitude of the bands II (AK3-7) and band III (AK3-11).

Table 4. Deconvolution parameters of the absorption spectra of trimethines in the presence of non-fibrillar proteins

		Non-fibrillar insulin				Non-fibrillar lysozyme			
Dye	Band	A_{max}	ν_c, cm^{-1}	FWHM, cm^{-1}	ρ, cm^{-1}	A_{max}	ν_c, cm^{-1}	FWHM, cm^{-1}	ρ, cm^{-1}
AK3-1	I	34000	16007	1784	0.43	43112	16024	1314	0.62
	II	7500	18100	910	1.02	11652	18056	1647	1.82
	III	19238	19300	3769	1.56	7895	18780	3378	1.48
AK3-3	I	77643	15900	1336	0.59	22100	15608	1436	0.81
	II	32698	17900	2334	0.92	9935	17025	1383	0.81
	III					11434	18858	4544	2.00
AK3-5	I	30000	15800	1000	0.5	70503	15858	891	0.66
	II	35497	17900	3619	1.23	52474	16930	2696	0.91
AK3-7	I	72000	15900	973	0.84	50720	15888	862	0.75
	II	39000	17000	1984	0.68	52102	16636	2243	1.21
	III	12857	19000	4324	2.23	2224	18921	10385	4.84
AK3-8	I	10935	15700	2202	0.61	27768	15871	863	0.72
	II	2028	17300	1750	0.75	24562	16622	2138	1.20
	III	37000	18950	2435	1.75	9627	18041	1461	0.70
AK3-11	I	12286	15234	1904	0.88	14355	15691	2256	1.69
	II	30000	17000	2380	0.96	5360	16934	922	0.77
	III	36042	18250	1667	1.03	30784	18289	2475	1.21

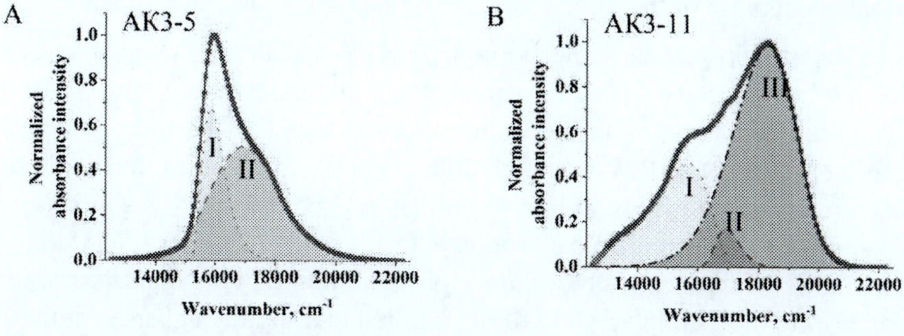

Figure 7. Deconvolution of the trimethines absorption spectra in the presence of non-fibrillized lysozyme into two (A) and three (B) bands, respectively.

It should be noted that the absorption spectra of AK3-3 in the non-fibrillar lysozyme were deconvoluted into three separate bands, where for the short- (dimeric dye form) and long-wavelength (monomeric dyes form) bands the 3.5-fold and 4.4-fold drop of the amplitude was observed. Besides, the incubation of AK3-3 with the non-fibrillar lysozyme resulted in the appearance of the III short-wavelength band of H-aggregates, the intensity of which even exceeded the amplitude of the band II. Furthermore, while comparing the ν_c values, it can be seen that monomeric lysozyme produced a hypsochromic shift of the monomer band I (AK3-3), dimer band II (all dyes excepting AK3-1), and the H-aggregates band III (AK3-1, AK3-7 and AK3-8). In turn, the position of the band I for AK3-8 та AK3-11 shifted to the higher values of wavenumber. Taken together, we can conclude that the trimethine dyes AK3-1, AK3 5, AK3-7 and AK3-8 bind to the native lysozyme mainly in the form of monomers and dimers, while AK3-3 and AK3-11 show a tendency to aggregate in the presence of the non-fibrillar lysozyme.

The incubation of AK3-1 and AK3-7 with the fibrillar insulin resulted in the drop of the amplitude of the band I coupled with an increase of the band III (Table 5). Interestingly, for these dyes the rise of the relative contribution of the band III to the overall spectrum was much more pronounced for fibrillar insulin compared to its non-aggregated counterpart. Taken into account that the bands I and III correspond to the absorption of monomers and H-aggregates, respectively, the decrease of amplitude of the band I, along with the increase of A_{max} values for band III indicate that the fibrillar insulin, similar to the control insulin, may also serve as a matrix for the aggregation of AK3-1 and AK3-7. However, the growth of FWHM of bands I and II, together with the decrease of the ν_c

aggregates there exists some portion of AK3-1 and AK3-7 bound to the fibrillar insulin in the form of monomers and dimers. Notably, similar results have been reported for heptamethine cyanine dyes in the presence of the fibrillar lysozyme [60]. Remarkably, the tendency to organize into card-pack assemblies on the fibrillar matrix of β-amyloid peptide was previously observed for the cationic pinacyanol dye [52].

In this context it should be noted, that the binding of the thiacarbocyanine dye 7514 to the insulin amyloid fibrils resulted in a slightly enhanced intensity of both H- and J- aggregate bands [53]. Similarly, the ability of Rhodamine 6G to aggregate on the amyloid template was found by Hanczyc et al., who demonstrated that the insulin fibrils promote the formation of the dye molecular assemblies [94]. Notably, the absorption spectra of AK3-3 in buffer solution or control insulin were deconvoluted into two separate bands, with the short- and long-wavelength spectral components seemingly corresponding to the dimeric and monomeric dye species, respectively. However, we did not succeed in decomposing the absorption spectrum of AK3-3 in the presence of the fibrillar insulin into two bands. Instead, it appeared that this spectrum can be represented as a sum of three separate bands. We assumed that the bands I and II correspond to the bound monomers and dimers, respectively. An additional argument in favor of the above assumption comes from the 24-nm shift of the peak position (decrease of the ν_c

monomers and dimers. The blue-shifted band III, which was not observed in the buffer and the control insulin, by analogy with AK3-1 and AK3-7, could be attributed to the formation of AK3-3 aggregates on the fibrillar matrix.

Table 5. Deconvolution parameters of the absorption spectra of trimethines in the fibrillar proteins

		Fibrillar insulin				Fibrillar lysozyme			
Dye	Band	A_{max}	ν_c, cm^{-1}	FWHM, cm^{-1}	ρ, cm^{-1}	A_{max}	ν_c, cm^{-1}	FWHM, cm^{-1}	ρ, cm^{-1}
AK3-1	I	39000	15500	1984	0.87	52327	15924	1405	0.66
	II	16952	17300	1673	0.76	13891	17943	1747	1.56
	III	20225	19300	3369	1.67	4918	18764	2931	1.73
AK3-3	I	62500	15500	1537	1.07	32383	15516	2077	1.00
	II	22000	17050	1446	1.23	6468	17115	1623	0.66
	III	29431	18200	2692	1.03	1757	18434	2183	2.30
AK3-5	I	124041	15500	1064	0.89	74203	15778	979	0.88
	II	79368	17000	2233	0.81	60928	16692	2471	0.85
AK3-7	I	41000	15400	1742	1.19	51437	15854	883	0.85
	II	23100	16600	2269	0.81	52887	16494	2185	1.27
	III	24032	18350	3253	1.30	39	19000	3448	3.95
AK3-8	I	28955	14850	1525	0.77	44261	15681	1052	0.84
	II	37667	17100	2979	1.33	25834	16717	2181	1.03
	III	6622	19100	2619	2.26	2091	18302	2054	1.35
AK3-11	I	88500	15000	1594	0.62	53443	15100	1353	0.83
	II	29169	16500	1267	0.81	22866	16500	1568	0.61
	III	17000	17800	2000	0.82	7296	17800	2356	1.59

The relative integral intensities of the bands I and II of AK3-5, AK3-8 and AK3-11 increase in the presence of fibrillar insulin. More specifically, the incubation of these dyes with the insulin fibrils resulted in 2.5, 1.4 and 5.8-fold increase of the monomer band, along with 2.1, 4.2 and 3.2-fold enhancement of the dimer band for AK3-5, AK3-8 and AK3-11, respectively (Table 5). At the same time, a marked drop in the amplitude of band III was observed for AK3-8 and AK3-11 in the presence of the aggregated protein. This effect was accompanied by the lowered ν_c

for AK3-5, AK3-8 and AK3-11 presumably arising from the reduced environmental polarity of the dye monomers and dimers associating with the fibrillar insulin. On the contrary, the insulin fibrils did not exert influence on the position of the band III for AK3-8 and AK3-11. While comparing the FWHM values, it can be seen that insulin fibrils produced a significant attenuation in FWHM of AK3-5 (band II) and AK3-11 (bands II and III), whereas in the case of AK3-8 this parameter showed a slight increase for the bands II and III. To summarize, being preferentially in the form of H-aggregates when bound to the control insulin, the dominating species of AK3-5, AK3-8 and AK3-11 in the presence of the fibrillar insulin are the dye monomers and dimers. Furthermore, the higher increase of amplitude for the band I of AK3-5 and AK3-11 indicates that the binding of monomers is more favorable for these dyes, whereas for AK3-8 the predominant binding mode is the dimerization on the fibrillar matrix. Similarly, the high affinity of Thiazole Orange – "TO" monomers for DNA was explained by the dye stacking with DNA bases, while low-affinity dimers (due to steric restrictions) were supposed to associate with the external surface of the DNA molecule [95].

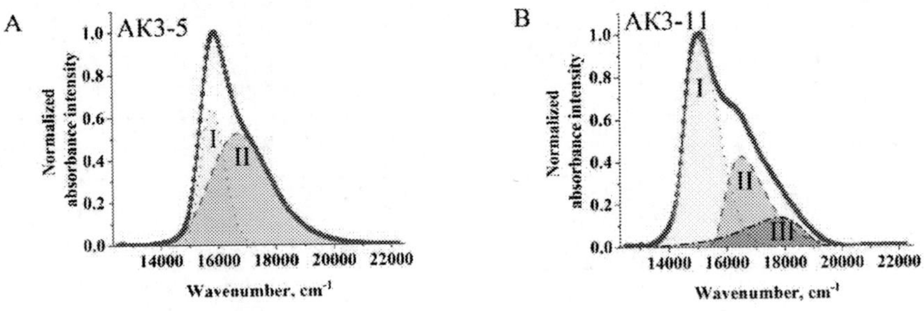

Figure 8. Deconvolution of the trimethine absorption spectra in the presence of fibrillar lysozyme into two (A) and three (B) bands, respectively.

The deconvolution analysis of the absorption spectra of the trimethine dyes in the presence of the fibrillar lysozyme demonstrated that the dominating dye species in the presence of the aggregated protein, similarly to the insulin amyloid fibrils, are the dye monomers and dimers (Figure 8, Table 5).

More specifically, the incubation of AK3-1, AK3-5, AK3-8 та AK3-11 with the fibrillar lysozyme resulted in a 1.3, 1.8, 2.1 and 3.5-fold increase of the relative integral intensities of the band I along with 1.6, 1.5, 2.9 and 2.4-fold enhancement of the amplitude of the band II, respectively. This effect was accompanied by the drop of the amplitude of the band III for AK3-1 (2.8

times), AK3-8 (13 times) та AK3-11 (2.6 times). At the same time, it can be seen that the lysozyme fibrils produced a significant attenuation in FWHM of the band I (AK3-1, AK3-5, AK3-8 та AK3-11) and band II (AK3-5, AK3-11), whereas in the case of AK3-1 and AK3-8 this parameter showed a slight increase for the bands II (AK3-1, AK3-8) and III (AK3-8). This effect was accompanied by the lowered ν_c

5, and AK3-11 presumably arising from the reduced environmental polarity of the dye monomers and dimers associating with the fibrillar lysozyme. On the contrary, the insulin fibrils did not exert influence on the position of the band III for AK3-1, AK3-8 and AK3-11. Remarkably, the higher increase of amplitude for the band I of AK3-5 and AK3-11 indicates that the binding of monomers is more favorable for these dyes. The complexation of AK3-3 with the fibrillar lysozyme, similar to the non-aggregated protein, led to the rise of the relative contribution of the bands III and II to the overall spectrum. Besides, the relative integral intensities of the bands II and III were 1.5 and 6.5 times lower that those observed in the presence of the non-aggregated lysozyme. Moreover, the incubation of AK3-3 with the lysozyme fibrils resulted in a 1.5-fold increase of the monomer band. This effect was accompanied by the lowered ν_c

above findings indicate that AK3-3 binds to the lysozyme amyloid fibrils presumably in the form of monomers. The incubation of AK3-7 with the fibrillar lysozyme resulted in the significant drop of the amplitude of the bands II and III in comparison with those observed in the buffer and control protein. The band III corresponding to the dye H-aggregates was almost indistinguishable. Furthermore, the enhancement of the amplitude of the band I along with the the lowered ν_c

monomers are dominating dye fraction in the presence of the lysozyme fibrils.

The decomposition analysis explains the fluorescence responses of the trimethine cyanine dyes to the nonfibrillar and fibrillar proteins. As can be seen from Figure 3, the chromophores AK3-11 and AK3-5 possess the largest ADF values in the presence of the fibrillar insulin, whereas in the case of fibrillar lysozyme the largest values of this parameter were observed for AK3-5, AK3-7 and AK3-11. The high amyloid-sensitivity of this dyes can be attributed to several factors: i) the formation of non-fluorescent H-aggregates of AK3-5 (only in the presence of InsF) and AK3-11(for both InsF and LzF) induced by the control protein; ii) significant fluorescence enhancement of AK3-11, AK3-5 and AK3-7 produced by the binding of the dye monomers to fibrillar insulin, with the low-order self-associated species playing a minor role in this process; iii) significantly lower binding affinity of the Ak3-7 monomers to the nonfibrillar lysozyme in comparison with the aggregated protein. It should be noted that aggregate formation, by analogy to AK3-11 and AK3-5, is also the predominant binding mode for the other dyes under investigation in the presence of the control insulin. Besides, the smaller ADF values for these dyes in the presence of InsF are attributed to the fact that dimerization and aggregation on the fibrillar insulin matrix prevailed over the binding of monomer species. In turn, significantly lower ADF values for AK3-3, AK3-1 and AK3-8 can be a consequence of the binding of dye monomers to the nonfibrillar lysozyme. Importantly, the AK3-5 and AK3-11 appeared to be highly sensitive to the both amyloid proteins. Besides, the AK3-11 seems to be the most effective amyloid reporter among the examined trimethines due to conversion of the non-emissive assemblies in aqueous solution and non-fibrillized proteins to the highly fluorescent monomeric species in the presence of the fibrillar insulin.

Potential Lysozyme and Insulin Binding Sites for the Trimethine Dyes

To identify the trimethine-protein binding sites and elucidate the nature of the interactions involved in the dye–protein complexation in the cases of non-fibrillized proteins and amyloid fibrils, we used the molecular docking technique. The examined trimethines were found to preferentially associate with the chain B of the native insulin. Similar tendencies in the dye binding to the native insulin were previously reported for Thioflavin T and cyanines with different length of polymethine bridge, observed using both PatchDock / FireDock and Autodock [55, 59].

As depicted in Figure 9 A, B, two binding sites are observed for the most energetically favorable complexes with InsN. Thus, AK3-1, AK3-3 and AK3-7 seem to interact with residues 14 B – 18 B, 13 A – 17 A and N-terminal fragment of the chain B, while the AK3-5, AK3-8 and AK3-11 bind to 16 B – 26 B amino acid residues. The free energy of the dye-InsN binding was found to follow the order: AK3-1 (-6.40 kcal/mol) > AK3-11 (-6.39 kcal/mol) > AK3-3, AK3-7 (-6.22 kcal/mol) > AK3-5 (-6.09 kcal/mol) > AK3-8 (-5.71 kcal/mol). To analyse the types of the AK3-protein complexation in more detail, the protein-ligand interaction profiler (PLIP, https://plip-tool.biotec.tu-dresden.de/plip-web/plip/index) was further used [69]. The hydrophobic interactions (the gray dashed lines on Figure 9, pannels C and D) were found to play a predominant role in the cyanine-insulin binding. The residues Leu 13, Glu 17 of the chain A as well as Leu 17 and Val 18 of the chain B are bound in a hydrophobic way with AK3-1, AK 3-3 and AK3-7. Specifically, the hydrophobic contacts were observed with Ala 14 B (AK3-1, AK3-7) and Phe 1 B, Gln 4 B, Leu 6 B (AK3-3) with the first potential binding site

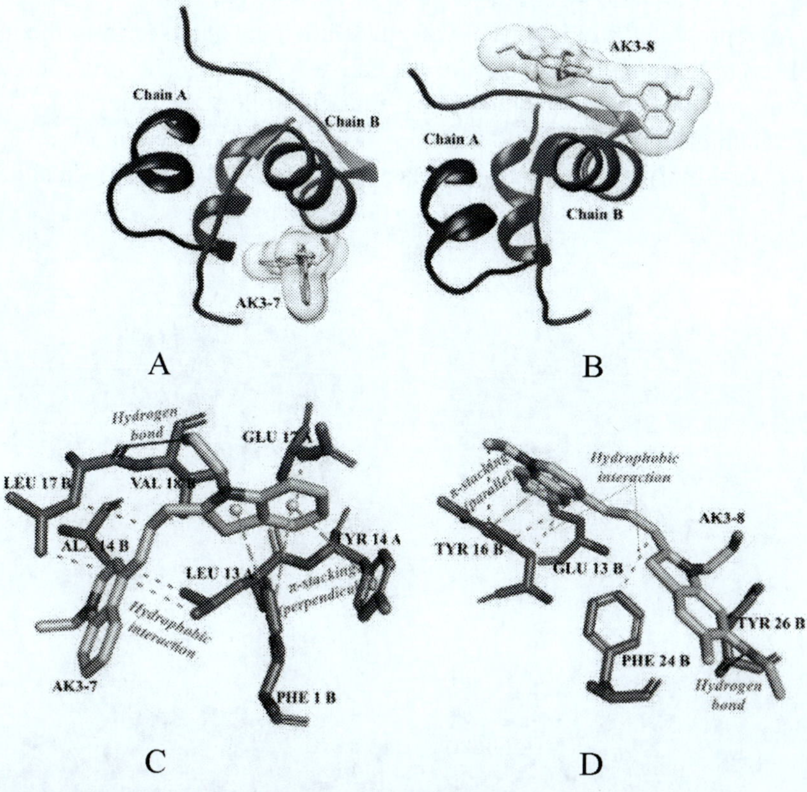

Figure 9. Representative most favorable modes of interactions between AK3-7 (A, C), AK3-8 (B, D) and native insulin, obtained using the AutoDock (panels A, B) and PLIP (panels c and d). The grey dashed lines on the panels c and d represent the hydrophobic interactions between the dye molecule and the insulin residues, while the green dashed line and blue solid line display the π-stacking contacts and hydrogen bonds, respectively.

(14 B – 18 B, 13 A – 17 A and N-terminal fragment of the chain B), and Val 12 (AK3-11), Glu 13 (AK3-8), Tyr 16, Phe 24, Phe 25 (AK3-5, AK3-11), Tyr 26 and Thr 27 (AK3-11) with the second binding site (16 B – 26 B amino acid residues). The dye-InsN complexes are additionly stabilized by π-stacking

interactions (green dashed lines in Figure 10, panels C and D) and hydrogen bonds (blue solid line). More specifically, the π-stacking contacts were observed between benzene and/or cyclopentane ring and Tyr 14 of the A-chain (AK3-1, AK3-3, AK3-7), Phe 1 (AK3-1, AK3-7) and Tyr 16 (AK3-8) of the B-chain, while the hydrogen bonds were formed between OH-groups of the aliphatic substituents on the nitrogen atom of AK3-7 and AK3-8, and Leu 17 (AK3-7), Tyr 26 (AK3-8) of the chain B of the native insulin.

As represented in Figure 10 A, B, the trimethines tend to form the most stable complexes with deep cleft of the native lysozyme. The free energy of cyanine – native lysozyme binding was found to follow the order: AK3-11 (-8.18 kcal/mol) > AK3-1 (-7.90 kcal/mol) > AK3-3 (-7.78 kcal/mol) > AK3-7 (-7.42 kcal/mol) > AK3-5 (-7.34 kcal/mol) > AK3-8 (-7.01 kcal/mol).

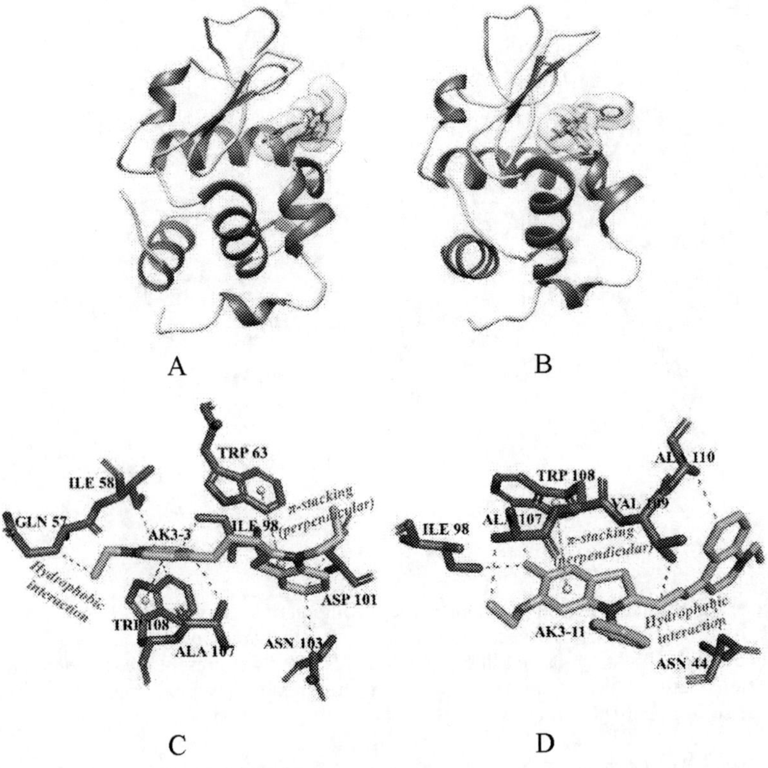

Figure 10. Representative most favorable modes of interactions between AK3-3 (A, C), AK3-11 (B, D) and native lysozyme obtained using the AutoDock (panels A, B) and PLIP (panels C and D). The grey dashed lines and green dashed line on the panels C and D represent the hydrophobic interactions and π-stacking contacts between the dye molecule and the lysozyme residues, respectively.

It is interesting to note that similar to the native insulin, the maximal AK3-LzN binding energy were observed for AK3-11 and AK3-1, while the minimal values were found for AK3-5 and AK3-11. As judged from the PLIP analysis, the hydrophobic interactions play a significant role in the binding of cyanines to the non-aggregated lysozyme and involve Asn 44 (only AK3-11), Gln 57 (except AK3-5, AK3-11), Ile 58 (except AK3-8, AK3-11), Asn 59 (AK3-1), Trp 62 (AK3-5), Trp 63 (AK3-5, AK3-7, AK3-8), Leu 75 (AK3-5), Ile 98 (except AK3-8), Asp 101 (except AK3-7, AK3-11), Asn 103 (except AK3-8, AK3-11), Ala 107 (except AK3-7, AK3-8), Trp 108 (except AK3-7, AK3-8), Val 109 (AK3-8, AK3-11) and Ala 110 (AK3-11). Moreover, the trimethines were found to form π-stacking contacts between aromatic or cyclopentane ring of the cyanine and Trp in position 63 (in the case of AK3-1,AK3-3, AK3-5) and 108 (except AK3-7, AK3-8). Similar to the dye-InsN binding, AK3-7, AK3-8 complexes with the native lysozyme are additionally stabilized by hydrogen bonds between the OH-groups of the trimethines and acceptor atoms of Asp 101 (AK3-7) and Gln 57 (AK3-8) of LzN.

It is known that amyloid fibril grooves represent specific binding sites for the classical amyloid markers ThT and Congo red [36]. The binding to fibril grooves was also previously observed for the mono-, tri- and heptamethine cyanine dyes, as well as for the DNA-intercalating dye YOYO-1 [55, 59, 60]. The width of the binding channels and the interstrand distance in β-sheet are equal to ~ 7 Å and 4 Å, respectively [36]. Thus, there are no spatial restrictions for the cyanines under study to interact with the fibril grooves since the trimethine lengths, widths and heights do not exceed 17.9 Å, 11.5 Å and 4.8 Å, respectively [55]. Therefore, by analogy to the other dyes we can expect that AK3 monomers tend to associate with at least 4 B-strands in such a way that their short axis is perpendicular to the fibril axis.

The results of the blind docking revealed that the examined cyanines tend to occupy the non-specific binding sites of the fibrillar insulin, namely the cross-strand ladders located on the dry steric zipper interface, formed by Leu 17 residues (Figure 11 A). The binding energy, which follows the order: AK3-3 (-11.21 kcal/mol) > AK3-1 (-10.63 kcal/mol) > AK3-5 (-9.77 kcal/mol) > AK3-11 (-9.39 kcal/mol) > AK3-7 (-9.28 kcal/mol) > AK3-8 (-7.20 kcal/mol), was found to be significantly higher than that for the native protein. The hydrophobic interactions between the dye molecules and nonpolar Ile 102 (except AK3-11), Leu in positions 217, 306, 317 and polar Gln 304 (AK3-8, AK3-11) residues were revealed by PLIP (Figure 11 B).

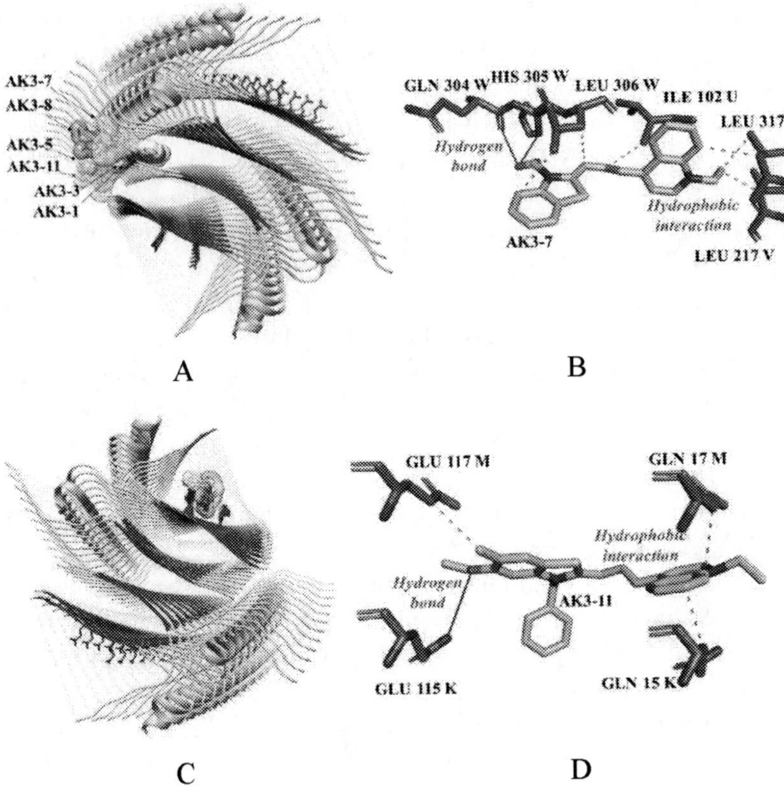

Figure 11. The binding of trimethines (AK3-1 – purple, AK3-3 – green, AK3-5 – orange, AK3-7 – grey, AK3-8 – magenta, AK3-11 – blue) to the fibrillar insulin obtained using the AutoDock at blind (A) and target (C) docking. Representative interactions between AK3-7 and cross-strand ladder (B), AK3-11 and groove Gln 15-Glu 17 (D) of the insulin fibril, revealed by PLIP. The grey dashed lines on the panels B and D represent the hydrophobic interactions between the dye molecule and InsF, while the blue solid lines display the hydrogen bonds between OH- or ether groups of the trimethines and Gln, His residues of InsF.

To estimate the trimethine affinity to the unique binding site such as the groove Gln 15-Glu 17 located on the wet fibril surface, the target docking was performed (Figure 11 C). The free energy of the trimetine binding was found to follow the order: AK3-5 (-5.44 kcal/mol) > AK3-1 (-5.21 kcal/mol) > AK3-7 (-4.83 kcal/mol) > AK3-11 (-4.25 kcal/mol) > AK3-3 (-4.02 kcal/mol) > AK3-8 (-3.07 kcal/mol). These dye-protein complexes were predominantly stabilized by hydrophobic contacts with the groove residues. Specifically, the hydrogen bonds between OH- or ether groups of AK3-7, AK3-8 and AK3-11 and Gln in position 15 (AK3-7), 204 (AK3-8) and 115 (AK3-11) were observed.

Figure 12 represents the energetically most favorable cyanine complexes with the groove Gly 2-Leu 4/Ser 8-Trp 10 of the lysozyme fibril core, which stabilised by hydrobhobic contacts with Leu 4, Trp 10 as well as π-stacking interaction between cyanine dyes and latter groove residue (Figure 4, panel B). The contribution of hydrogen bonds in AK3-7 (Ile 3), AK3-8 (Ile 3, Arg 9) complexation with LzF was detected. The values of binding energies appeared to be comparable with the native lysozyme and follow the order: AK3-11 (-7.78 kcal/mol) > AK3-3 (-7.11 kcal/mol) > AK3-1 (-6.74 kcal/mol) > AK3-5 (-6.59 kcal/mol) > AK3-7 (-6.43 kcal/mol) > AK3-8 (-6.13 kcal/mol).

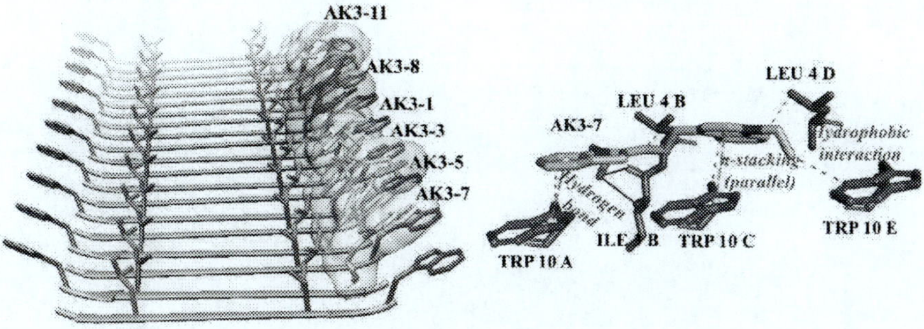

A B

Figure 12. The binding of trimethines (AK3-1 – purple, AK3-3 – green, AK3-5 – orange, AK3-7 – grey, AK3-8 – magenta, AK3-11 – blue) to the fibrillar lysozyme, obtained using the AutoDock (A). Representative interactions between AK3-11 and groove Gly 2-Leu 4 / Ser 8-Trp 10 of the lysozyme fibril revealed by PLIP (B). The grey dashed lines on the panel B represent the hydrophobic interactions between the dye molecule and the groove, while the green dashed line and blue solid line display the π-stacking contacts and hydrogen bonds, respectively.

The Amyloid-Sensing Potential of the Donor-Acceptor Pair ThT – Cyanine Dye

During the last decades the concept of selective recognition and characterization of the fibrlillar protein aggregates using the Förster resonance energy transfer (FRET) between extrinsic fluorophores has become very useful for the detection of amyloid fibrils [46, 61-63, 96-98]. By exploring the FRET between the benzanthrone dye ABM and the squaraine dye SQ-1 or the polymethine dye V2 the limits for the fractal-like dimension of the lysozyme fibrils were determined [97]. Besides, the FRET between bis-ANS and a styrylquinoxalin dye was employed to monitor simultaneously the presence of oligomers and fibrils of the Aβ-peptides [96]. Furthermore, the advantages of multistep FRET have been demonstrated while exploring the amyloid transformation of the N-terminal fragment of apolipoprotein A-I [61] and insulin [62]. In view of this, we assessed the amyloid-sensing potential of the donor-acceptor pair ThT – cyanine dye using a series of trimethine cyanine compounds and fibrillar insulin as a model protein [63]. Due to the overlap between the ThT emission and cyanine absorption spectra (Figure 13), the energy can be transferred between these fluorophores by a distance-dependent Förster mechanism.

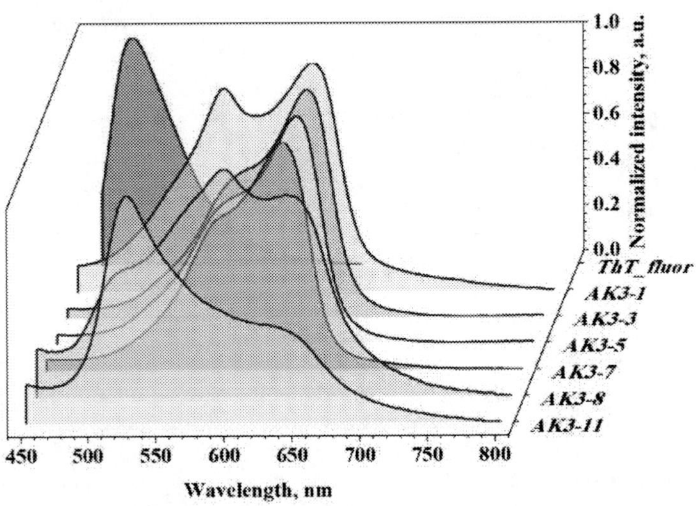

Figure 13. The overlap between the donor emission and acceptor absorption spectra for the ThT/AK3 dye donor-acceptor pairs.

As seen in Figure 14, the addition of the trimethine dyes to the ThT-InsF systems resulted in the significant quenching of ThT fluorescence coupled with the slight intensity increase of the peak at 654-665 nm (depending on the dye chemical structure) with increasing trimethine dye concentration up to 0.13 μM. Further increase of acceptor concentration (up to 1.33 μM) led to the subsequent attenuation of the ThT emission band along with the simultaneous decrease of the long-wavelength peaks (except AK3-11 for which this peak remained unchanged) corresponding to the bound AK3-dye. The decrease in the fluorescence intensity of the peaks at 654-665 nm with increasing the trimethine concentration most probably arises from the concentration-dependent ability of cyanine dyes to aggregate into non-fluorescent H-aggregates [59].

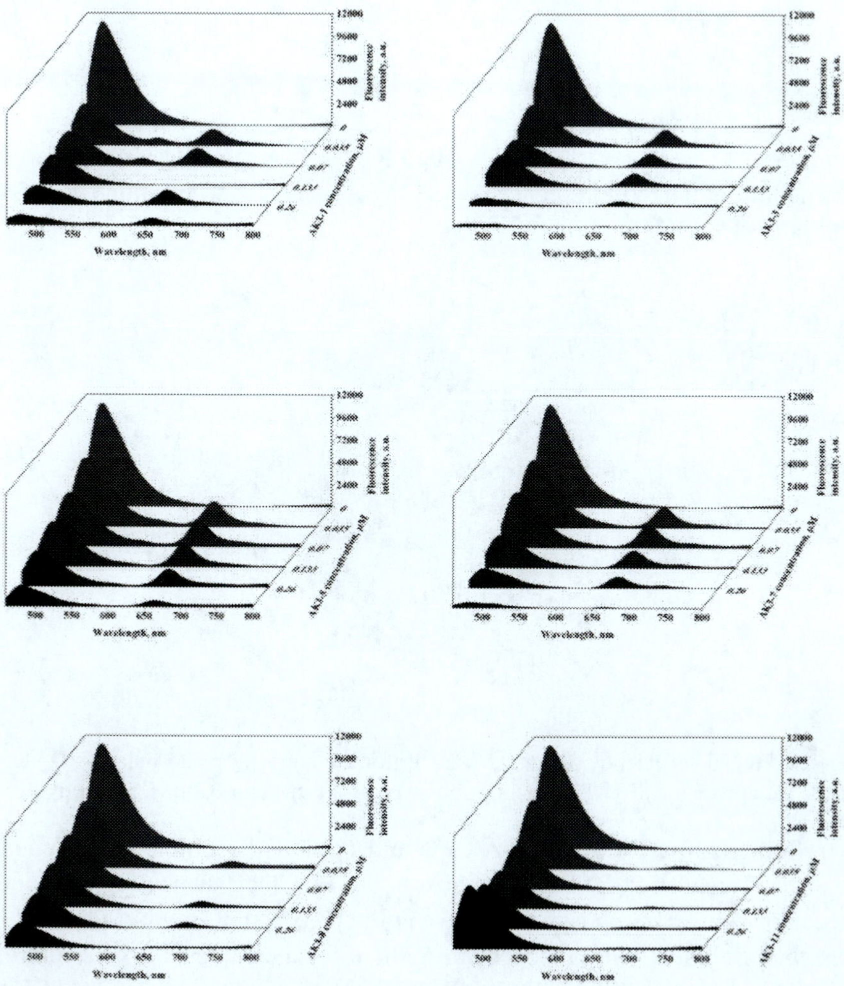

Figure 14. The changes in the fluorescence spectra of ThT bound to the insulin amyloid fibrils upon addition of the trimethines. The protein and ThT concentrations were 1.44 μM and 0.2 μM, respectively.

It should be noted that the quenching of ThT fluorescence in the presence of acceptors was observed only in the fibrillar insulin indicating that the insulin amyloid fibrils serve as a scaffold for the positioning of ThT and cyanine dyes. The efficiency of energy transfer between ThT as a donor and AK3-dye as an acceptor was calculated using the quenching of the donor fluorescence in the presence of acceptor [99]:

$$E = 1 - \frac{I_{DA}}{I_D}, \tag{6}$$

where I_D and I_{DA}

presence of acceptor, respectively. The FRET efficiencies calculated from the quenching of ThT fluorescence at varying acceptor concentrations are given in Figure 15.

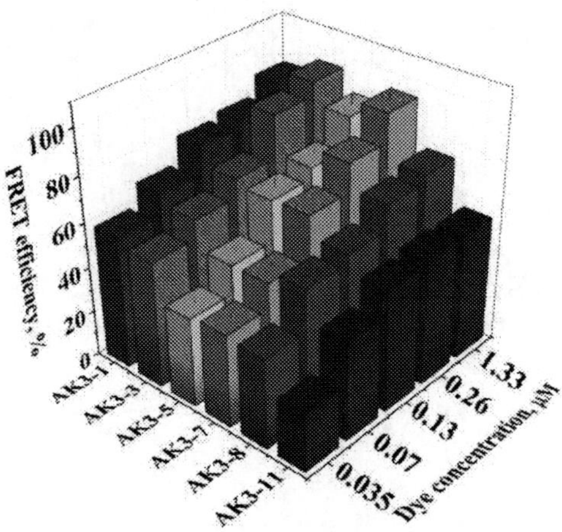

Figure 15. The efficiency of energy transfer for the donor-acceptor pairs ThT-AK3-dyes. The protein and ThT concentrations were 1.44 µM and 0.2 µM, respectively.

Remarkably, for AK3-1, AK3-3 and AK3-8 the efficiency of energy transfer exceeded 90%. For comparison, the FRET efficiency for the donor-acceptor pair ThT – phosphonium dye in the presence of amyloid fibrils of N-terminal fragment of apolipoprotein A-I and for the pair ThT – benzanthrone dye in the presence of fibrillar insulin did not exceed 70% [61] and 82% [62], respectively. The recent studies indicate that a range of parameters can influence on the FRET efficiency between fluorophores, among which are: the donor-acceptor distance, the orientation of the donor and acceptor transition dipoles, the donor quantum yield, the refractive index of the medium, the overlap between the donor emission and acceptor absorption spectra [99, 100]. To understand the possible factors contributing to the observed extraordinary strong energy transfer, the FRET efficiencies have been calculated using the stretched exponential model that describes energy transfer in the spatially confined molecular systems [101]:

$$Q_r = \int_0^{\infty} \exp\left(-\lambda - C_A V_d \Gamma \left(1 - d/6\right) \lambda^{\frac{d}{6}} \right) d\lambda \qquad (7)$$

where $\lambda = t / \tau_D$, τ_D

acceptor; $V_d = \pi^{\frac{d}{2}} R_0^d / \Gamma\left(\frac{d}{2} + 1\right)$ *d*-dimensional sphere of radius

R_0, d is the dimensionality of fluorophore distribution (fractal-like dimension); $C_A = C_B / C_P V_{PF}$, C_P is the total protein concentration; V_{PF} volume of insulin molecule in a fibrillar state taken as 5.8 nm^3 [14], C_B molar concentration of bound acceptor. To obtain the best agreement between theory (Eq. 7) and experiment we varied the set of parameters: the fraction of bound acceptor; the dimensionality of fluorophore distribution d; and the orientation factor (κ^2

First, we made an attempt to approximate the experimental data by equation (7) assuming that overall volume of the insulin fibrils is accessible to the acceptors (coefficient of accessibility $\alpha = 1$) and varying the orientation factor in almost widest possible limits (0.01 – 4). It appeared that for AK3-1, AK3-3, AK3-5 and AK3-7 a good agreement between theory and experiment can be achieved under assumption of one-dimensional acceptor distribution ($d = 1$) and only partial accessibility of fibril volume to acceptors (coefficient α varied from ~0.1 to ~0.7) (Figure 16). Notably, these α estimates represent the upper limits for this parameter since the fraction of bound acceptor (f_b was taken as equal to 1. For AK3-8 and AK3-11 the theoretical estimates were close to the experimental FRET efficiencies only at $\kappa^2 \leq 0.67$ (AK38) or κ^2 ~ 0.01 (AK3-11). Another finding noteworthy is that in the case of three-dimensional acceptor distribution ($d = 3$) the agreement between theoretical and experimental E values can be reached only for low fractions of bound acceptor (f_b $\alpha = 1$).

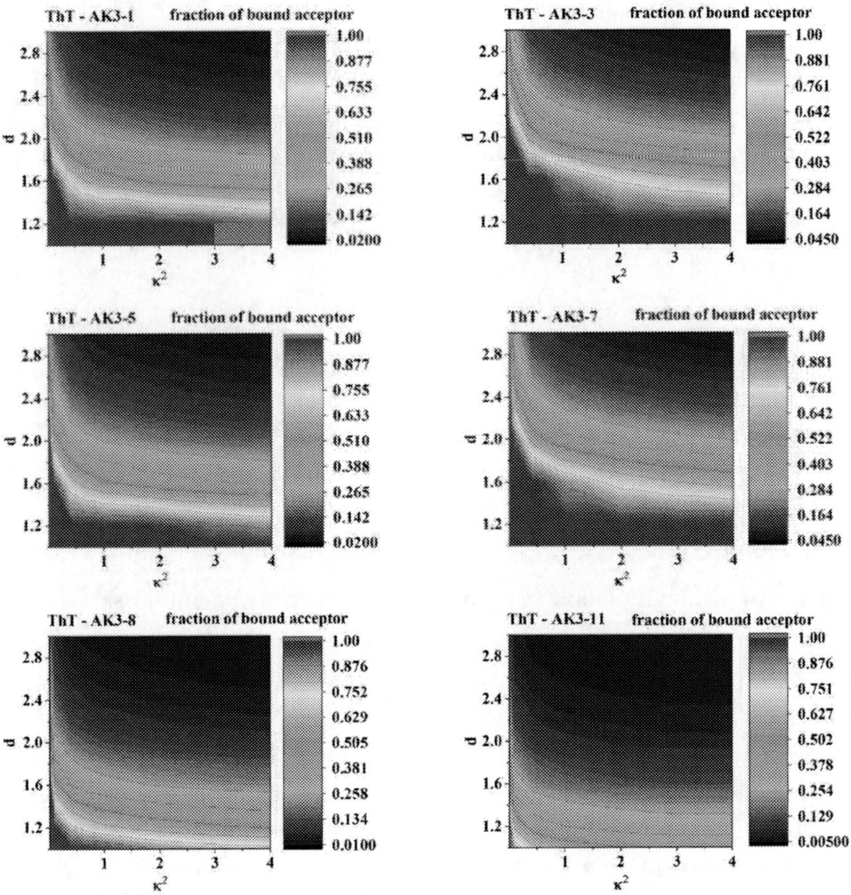

Figure 16. Three-dimensional plots illustrating the relationship between the fraction of bound acceptor, the orientation factor and the dimensionality of fluorophore distribution for the donor-acceptor pairs.

To obtain approximate estimates for the donor-acceptor separation R, the FRET data were analyzed in terms of the classical expression for the distance dependence of FRET efficiency [99]:

$$R = R_0 \left(1/E - 1\right)^{1/6} \tag{8}$$

The critical distance of energy transfer was calculated as [1]:

$$R_0 = 979(k^2 n_r^{-4} Q_D J)^{1/6},\ J = \int_0^\infty F_D(\lambda)\varepsilon_A(\lambda)\lambda^4 d\lambda \Big/ \int_0^\infty F_D(\lambda) d\lambda \qquad (9)$$

where κ^2 is the orientation factor, n_r

($n_r = 1.4$), Q_D J is the overlap integral derived from numerical integration, $F_D(\lambda)$ is the donor fluorescence intensity, $\varepsilon_A(\lambda)$ is the acceptor molar absorbance at the wavelength λ. aBy introducing into Eq. 8 the R_0

9) and the highest measured FRET efficiencies (E) corresponding to the maximum AK3-dye concentration (1.33 μM), we obtained the possible limits for the donor-acceptor distances (Table 6). Along with the isotropic value of orientation factor ($\kappa^2 = 0.67$) corresponding to the unrestricted rotation of fluorophores, we considered the cases of nearly perpendicular ($\kappa^2 = 0.01$); parallel and oppositely directed ($\kappa^2 = 1$), and parallel and identically directed transition dipoles ($\kappa^2 = 4$).

The insulin amyloid fibrils obtained under orbital rotation at neutral pH seem to form the tangles of intertwined fibrils with a width ~ 10-50 nm [81, 102]. These fibrils consist of several protofilaments with a width ~ 5 nm and length ~ 50 - 400 nm [81] and exhibited twisted morphology with a pitch of ~ 66 nm [81,102]. Given the above thickness of the insulin amyloid fibrils, and the preference of all dyes under study to the wet surface groove Gln15-Glu17 mentioned above while discussing the molecular docking results, the obtained distance (R) estimates suggest the location of ThT and AK3-dyes within the surface groove on one side of insulin amyloid core.

Table 6. FRET parameters derived for the employed donor-acceptor pairs

Donor-acceptor pair	R_0, nm	J, $M^{-1} cm^{-1} nm^4$	R, nm*			
			$\kappa^2 = 0.01$	$\kappa^2 = 0.67$	$\kappa^2 = 1$	$\kappa^2 = 4$
ThT-AK3-1	3.8	4.38·1015	1.4	2.9	3.9	3.1
ThT-AK3-3	3.7	3.75·1015	1.1	2.1	2.9	2.3
ThT-AK3-5	3.6	3.49·1015	1.3	2.7	3.6	2.8
ThT-AK3-7	3.4	2.75·1015	1.1	2.2	3.0	2.3
ThT-AK3-8	3.8	4.23·1015	1.6	3.1	4.2	3.3
ThT-AK3-11	3.8	4.26·1015	1.9	3.7	4.9	4.0

*The uncertainty in R value does not exceed 0.3 nm.

Remarkably, a linear chromophore arrangement within the same surface groove ($d = 1$) can, in principle, account for the extremely high FRET efficiencies revealed for ThT-AK3 dye donor-acceptor pairs (Figure 15). Likewise, we cannot rule out that the observed strong energy transfer can be brought about by the accumulation of the donors and acceptors in specific cavities within the insulin fibril structure. These cavities can emerge from: i) fibril clustering [103, 104] and ii) coiling of twisted protofilaments around each other [104]. Recently, Kuznetsova et al. demonstrated the existence of two possible modes of ThT binding to the insulin fibrils [103]. The first mode involves the dye incorporation into the solvent-exposed grooves of fibril spine, whereas the second one implicates ThT localization in the clustered areas of fibril network [104]. Due to the preference of ThT and cyanine dyes to the same binding sites, the accumulation of the dyes in the cavities formed between the intertwined insulin fibrils seems highly probable. Besides, amyloid fibrils grown under agitation at neutral pH exhibited twisted morphology [103] and had a more loosely packed β-sheet structure compared to fibrils formed at acidic pH [81]. While examining the protofilament packing in the insulin amyloid fibrils, Jimenez et al. showed that twisting of a pair or several protofilaments around each other leads to the appearance of interacting regions between the protofilaments which are fixed relative to the cross-β-structure and regions free of interactions containing loops and/or domains that do not interfere with the protofilament packing [81]. These regions emerge due to a shorter twist between β-strands in the insulin protofilament in comparison with a twist of the whole fibril [81]. It can also be assumed that the presence of alternating interacting and non-interacting regions between protofilaments enables the formation of cavities suitable for accommodation of the FRET donors and acceptors. However, we can not rule out the possibility that such a striking decrease of ThT fluorescence in the presence of AK3-dyes results from the energy transfer between ThT and H-aggregates of cyanines. Specifically, we demonstrated above that for AK3-3, AK3-7 and AK3-1 the aggregation of acceptors on the fibrillar insulin matrix prevailed over the binding of monomeric dye species. In turn, the dominating species of AK3-5, AK3-8 and AK3-11 in the presence of insulin fibrils were represented by the dye monomers. Notably, for the ThT-AK3-dye donor-acceptor pairs the maximum FRET efficiencies were found to follow the order: AK3-3 (96%) > AK3-7 (95%) > AK3-1 (90%) > AK3-5 (87%) > AK3-8 (74%) > AK3-11 (56%). Taken together, the above findings indicate that the higher efficiencies of energy transfer were observed for the dyes tending to aggregate in the presence of fibrillar insulin.

Conclusion

In conclusion, using the optical spectroscopy and molecular docking techniques, a series of the trimethine cyanine dyes have been tested as potential probes for selective detection and characterization of fibrillar protein aggregates. The comprehensive analysis of the spectral characteristics of the trimethine cyanine dyes in the buffer solution and in the presence of control and fibrillar insulin and lysozyme allowed us to demonstrate that all dyes under study are capable of distinguishing between the non-aggregated and fibrillar proteins. The quantitative characterization of the complexation between the trimethines and the native or fibrillar proteins was performed. The comparison of the fluorescence responses and the binding parameters in the presence of the fibrillar insulin and lysozyme led us to hypothesize that the novel trimethine cyanine dyes are sensitive to the fibril morphology. Based on the results of absorption studies, it was concluded that the trimethine association with non-fibrilar and fibrillar proteins is accompanied by the changes in dye aggregation extent. Among our main findings are: i) the trimethine cyanine dyes under study tend to aggregate in the presence of the non-fibrillized insulin (all dyes) and lysozyme (AK3-3 and AK3-11), forming the non-fluorescent H-aggregates; ii) the trimethines AK3-8, AK3-1, AK3-5 and AK3-7 bind to the native lysozyme mainly in the form of monomers and dimers, but the magnitude of the dye-protein association (excepting AK3-8) is significantly lower than that in the presence of the fibrillar lysozyme; iii) the fibrillar proteins led to a strong increase of the trimethine fluorescence as compared to the control protein (excepting AK3-8) and the buffer solution, along with the 4-22 nm bathochromic shifts of the emission maxima of the dyes being more pronounced in the presence of the fibrillar insulin. The dye AK3-11 was selected as the most effective probe among the examined trimethines due to conversion of the non-emissive assemblies in aqueous solution and non-fibrillized proteins to the highly fluorescent monomeric species in the presence of the fibrillar insulin. The applicability of the Förster resonance energy transfer between the classical amyloid marker ThT as a donor and trimethine cyanine dyes as acceptors for the detection of the insulin amyloid fibrils was evaluated. It was suggested that the observed extremely high FRET efficiencies for the examined donor-acceptor pairs result from the dye accumulation in the cavities within suprafibrillar structures and/or energy transfer from ThT to the H-aggregates of cyanine dyes formed on the fibrillar matrix. To summarize, we successfully demonstrated that trimethine cyanine dyes are prospective amyloid probes capable of operating in both fluorescence and absorption modes.

References

[1] Chiti, F., Dobso, C. (2017). Protein misfolding, amyloid formation and human disease: a summary of progress over the last decade. *Annu Rev Biochem*, 27-68.

[2] Murph, R. M. (2002). Peptide aggregation in neurodegenerative disease. *Annu Rev Biomed Eng*, 174-155.

[3] Petrou, M., Dwamena, B. A., Foerster, B. R., MacEachern, M. P., Bohnen, N. I., Müller, M. L., Albin, R. L., & Frey, K. A. (2015). Amyloid deposition in Parkinson's disease and cognitive impairment: a systematic review. *Mov Disord*, 928-935.

[4] Drombosky, K. W., Rode, S., Kodali, R., Jacob, T. C., Palladino, M. J., & Wetzel, R. (2018). Mutational analysis implicates the amyloid fibril as the toxic entity in Huntington's disease. *Neurobiol Dis*, 126-138.

[5] Knowles, T. P. J., Vendruscolo, M., Dobson, C. M. (2014). The amyloid state and its association with protein misfolding diseases. *Nat Rev Mol Cell Biol*, 384-396.

[6] Chiti, F., Dobson, C. M. (2017). Protein misfolding, amyloid formation and human disease: a summary of progress over the last decade. *Annu Rev Biochem*, 27-68.

[7] Gallardo, R., Ranson, N. A., Radford, S. E. (2020). Amyloid structure: much more than just a cross-β-fold. *Curr Opin Struct Biol*, 7-16.

[8] Nguyen, P. H., Ramamoorthy, A., Sahoo, B. R., Zheng, J., Faller, P., Straub, J. E., Dominguez, L., Shea, J. E., Dokholyan, N. V., De Simone, A., Ma, B., Nussinov, R., Najafi, S., Ngo, S. T., Loquet, A., Chiricotto, M., Ganguly, P., McCarty, J., Li, M. S., Derreumaux, P. (2021). Amyloid oligomers: a joint experimental/ computational perspective on Alzheimer's disease, Parkinson's disease, type II diabetes, and amyotrophic lateral sclerosis. *Chem Rev*, 2545-264.

[9] Dobson, C. M. (2017). The amyloid phenomenon and its links with human disease. *Cold Spring Harb Perspect Biol*, a023648.

[10] Knowles, P. J., Mezzenga, R. (2016). Amyloid fibrils as building blocks for natural and artificial functional materials. *Adv Mater,* 6546-6561.

[11] Zheyu Wang, Saewon Kang, Sisi Cao, Michelle Krecker, Vladimir V. Tsukruk and Srikanth Singamaneni (2020). Protein-based functional nanomaterials. *MRS Bulletin*, 1017-1026.

[12] Trusova, V., Vus, K., Tarabara, U., Zhytniakivska, O., Deligeorgiev, T., & Gorbenko, G. (2020). Liposomes integrated with amyloid hydrogels: a novel composite drug delivery platform. *BioNanoSci*, 446-454.

[13] Li, Y., Zhao, C., Luo, F., Liu, Z., Gui, X., Luo, Z., Zhang, X., Li, D., Liu, C., & Li, X. (2018). Amyloid fobril structure of α-synuclein determined by cryo-electron microscopy. *Cell Res,* 897-903.

[14] Rambaran, R., Serpell, L. (2008). Amyloid fibrils. *Prion,* 112-117.

[15] Jimenes, J. L., Tennent, G., Perys, M., Saibil, H. R. (2001). Structural diversity of ex vivo amyloid fibrils studied by cryo-electron microscopy. *J Mol Biol*, 241-247.

[16] Iadanza, M. G, Jackson, M. P., Hewitt, E. W., Ranson, N. A., Radfordm S. E. (2018). A new era for understanding amyloid structures and disease. *Nat Rev Mol Cell Biol,* 755-773.

[17] Adamcik, J., Mezzenga, R. (2012). Study of amyloid fibrils via atomic force microscopy. *Curr Opin Colloid Interface Sci,* 369-376.

[18] Ruggeri, F. S., Sneideris, T., Vendruscolo, M., Knowles, T. P. J. (2019). Atomic force microscopy for single molecule characterization of protein aggregation. *Arch Biochem Biophys,* 134-148.

[19] Visser, M. J. E., Pretorius, E. (2019). Atomic force microscopy: the characterisation of amyloid protein structure in pathology. *Curr Top Med Chem,* 2958-2973.

[20] Adamcik, J., Jung, J.-M., Flakowski, J., De Los Rios, P., Dietler, G., & Mezzenga, R. (2010). Understanding amyloid aggregation by statistical analysis of atomic force microscopy images. *Nat Nanotechnol,* 423-428.

[21] Kaur, A., New, A. J., Sunde, M. (2020). Strategies for the molecular imaging of amyloid and the value of a multimodal approach. *ACS Sens,* 2268-2282.

[22] Lv, Z., Li, L., Man, Z., Xu, Z., Cui, H., Zhan, R., He, Q., Zheng, L., & Fu, H. (2020). Polarity-activated super resolution imaging probe for the formation and morphology of amyloid fibrils. *Nano Res,* 2556-2563.

[23] Yang, H. L., Fang, S. Q., Tang, Y. W., Wang, C., Luo, H., Qu, L. L., Zhao, J. H., Shi, C. J., Yin, F. C., Wang, X. B., & Kong, L. Y. (2019). A hemicyanine derivative for near-infrared imaging of β-amyloid plaques in Alzheimer's disease. *Eur J Med Chem,* 736-743.

[24] Mu, X., Wu, F., Wang, R., Huang, Z., Lv, T., Lu, Y., Liu, B., & Zhou, X. (2021). A cyanine-derived NIR molecular rotor for ratiometric imaging of amyloid-β aggregates. *Sens Actuators B:Chemical,* 29842.

[25] Duboisset, J., Ferrand, P., He, W., Wang, X., Rigneault, H., & Brasselet, S. (2013). Thioflavine-T and Congo Red reveal the polymorphism of insulin amyloid fibrils when probed by polarization-resolved fluorescence microscopy. *J Phys Chem B,* 784-788.

[26] Smith, J. F., Knowles, T. P. J., Dobson, C. M., MacPhee, C. E., & Welland, M. E. (2006). Characterization of the nanoscale properties of individual amyloid fibrils. *PNAS*, 15806-15811.

[27] Jurado, R., Adamcik, J., Sánchez-Ferrer, A., Bolisetty, S., Mezzenga, R., & Gálvez, N. (2021). Understanding the formation of apoferritin amyloid fibrils. *Biomacromolecules,* 2057-2066.

[28] Shen, C. L., Murphy, R. M. (1995). Solvent effects on self-assembly of beta-amyloid peptide. *Biophys J,* 640-651.

[29] Dzwolak, W., Pecul, M. (2005). Chiral bias of amyloid fibrils revealed by the twisted conformation of Thioflavin T: an induced circular dichroism/DFT study. *FEBS Letters,* 6601-6603.

[30] Sadqi, M., Hernández, F., Pan, U., Pérez, M., Schaeberle, M. D., Ávila, J., & Muñoz, V. (2002). α,-Helix structure in Alzheimer's disease aggregates of tau-protein. *Biochemistry,* 7150-7155.

[31] Stroud, J. C., Liu, C., Teng, P. K., Eisenberg, D. (2012). Toxic fibrillar oligomers of amyloid-β have cross-β structure. *PNAS,* 7717-7722.

[32] Torra, J., Viela, F., Megías, D., Sot, B., & Flors, C. (2022). Versatile near-infrared super-resolution imaging of amyloid fibrils with the fluorogenic probe CRANAD-2. *Eur J Chem,* e202200026.

[33] Squires, A. M., Devlin, G. L., Gras, S. L., Tickler, A. K., MacPhee, C. E., & Dobson, C. M. (2006). X-ray scattering study of the effect of hydration on the cross-β structure of amyloid fibrils. *J Am Chem Soc,* 11738-11739.

[34] Shinpei Tanaka, Tomomi Kubota, KatsuzoWakabayashi, Katsuhide Yutani, SatoruFujiwara (2002). An insight into the pathway of the amyloid fibril formation of hen egg white lysozyme obtained from a small-angle X-ray and neutron scattering study. *J Mol Biol,* 237-251.

[35] Hiramatsu, H., Kitagawa, T. (2005). FT-IR approaches on amyloid fibril structure. *Biochim Biophys Acta Proteins Proteom,* 100-107.

[36] Krebs, M., Bromley, E., Donald, A. (2005). The binding of thioflavin T to amyloid fibrils: localization and implications. *J Struct Biol,* 30-37.

[37] Mudliar, N. H., Pettiwala, A. M., Awasthi, A. A., Singh, P. K. (2016). On the molecular form of amyloid marker, Auramine O in human insulin fibrils. *J Phys Chem B,* 12474-12485.

[38] Smith, N. W., Alonso, A., Brown, S. M., Dzyuba S. V. (2010). Triazole-containing BODIPY dyes as novel fluorescent probes for soluble oligomers of amyloid Aβ1-42 peptide. *Biochem Biophys Res Commun,* 1455-1458.

[39] Kovalska, V., Chernii, S., Losytskyy, M., Tretyakova, I., Dovbii, Y., Gorski, A., Chernii, V., Czerwieniec, R., & Yarmoluk, S. (2018). Design of functionalized β-ketoenole derivatives as efficient fluorescent dyes for detection of amyloid fibrils. *New J Chem,* 13308-13318.

[40] Lindgren, M., Sorgjerd, K., Hammarstrom, P. (2005). Detection and characterization of aggregates, prefibrillar amyloidogenic oligomers, and protofibrils using fluorescencespectroscopy. *Biophys J,* 4200-4212.

[41] Kitts, C. C., Beke-Somfai, T., Norden, B. (2011). Michler's hydrol blue: a sensitive probe for amyloid fibril detection. *Biochemistry,* 3451-3461.

[42] Ryzhova, O., Vus, K., Trusova, V., Kirilova, E., Kirilov, G., Gorbenko, G., & Kinnunen, P. (2016). Novel benzanthrone probes for membrane and protein studies. *Methods Appl Fluoresc,* 034007.

[43] Li, Q., Lee, J. S., Ha, C., Park, C. B., Yang, G., Gan, W. B., & Chang, Y. T. (2004). Solid-phase synthesis of styryl dyes and their application as amyloid sensors. *Angew Chem Int Ed,* 6331-6335.

[44] Celej, M. S., Jares-Erijman, E. A., Jovin, T. M. (2008). Fluorescent N-arylaminonaphthalene sulfonate probes for amyloid aggregation of alpha-synuclein. *Biophys J,* 4867-4879.

[45] Åslund, A., Sigurdson, C. J., Klingstedt, T., Grathwohl, S., Bolmont, T., Dickstein, D. L., Glimsdal, E., Prokop, S., Lindgren, M., Konradsson, P., Holtzman, D. M., Hof, P. R., Heppner, F. L., Gandy, S., Jucker, M., Aguzzi, A., Hammarström, P., & Nilsson, K. P. R. (2009). Novel pentameric thiophene derivatives for *in vitro* and

in vivo optical imaging of a plethora of protein aggregates in cerebral amyloidoses. *ACS Chem Biol,* 673-684.

[46] Mishra, R., Sjolander, D., Hammarstrom, P. (2011). Spectroscopic characterization of diverse amyloid fibrils *in vitro* by the fluorescent dye Nile red. *Mol Biosyst*, 1232-1240.

[47] Clement, C. G., Truong, L. D. (2014). An evaluation of Congo red fluorescence for the diagnosis of amyloidosis. *Hum Pathol,* 1766-1772.

[48] Cui, M., Ono, M., Watanabe, H., Kimura, H., Liu, B., & Saji, H. (2014). Smart near-infrared fluorescence probes with donor-acceptor structure for *in vivo* detection of beta-amyloid deposits. *J Am Chem Soc,* 3388-3394.

[49] Wang, X., Chan, H. N., Desbois, N., Gros, C. P., Bolze, F., Li, Y., Li, H. W., & Wong, M. S. (2021). Multimodal theranostic cyanine-conjugated Gadolinium(III) complex for *in vivo* imaging of amyloid-β in an Alzheimer's disease mouse model. *ACS Appl Mater Interfaces*, 18525-18532.

[50] Yan, J., Zhu, J., Zhou, K., Wang, J., Tan, H., Xu, Z., Chen, S., Lu, Y., Cui, M., & Zhang, L. (2017). Neutral merocyanine dyes: for *in vivo* NIR fluorescence imaging of amyloid β-plaques. *Chem Commun,* 9910-9917.

[51] Smidlehner, T., Bonnet, H., Chierici, S., Piantanida, I. (2020). Fluorescently-labelled amyloid paired helical filaments (PHF) in monitoring its fibrillation kinetics. *Bioorg Chem,* 104196.

[52] Sabate, R., Estelrich, J. (2003). Pinacyanol as effective probe of fibrillar β-amyloid peptide: Comparative study with Congo Red. *Biopolymers,* 455-463.

[53] Kovalska, V. B., Losytskyy, M. Y., Tolmachev, O. I., Slominskii, Y. L., Segers-Nolten, G. M. J., Subramaniam, V., & Yarmoluk, S. M. (2012). Tri- and pentamethine cyanine dyes for fluorescent detection of α-synuclein oligomeric aggregates. *J Fluoresc,* 1441-1448.

[54] Gao, G. Q., & Xu, A. W. (2013). A new fluorescent probe for monitoring amyloid fibrillation with high sensitivity and realibility. *RSC Adv,* 21092-21098.

[55] Vus, K., Girych, M., Trusova, V., Gorbenko, G., Kurutos, A., Vasilev, A., Gadjev, N., & Deligeorgiev, T. (2019). Cyanine dyes derived inhibition of insulin fibrillization. *J Mol Liq,* 541-552.

[56] Kovalska, V., Losytskyy, M., Chernii, V., Volkova, K., Tretyakova, I., Cherepanov, V., Yarmoluk, S., & Volkov, S. (2012). Studies of anti-fibrillogenic activity of phthalocyanines of zirconium containing out-of-plane ligands. *Bioorg Med Chem*, 330-334.

[57] Chernii, S., Gerasymchuk, Y., Losytskyy, M., Szymański, D., Tretyakova, I., Łukowiak, A., Pekhnyo, V., Yarmoluk, S., Chernii, V., & Kovalska, V. (2021). Modification of insulin amyloid aggregation by Zr phthalocyanines functionalized with dehydroacetic acid derivatives. *PLOS ONE,* e0243904.

[58] Kovalska, V., Chernii, S., Losytskyy, M., Ostapko, J., Tretyakova, I., Gorski, A., Chernii, V., & Yarmoluk, S. (2018). Activity of Zn and Mg phthalocyanines and porphyrines in amyloid aggregation of insulin. *J Mol Recognit*, e2660.

[59] Zhytniakivska, O., Kurutos, A., Tarabara, U., Vus, K., Trusova, V., Gorbenko, G., Gadjev, N., & Deligeorgiev, T. (2020). Probing the amyloid protein aggregates with unsymmetrical monocationic trimethine cyanine dyes. *J Mol Liq*, 113287.

[60] Vus, K., Tarabara, U., Kurutos, A., Ryzhova, O., Gorbenko, G., Trusova, V., Gadjev, N., & Deligeorgiev, T. (2017). Aggregation behavior of novel heptamethine cyanine dyes upon their binding to native and fibrillar lysozyme. *Mol Biosyst,* 1970-980.

[61] Gorbenko, G., Trusova, V., Deligeorgiev, T., Gadjev, N., Mizuguchi, C., & Saito, H. (2019). Two-step FRET as a tool for probing the amyloid state of proteins. *J Mol Liq,* 111675.

[62] Tarabara, U., Kirilova, E., Kirilov, G., Vus, K., Zhytniakivska, O., Trusova, V., & Gorbenko, G. (2021). Benzanthrone dyes as mediators of cascade energy transfer in insulin amyloid fibrils. *J Mol Liq,* 115102.

[63] Zhytniakivska, O., Kurutos, A., Shchuka, M., Vus, K., Tarabara, U., Trusova, V., & Gorbenko, G. (2021). Forster resonance energy transfer between Thioflavin T and unsymmetrical trimethine cyanine dyes on amyloid fibril scaffold. *Chem Phys Lett,* 139127.

[64] Gorbenko, G., Zhytniakivska O., Vus K., Tarabaraa U. and Trusovaa V. (2021). Three-step Förster resonance energy transfer on an amyloid fibril scaffold. *Phys Chem Chem Phys,* 14746-14754.

[65] Iannuzzi, C., Borriello, M., Portaccio, M., Irace, G., & Sirangelo, I. (2017). Insights into insulin fibril assembly at physiological and acidic pH and related amyloid intrinsic fluorescence. *Int J Mol Sci,* E2551.

[66] Bacalum, M., Zorila, B., Radu, M. (2013). Fluorescence spectra decomposition by asymmetric functions: Laurdan spectrum revisited. *Anal Biochem,* 123-129.

[67] Vus, K., Trusova, V., Gorbenko, G., Kirilova, E., Kirilov, G., Kalnina, I., & Kinnunen, P. (2012). Novel aminobenzanthrone dyes for amyloid fibril detection. *Chem Phys Lett*, 110-115.

[68] Dallakyan, S., Olson, A. J. (2015). Small-molecule library screening by docking with PyRx. *Methods Mol Biol*, 243-250.

[69] Salentin, S., Schreiber, S., Haupt, V. J., Adasme, M. F., & Schroeder, M. (2015). PLIP: fully automated protein-ligand interaction profiler. *Nucleic Acids Res*, W443-W447.

[70] Trusova, V. (2015). Modeling of amyloid fibril binding to the lipid bilayer. *East Eur J Phys,* 51-58.

[71] Csizmadia, P. (1999). MarvinSketch and MarvinView: Molecule Applets for the World Wide Web. In: *Proceedings of ECSOC-3, the third international electronic conference on synthetic organic chemistry*, 367-369.

[72] Hanwell, M. D., Curtis, D. E., Lonie, D. C., Vandermeersch, T., Zurek, E., & Hutchison, G. R. (2012). Avogadro: An Advanced Semantic Chemical Editor, Valization, and Analysis Platform. *J Cheminform*, 17.

[73] Pettersen, E. F., Goddard, T. D., Huang, C. C., Couch, G. S., Greenblatt, D. M., Meng, E. C., & Ferrin, T. E. (2004). UCSF Chimera – a visualization system for exploratory research and analysis. *J Comput Chem,* 1605-1612.

[74] Cui, F., Shi, K., Zhang, L., Kawashima Y. (2006). Biodegradable nanoparticles loaded with insulin–phospholipid complex for oral delivery: Preparation, *in vitro* characterization and *in vivo* evaluation. *J Control Release*, 242–250.

[75] Scott, D. A. (1934). Crystalline insulin. *Biochem J.* 1592–1602.

[76] Vus, K., Trusova, V., Gorbenko, G., Sood R., Kinnunen, P. (2015). Thioflavin derivatives for the characterization of insulin and lysozyme amyloid fibrils *in vivo*:Fluorescence and quantum-chemical studies. *J Luminesc, 284-293.*

[77] Murali, J., Jayakumar, R. (2005). Spectroscopic studies of native and protofibrillar insulin. *J Struct Biol*, 180-189.

[78] Donabedian, P. L., Evanoff, M., Monge, F. A., Whitten, D. G., & Chi, E. Y. (2017). Substituent, charge, size effects on the fluorogenic performance of amyloid ligands: a small-library screening study. *ACS Omega,* 3192-3200.

[79] Hill, S. E., Miti, T., Richmond, T., Muschol, M. (2011). Spatial extent of charge repulsion regulates assembly pathways for lysozyme amyloid fibrils. *PLoS ONE,* e18171.

[80] Wang, S., Lium K.-N., Lum Y.-C. (2009). Amyloid fibrillation of hen egg-white lysozyme is inhibited by TCEP. *Biophys Biochem Res Comunn*, 639-642.

[81] Jiménez, J. L., Nettleton, E. J., Bouchard, M., Robinson, C. V., Dobson, C. M., & Saibil, H. R. (2002). The protofilament structure of insulin amyloid fibrils, *PNAS,* 9196-9201.

[82] Sulatskaya, A. I., Volova, E. A., Komissarchik, Ya. Yu., Snigirevskaya, E. S., Maskevich, A. A., Drobchenko, E. A., Kuznetsova, I. M., & Turoverov, K. K. (2014). Investigation of the kinetics of insulin amyloid fibrils formation. *Cell and Tissue Biology,* 186-191.

[83] Knowles, T. P. J., Oppenheim, T. W., Buell, A. K., Chirgadze, D. Y., & Welland, M. E. (2010). Nanostructured films from hierarchical self-assembly of amyloidogenic proteins. *Nature Nanotech,* 204-207.

[84] Kuperman, M. V., Chernii, S. V., Losytskyy, M. Yu., Kryvorotenko, D. V., Derevyanko, N. O., Slominskii, Y. L., Kovalska, V. B., & Yarmoluk, S. M. (2015). Trimethine cyanine dyes as fluorescent probes for amyloid fibrils:the effect of *N,N'*-substitutients. *Anal Biochem,* 9-17.

[85] Spano, F. C. (2006). Excitons in conjugated oligomer aggregates, films, and crystals. *Annu Rev Phys Chem*, 217–243.

[86] Beckford, G., Owens, E., Henary, M., Patonay, G. (2012). The solvatochromic effects of side chain substitution on the binding interactions of novel tricarbocyanine dyes with human serum albimin. *Talanta,* 45-52.

[87] Guralchuk, G. Ya., Katrunov, I. K., Grynyov, R. S., Sorokin, A. V., Yefimova, S. L., Borovoy, I. A., & Malyukin, Y. V. (2008). Anomalous surfactant-induced enhancement of luminescence quantum yield of cyanine dye J-aggregates. *J Phys Chem C,* 14762-14768.

[88] Volkova, K. D., Kovalska, V. B., Balanda, A. O., Vermeij, R. J., Subramaniam, V., Slominskii, Yu. L., & Yarmoluk, S. M. (2007). Cyanine dye-protein interactions: looking for fluorescent probes for amyloid structures. *J Biochem Bioph Methods,* 727-733.

[89] Kurutos A., Ryzhova O., Trusova V., Gorbenko G., Gadjev N., Deligeorgiev T. (2016). Synthesis of meso-chloro-substituted pentamethine cyanine dyes containing benzothiazolyl/benzoselenazolyl chormophores Novel synthetic approach and studies on photophysical properties upon interaction with bio-objects. *J Fluor,* 177-187.

[90] Bacalum, M., Zorila, B., Radu, M. (2013). Fluorescence spectra decomposition by asymmetric functions: Laurdan spectrum revisited. *Anal Biochem,* 123-129.

[91] Kurutos A., Ryzhova O., Trusova V., Gorbenko G., Gadjev N., Deligeorgiev T. (2016). Synthesis of meso-chloro-substituted pentamethine cyanine dyes containing benzothiazolyl/benzoselenazolyl chormophores Novel synthetic approach and studies on photophysical properties upon interaction with bio-objects. *J Fluor,* 177–187.

[92] Slavnova, T. D., Görner, H., Chibisov, A. K. (2007). J-aggregation of anionic ethyl meso-thiacarbocyanine dyes induced by binding to proteins. *J Phys Chem B,* 10023-10031.

[93] Slavnova, T. D., Görner, H., Chibisov, A. K. Cyanine-based J-aggregates as a chirality-sensing supramolecular. *J Phys Chem B*, 3379-3384.

[94] Hanczyc, P., Sznitko, L., Zhong, C., Heeger, A. J. (2015). Stimulated emission from rhodamine 6G aggregates self-assembled on amyloid protein fibrils. *ACS Photonics,* 1755-1762.

[95] Nygren, J., Svanvik, N., Kubitsa, M. (1998). The interactions between the fluorescent dye Thiazole orange and DNA. *Biopolymers*, 39-51.

[96] Alies, B., Eury, H., Essassi, E. M., Pratviel, G., Hureau, C., & Faller, P. (2014). Concept for simultaneous and specific in situ monitoring of amyloid oligomers and fibrils via Förster resonance energy transfer. *Anal Chem* 11877-11882.

[97] Gorbenko, G., Trusova, V., Kirilova, E., Kirilov, G., Kalnina, I., Vasilev, A., Kaloyanova, S., & Deligeorgiev, T. (2010). New fluorescent probes for detection and characterization of amyloid fibrils. *Chem Phys Lett*, 275-279.

[98] Lee, J. H., Lee, I. H., Choe, Y. J., Kang, S., Kim, H. Y., Gai, W. P., Hahn, J. S., & Paik, S. R. (2009). Real-time analysis of amyloid fibril formation of alpha-synuclein using a fibrillation-state-specific fluorescent probe of JC-1. *Biochem J,* 311-323.

[99] Lakowicz, J. R. (2006). *Principles of fluorescence spectroscopy,* Springer, New York,

[100] Drake, J. M., Klafter, Levitz, J., P. (1991). Chemical and biological microstructures as probed by dynamic processes. *Science,* 1574-1579.

[101] Yoshihara, H., Saito, J., Tanabe, A., Amada, T., Asakura, T., Kitagawa, K., & Asada, S. (2016). Characterization of novel insulin fibrils that shows strong cytotoxicity under physiological pH. *J Pharm Sci* 1419-1426.

[102] Mawhinney, M. T., Williams, T. L., Hart, J. L., at al. (2017). Elucidation of insulin assembly at acidic and neutral pH: characterization of low molecular weight oligomers. *Proteins,* 2096-2110.

[103] Sulatskaya, A., Rodina, N., Sulatsky, M., Povarova, O., Antifeeva, I., Kuznetsova, I., & Turoverov, K. (2018). Investigation of α-Synuclein Amyloid Fibrils Using the Fluorescent Probe Thioflavin T. *Int J Mol Sci*, 2486.

[104] Kuznetsova, I. M., Sulatskaya, A. I., Uversky, V. N., Turoverov, K. K. (2012). A new trend for the analysis of the Thioflavin T binding to amyloid fibrils. *Mol Neurobiol*, 488-498.

Chapter 3

Membrane Effects of Amyloid Fibrils: Fluorescence Study and Computer Modeling

V. Trusova*, O. Zhytniakivska, U. Tarabara, K. Vus and G. Gorbenko
Department of Medical Physics and Biomedical Nanotechnologies, V.N. Karazin Kharkiv National University, Kharkiv, Ukraine

Abstract

Amyloid fibrils, highly ordered elongated protein aggregates with a characteristic cross-β-sheet molecular architecture, have received growing research attention throughout the past decades. These nanoassemblies are explored from three main perspectives concerning the pathological, functional and artificial amyloids. Pathological amyloids are proved to be associated with a range of human diseases, among which are neurodegenerative disorders, type II diabetes, systemic amyloidoses, etc. Ongoing research efforts are directed at elucidating the nature of protein misfolding and amyloid conversion, search for therapeutic molecules capable of preventing these processes, increasing the sensitivity and specificity of amyloid detection. Functional amyloids, found in bacteria, fungi, plants and mammals, are involved in diverse physiological activities, such as biofilm formation, storage of peptide and proteins, information carriage, signaling, etc. Artificial amyloids are scrutinized as building blocks for bioinspired functional nanomaterials with versatile applications, including tissue engineering, drug delivery, biosensing, nanowire production, etc. Major research efforts are currently focused on pathological amyloid with the emphases on the molecular-level details of cell damage by pathogenic protein assemblies.

* Corresponding Author's Email: valerija.trusova@karazin.ua.

In: Advances in Biology. Volume 2
Editor: Charles D. Grant
ISBN: 979-8-88697-422-5

Accumulating evidence supports the idea that cytotoxic action of amyloid fibrils and their oligomeric intermediates is primarily targeted at cellular membranes. The cytotoxicity of oligomeric aggregates is displayed in the formation of non-specific ion channels, the alterations in calcium homeostasis and activity of membrane enzymes, and the loss of membrane integrity, while the mature fibrils are capable of producing the increase in membrane permeability, lipid bilayer thinning, membrane fragmentation and the uptake of lipids, etc. The membrane effects of aggregated proteins essentially depend on the basic physicochemical characteristics of a lipid bilayer, such as chemical nature of phospholipid headgroups, electrostatic surface potential, degree of hydration, lipid packing density, phase state, bilayer curvature, etc. which, in turn, are determined by the exact lipid composition. This emphasizes the importance of systematic investigation of the model protein-lipid systems with controllable structural and physicochemical properties. The present contribution is a quintessence of our research efforts focused on elucidating the membrane effects of amyloid fibrils formed from structurally different proteins – lysozyme, insulin, N-terminal fragment of apolipoprotein A-I and Aβ-peptide. The lipid component of the model protein-lipid systems was represented by the unilamellar liposomes composed of zwitterionic (phosphatidylcholine, PC), anionic (cardiolipin, CL) phospholipids and cholesterol (Chol). The following main questions have been addressed: i) whether the membranotropic action of the examined fibrillar proteins depends on the electrostatic potential of lipid bilayer; ii) how the amyloid-induced structural alterations of the model membranes are modulated by cholesterol in the two- (PC/Chol) and three-component (PC/CL/Chol) lipid systems; iii) what the membrane disposition of amyloid fibrils is. To answer these questions, several fluorescence-based approaches have been used, involving the fluorescent probes located in the polar (Laurdan) and nonpolar (pyrene, diphenylhexatriene) regions of a lipid bilayer, the intrinsic protein fluorescence and its quenching by a polar quencher acrylamide. To uncover the preferable modes of the fibril-membrane binding, the computer modeling tools, such as CreateFibril, AQUASOL, PPM and molecular docking have been employed. The results obtained provide evidence for i) different modes of interactions of monomeric and fibrillar proteins with the model membranes; ii) more pronounced influence of fibrillar proteins on the degree of hydration and free volume of the lipid bilayer; iii) essential roles of electrostatic, hydrophobic protein-lipid interactions and amyloid twist in determining the bilayer location of amyloid fibrils; iv) strong dependence of the membrane-modifying effects of fibrillar proteins on the lipid bilayer composition. Overall, our findings support the notion that the development of effective anti-amyloid strategies should involve the search for compounds capable of preventing the changes in membrane structure and physicochemical properties.

Keywords: amyloid fibrils, lipid bilayer, membranotropic activity, fluorescence, computer modeling, molecular docking

Introduction

During its lifecycle between the synthesis on ribosome and eventual degradation through the proteolysis, protein molecule can adopt a bewildering collection of conformational states, interconverting over a wide range of timescales [1, 2]. The number and activity of different protein intermediates are controlled by intrinsic amino acid sequence of the polypeptide as well as by interactions with molecular chaperons, degradation processes and other complicated quality control mechanisms [3, 4]. The intermediate states are generally characterized by the exposed patches of hydrophobic amino acid residues which are prone to self-association. Hydrophobic interactions between partially folded states force the polypeptide chain to leave the native folding pathway and enter the so-called off-folding route. One of the possible off-folding pathways involves the transition of partially unfolded intermediates into the extended highly ordered fibrillar structures, amyloid fibers [2-4]. The transformation of normally soluble proteins into insoluble mechanically stable fibrous aggregates has become a focus of considerable attention due to recognition of these non-covalent protein linear homopolymers as the main identifiers of devastating neurodegenerative and metabolic disorders including Parkinson's, Alzheimer's, Huntington's, Creutzfeld-Jakob diseases, type II diabetes, bovine spongiform encephalopathy, etc. [5].

Due to small difference between stabilizing and destabilizing forces, native protein structure is only marginally stable, so that any variation in physicochemical properties of polypeptide surroundings may appear critical for protein transition from monomeric to aggregated state. Accumulating evidence suggests that in biological environment protein self-associating tendency can be enhanced by cell membranes [6, 7]. Lipid bilayer, the basic structural element of biological membranes, is commonly regarded as two-dimensional liquid providing a variety of environments which can affect protein structure and dynamics via both specific and non-specific interactions. These interactions are controlled not only by general physicochemical characteristics of a membrane, such as surface charge, degree of hydration, hydrophobicity, bilayer curvature, but also by the exact chemical nature of membrane lipids, conformation and dynamics of lipid headgroups and acyl chains, and protein-lipid selectivity arising from hydrophobic matching at the protein-lipid interface [8].

A wealth of recent data corroborates the hypothesis that protein oligomers formed on a membrane template, serve as nuclei for amyloid fibril growth on the one hand, while on the other hand they represent toxic species producing membrane destabilization and subsequent cell death [9, 10]. Due to fundamental similarity between intra- and interchain interactions, protein fibrillization represents an alternative folding pathway of partially unfolded or misfolded proteins that acquire a stable structure by reaching the minimum on the energy landscape [1]. Fibril formation is nucleation-dependent and proceeds through thermodynamically unfavorable monomer association into a critical oligomeric nucleus followed by its subsequent energetically favorable elongation and the exponential fibril growth [11].

Lipid bilayer provides a variety of environments in which aggregated state of polypeptide chain appears to be more thermodynamically favorable than its monomeric form. The principal membrane-related factors promoting protein aggregation have been tentatively outlined as follows [12, 13]: i) structural transformation of polypeptide chain into aggregation-prone conformation featured by the exposed patches complementary to each other in terms of hydrophobicity, charge distribution and H-bonding propensity; ii) increase of protein local concentration in a lipid phase; iii) specific aggregation-favoring orientation of a protein molecule in a lipid environment; iv) aggregation-competent arrangement of the solvent exposed and bilayer-buried portions of polypeptide chain; v) electrostatic attraction between the dipoles of antiparallel α-helixes; vi) mismatch between hydrophobic thicknesses of the lipid bilayer and the embedded protein fragment.

Furthermore, the modern theories of proteinopathy in membrane environment postulate the existence of two pathways by which lipids can affect the conversion of native proteins into the amyloid state [14-16]. According to the first, the so called desposition pathway, monomeric protein species initially self-assemble into disordered oligomeric species coming through the set of nucleation events with subsequent conversion of these soluble aggregates into grown highly ordered amyloid structures [15]. Second route referred to as condensation pathway, assumes that protein monomers initially phase separate into liquid proplets which facilitates the formation of disordered oligomers. These early oligomers further convert into the extended fibrillar structures [14].

Regardless of the exact type of mechanism realized in any particular system, lipid molecules are considered as potent promoters of protein fibrillzation. Furthermore, a strong causative link between the impairment of lipid bilayer homeostasis and the onset and progression of amyloid diseases is emerged [16]. Accordingly, a vast majority of theoretical and experimental reports suggest that the toxic potential of amyloid aggregates correlates with their ability to damage the cell membranes [17]. Several factors have been reported to regulate the mechanisms of membrane disruption induced by amyloid species [10]. Among these are lipid bilayer composition, phase state, curvature, elasticity, electrostatic surface charge, hydrophilicity and lipid packing density, presence/absence of metal ions, etc. This highlights the importance of systematic studies of the model protein-lipid systems with controllable structural and physicochemical characteristics.

The present contribution is a quintessence of our research efforts focused on elucidating the membrane effects of amyloid fibrils formed from structurally different proteins – lysozyme, insulin, N-terminal fragment of apolipoprotein A-I and Aβ-peptide. The lipid component of the model protein-lipid systems was represented by the unilamellar liposomes composed of zwitterionic (phosphatidylcholine, PC), anionic (cardiolipin, CL) phospholipids and cholesterol (Chol). The key issues which will be discussed here are as follows: i) to evaluate the effect of lipid bilayer surface potential on the membranotropic activity of the fibrillar proteins, ii) to explore the role of cholesterol in amyloid-induced membrane disruption, iii) to analyze the spatial distribution of fibrillar proteins with respect to a membrane surface. To answer these questions, several fluorescence-based approaches have been used, involving the fluorescent probes located in the polar (Laurdan) and nonpolar (pyrene, diphenylhexatriene) regions of a lipid bilayer, the intrinsic protein fluorescence and its quenching by a polar quencher acrylamide. To uncover the preferable modes of the fibril-membrane binding, the computer modeling tools, such as CreateFibril, AQUASOL, PPM and molecular docking have been employed.

Materials and Methods

Materials

The hen egg white lysozyme, pyrene, diphenylhexatriene (DPH) and cholesterol were purchased from Sigma (St. Louis, MO, USA). 1-palmitoyl-2-oleoyl-sn-glycero-3-phosphocholine and cardiolipin were from Avanti Polar Lipids (Alabaster, AL). Laurdan (6-Lauroyl-2-dimethylaminonaphthalene) was from Invitrogen Molecular Probes (Eugene, OR, USA). 3-methoxybenzanthrone (MBA) was from Zonde (Latvia). The N-terminal 1-83 fragment of human apolipoprotein A-I (apoA) 1-83/G26R and its single tryptophan variants G26R/W@8, G26R/W@50 and G26R/W@72 were expressed and purified as described elsewhere [18]. The apoA preparations were at least 95% pure as assessed by SDS-PAGE.

Preparation of Amyloid Fibrils

The lysozyme amyloid fibrils were grown by the incubation of the protein solution (10 mg/ml) in 10 mM glycine buffer (pH 2) at 60°C for 14 days. The working solutions of the native and fibrillar lysozyme were prepared in 5 mM Na-phosphate buffer (pH 7.4). ApoA variants were freshly dialyzed from 6M guanidine hydrochloride solution into 10 mM Tris buffer, 150 mM NaCl, 0.01% NaN_3, pH 7.4 before use. The reaction of protein fibrillization was conducted at 37°C in the above buffer with constant agitation on an orbital shaker. The amyloid nature of fibrillar aggregates was confirmed in Thioflavin T (ThT) assay [19].

Preparation of Lipid Vesicles

The large unilamellar lipid vesicles were prepared by the extrusion technique from PC and its mixtures with CL (5, 10 or 20 mol%) and/or Chol (30 mol%). The thin lipid films were obtained by the evaporation of lipids' ethanol solutions and then hydrated with 1.2 ml of 5 mM Na-phosphate buffer (pH 7.4) to yield the final lipid concentration 2 mM. Afterwards, the lipid suspensions were extruded through a 100 nm pore size polycarbonate filter. Hereafter, the liposomes composed of binary mixtures of PC with 30 mol%

Chol, 5, 10 or 20 mol% CL were referred to as Chol30, CL5, CL10 and CL20, while the lipid vesicles composed of ternary mixtures of PC with CL (5 or 20 mol%) and Chol (30 mol%) were referred to as CL5/Chol30 and CL20/Chol30, respectively.

Fluorescence Measurements

The steady-state fluorescence spectra were recorded with RF6000 spectrofluorimeter (Shimadzu, Japan). The fluorescence measurements were performed at 20°C using 10 mm path-length quartz cuvettes. The excitation wavelengths were 340 nm for pyrene and 364 nm for Laurdan. The ratio of vibronic bands in the pyrene emission spectra (I_I / I_{III}
the fluorescence intensities at 371 nm (I_I) and 382 nm (I_{III}). The excimer-to-monomer fluorescence intensity ratio (*E/M*) was determined by measuring the fluorescence intensities at the monomer (391 nm) and excimer (466 nm) peaks. The generalized polarization (GP) of Laurdan fluorescence was determined as [20]:

$$GP = \frac{I_B - I_R}{I_B + I_R} \quad (1)$$

where I_B and I_R are the fluorescence intensities of the blue (440 nm) and red (490 nm) spectral components, respectively.

The fluorescence quenching experiments were carried out with the neutral water-soluble quencher acrylamide. The emission spectra of lysozyme were recorded with excitation at 296 nm, using 5 nm band passes for both excitation and emission. Small aliquots (10 μl) of the acrylamide stock solution (4 M) were added to a stirred and temperature-controlled protein solution or protein-lipid mixtures. Fluorescence intensity measured in the presence of quencher was corrected for the reabsorption and inner filter effects using the following coefficient:

$$k = \frac{\left(1 - 10^{-A}\right) A_s}{\left(1 - 10^{-A_s}\right) A} \quad (2)$$

where A is the protein absorbance in the absence of a quencher, and A_s is the total absorbance of the sample at excitation or emission wavelengths. The fluorescence quenching data were analyzed in terms of the Stern-Volmer equation [19]:

$$I_0/I = 1 + k_q\tau_0 = 1 + K_{SV}[Q] \quad (3)$$

Where I_0 and I are fluorescence intensities recorded in the absence and presence of a quencher, respectively, k_q is the bimolecular quenching rate constant, τ_0 is the fluorophore lifetime in the absence of the quencher, K_{SV} is the Stern-Volmer constant, and [Q] is the quencher concentration.

Steady-state DPH anisotropy was measured at the excitation and emission wavelengths 0f 350 and 450 nm, respectively. Probe stock solution was prepared in tetrahydrofuran. To incorporate DPH into the lipid bilayer, liposome suspensions were incubated with the probe during 30 minutes at room temperature.

Transmission Electron Microscopy

For electron microscopy assay, a 10 μl drop of the protein solution was applied to a carbon-coated grid and blotted after 1 min. A 10 μl drop of 2% (w/v) uranyl acetate solution was placed on the grid, blotted after 30 sec, and then washed 3 times by deionized water and air dried. The resulting grids were viewed at Jeol JSM-840 or Tecnai 12 BioTWIN electron microscope.

Molecular Docking Calculatuions

To predict the most favorable modes of interactions between amyloid fibrils and lipid molecules, the molecular docking studies were conducted using the PatchDock server with futher refinement with FireDock server. In brief, PatchDock server is based on shape complementarity principles and follows a geometry-based molecular docking algorithm to find the docking transformations with good molecular shape complementarity. PatchDock algorithm separates the Connolly dot surface representation [21-23] of the molecules into concave, convex and flat patches. These divided complementary patches are matched to generate candidate transformations and evaluated by geometric fit and atomic desolvation energy scoring [24] function. RMSD (root mean square deviation) clustering is applied to the docked solutions to select the non-redundant results and to discard redundant docking structures. The best ten solutions obtained with PatchDock were submitted to the following refinement by FireDock [25, 26] yielding the results ranked on the basis of global energy of binding (GBE). The refined complex with the lowest GBE was used for further analysis. Optimized three-dimensional coordinates of lipid molecules were generated from LIPID MAPS web-resource (https://www.lipidmaps.org/). The lipids were docked with the model amyloid fibrils of insulin provided by M. Sawaya (https://www.molnac.unisa.it/BioTools/cocomaps/) or prepared by our models as described in the section "Construction of amyloid fibrils section". The three-dimensional structure of Aβ (1-42)-peptide fibrils was obtained from Protein Data Bank, PDB entry 2BEG.

Molecular Characterization of Lysozyme and Apolipoprotein A-I Amyloid Structures

Lysozyme is a multifunctional cationic protein with extensive bactericidal, antitumor and immunomodulatory activities. The mutants of human lysozyme are prone to pathological fibrillization associated with hereditary nonneuropathic systemic amyloidosis, a disease in which amyloid deposits accumulate in a range of tissues including the kidney, liver and spleen [27, 28]. The ortholog of human lysozyme, hen egg white lysozyme (Lz) sharing ~60% of sequence homology, readily undergoes fibrillization *in vitro* and is extensively employed in the amyloid studies.

Apolipoprotein A-I (apoA), a major component of plasma high density lipoproteins, plays a crucial role in a wide diversity of physiological processes, including lipid metabolism, reverse cholesterol transport, etc. [29]. Different mutations in apoA gene significantly affect the normal physiological functioning of the native protein by converting it to the precursor for amyloid fibril formation. Specifically, *in vivo* formation of amyloid deposits of sixteen variants of apoA was shown to be directly associated with the development of familial hypercholesterolemia and autosomal dominant systemic amyloidosis [30]. The amyloidogenic mutations of apoA are generally divided into two groups – those located within the N-terminal fragment of the protein (residues 1-90) and associated with hereditary amyloidosis, and those occurring in the central region (residues 140-170) and inducing the defective activation of lecithin-cholesterol acyltransferase [31]. One of the most important N-terminal amyloidogenic mutations of apoA is known as Iowa and involves the substitution of Gly residue in the position 26 with Arg (G26R mutation). This destabilizes the N-terminal α-terminal helix bundle domain of the polypeptide and greatly enhances the formation of amyloid fibrils.

Figure 1. TEM images of Lz and apoA amyloid fibers.

Figure 1 represents the transmission electron images of amylid fibrils of hen egg white lysozyme and different mutants of apoA referred to here as G26R, G26R/W@8, G26R/W@50 та G26R/W@72, formed in the absence of lipid vesicles.

As seen from this figure, amyloid assemblies were visuzled as filamentous structures of different length with the diameter 15-20 nm in the case of Lz, and 5-10 nm in the case of apoA mutants.

In our previous studies [32] we proposed the structural model of apoA amyloid fibers. According to this model, the fibrillar aggregates have U-like shape in which two β-hairpins, embracing the segments Leu14-Val31 and Gln41-Ser58, form the self-complimentary zippers, stabilized by van der Waals and hydrophobic interactions as well as by salt bridge between Arg27 and Asp48.

A number of works were devoted also to deciphering the structural peculiarities of Lz amyloid fibrils [33-35]. More specifically, based on the combined results from proteolysis and Fourier-transform infrared spectroscopy studies, Frare et al. showed that amyloid fibers are formed by the protein fragment involving the amino acid residues 32-108 [34]. this fragment is characterized by cross-β-conformation, while the rest of the residues are in a disordered configuration. In addition, it was shown that lysozyme nanopeptide $_{54}$GILQINSRW$_{62}$ plays a key role in protein fibrillization and represents the hydrophobic core of amyloid assemblies. Finally, utilizing the atomic force microscopy as an analytical tool, Lara et al. examined the mechanical properties of Lz amyloids and calculated the Young's modulus which was equal 2.8 ± 0.9 GPa [36].

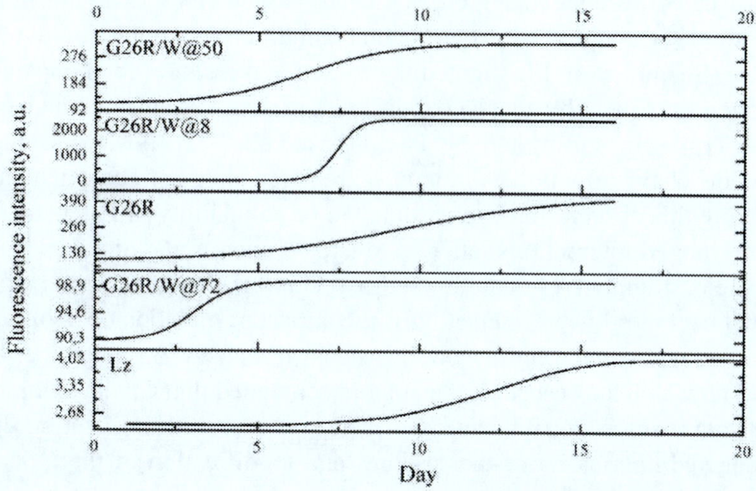

Figure 2. ThT kinetics profiles in the presence of different amylod fibrils measured at 484 nm.

The protein amyloidogenic potential was evaluated using the amyloid-specific fluorophore Thioflavin T (ThT). Kinetic profiles of this dyes are characterized by sigmoidal pattern, the hallmark of protein fibrillization process (Figure 2). The unique photophysical properties of ThT are believed to underly a marked increase of the dye emission in the amyloid-bound state. ThT belongs to the so-called molecular rotors whose spectral properties are governed by environmental viscosity. In low viscosity solutions benzyl-amine and -thiazole rings rotate freely around the central C-C bond resulting in the quenching of ThT fluorescence [37]. Upon formation of the dye-amyloid complexes, ThT immobilization within the fiber core suppresses the rotation of the probe giving rise to the increase in its fluorescence. ThT affinity to the fibrillar structures is generally explained by specific location of the probe within fibril matrix. According to the channel hypothesis, ThT resides in the grooves formed by the side amino acid residues with the principal axis of the dye being parallel to the fibril axis. Such orientation hinders the probe rotation and enhances its emission.

As seen from Figure 2, the kinetic profiles of ThT fluorescence were dependent on the type of apoA mutation. Specifically, the degree of fibrillization was found to increase in the order: apoA – G26R/W@72 – G26R/W@50 – G26R – G26R/W@8. In this context it should be mentioned that it seems impossible to draw an umbiguous conclusion about the reason of the observed difference in aggregation potential of different variants of apoA – the subtle variations in experimental conditions or mutation type. To illustrate, Sechini et al. highlighted the role of Trp residues in apomyoglobin fibrillogenesis [38]. The analysis of different protein variants in which one or several Trp were substituted by Phe, showed that Trp position along the polypetide chain affects the protein aggregation degree. The phenomenon underlying this observation was assumed to be an inability of Phe to maintain the long-ranged interactions inherent to Trp. If such mutation takes place in the protein fragment critical for self-association, the general aggregation potential becomes lower. Interestingly, the impact of mutation on apoA ability to fibrillize was shown in the work of Adachi et al. [39]. Based on the results of ANS and ThT fluorescence, the authors concluded that Gly substitution by Lys at the position 26, reduces the kinetics of peptide fibrillization due to different hydrophobicity of the residues and modification of the monomer-monomer intermolecular contacts.

Fluorescence Studies of Membranotropic Activity of Amyloid Fibers

It is becoming increasingly apparent that although mature amyloid fibrils bind to a membrane relatively weakly, they are far from being chemically inert nontoxic species. Numerous studies indicate substantial cytotoxic potential of fibrillar aggregates [40-42]. The integral picture of membrane damage by toxic protein aggregates have not been emerged yet because of three main disagreements discussed in a scientific community, viz.: i) whether mature amyloid fibrils or their precursors (oligomeric prefibrillar aggregates) are the most cytotoxic species; ii) whether amyloids increase lipid bilayer conductance or form discrete pores and channels; iii) what kind of cell death, apoptosis or necrosis is induced by the protein aggregates. The alterations in membrane integrity initiated by mature fibrils may result in the enhancement of lipid bilayer permeability [43], lipid loss [44], receptor activation [45], membrane fragmentation [46] and oxidation of membrane lipids [47]. For instance, Novitskaya et al. showed that both the β-oligomers and amyloid fibrils of recombinant mammalian prion protein are highly toxic to cultured cells, hippocampal and cerebellar neurons. The percent of cell death in cultured cells treated with β-oligomers was lower than that in the cells treated with fibrils [48]. In this study amyloid fibrils and the β-oligomers of prion protein were found to induce apoptosis. However, Gharibyan and coworkers reported that lysozyme fibrils produce necrosis-like cell death, while the cells treated with protein oligomers undergo the apoptotic changes [49]. Another pathway for the cytotoxicity of amyloid fibrils was proposed by Huang and coworkers, who observed the hemolysis and aggregation of erythrocytes accompanied by the disulfide cross-linking under the influence of lysozyme protofibrils and fibrils [50]. Recent studies revealed that fibrillar lysozyme induced the liposome leakage resembling the effect of carbon nanotube conjugates with adsorbed lysozyme [51]. Along with the direct membrane-disrupting action of mature fibrils, their cytotoxicity has been proposed to be associated with fibril growth at the lipid matrix [52]. Another possibility involves fibril fragmentation leading to the production of shorter fibrillar aggregates, which possess a higher cytotoxic activity than their longer counterparts [53]. Nevertheless, despite the rapidly growing amount of experimental data on the role of aggregated proteins in the amyloid related diseases, the nature of the toxic species still remains obscure.

In the present study we focused our efforts on the investigation of the influence of fibrillar aggregates of lysozyme on the structural and physical properties of the model membranes composed of phosphatidylcholine (PC) and its mixtures with cardiolipin (CL) (5, 10, 20 mol%) and/or cholesterol (Chol) (30 mol%). To evaluate the extent of lipid bilayer modification, we used fluorescence spectroscopy, one of the most accurate and powerful tools for monitoring the changes in membrane structure. Our goal was to ascertain what kind of structural alterations the fibrillar protein species can produce in the lipid bilayer. To achieve this purpose, four fluorescent probes with different spectral properties and bilayer location have been employed, viz. pyrene and 1,6-diphenyl-1,3,5-hexatriene (DPH), distributing in the acyl chain region, as well as Laurdan and 3-methoxybenzantrone (MBA) residing at the lipid-water interface.

Impact of Protein Amyloid Fibrils on Lipid Bilayer Nonpolar Region

To explore the membrane modifications at the level of lipid acyl chains, induced by amyloid fibers, classical membrane nonpolar probes pyrene, DPH and Laurdan were utilized. Pyrene is a polycyclic aromatic compound which localizes predominantly in the region of 4-13 lipid carbons with the long axis of the probe being parallel to the bilayer normal [54]. As seen in Figure 4, pyrene fluorescence spectrum is characterized by the existence of five vibronic bands of monomer emission, numbered from I to V starting from the lowest wavelength. Vibronic bands reflect the π-π* transitions in pyrene molecule. Specifically, the first vibronic band (peak at 372 nm) corresponds to 0-0 transition, while the third band (peak at 383 nm) describes the 0-2 transition [55]. Due to the strong relationship between the electronic and vibronic states of the fluorophore molecule, the third monomer peak appears to be extremely sensitive to the changes in environmental polarity, the effect manifesting itself in the stronger emission in hydrophobic solvents as compared to the first vibronic band, whose intensity dominates in the polar surroundings. The ratio of the first-to-third pyrene peaks is known as the "Ham effect" and represents the indicator of the probe microenvironmental polarity [56]. Along with monomer emission, pyrene fluorescence spectra are featured also by excimer emission. The collisions between the excited and non-excited pyrene monomers results in the formation of the excited dimer, or excimer, emitting in the range of 470-480 nm. The close proximity of excited and nonexcited pyrene species and their steric conformity are reported to be the main prerequisites for excimer formation [57]. These rationales suggest that pyrene excimerization is a diffusion-controlled process, which is governed by the dynamic properties of the lipid bilayer. The formation of excited dimers can be described by the following scheme:

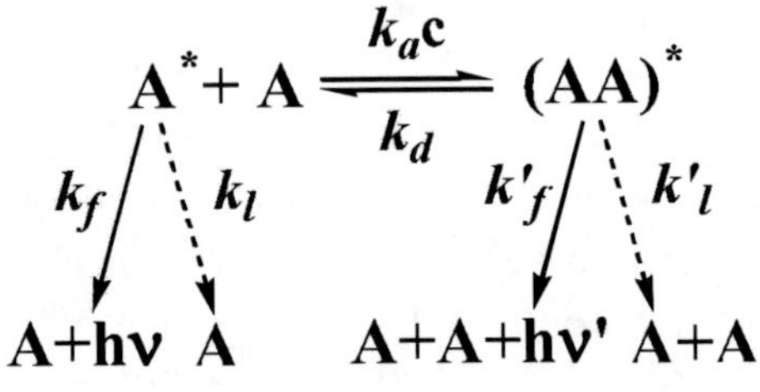

where A and A^* denote the pyrene monomers in the ground and excited states, respectively, k_a – the constant of pyrene formation, $k_f, k_f^{'}$ probabilities of radiative decay of excited monomers and excimers, respectively, $k_l, k_l^{'}$ – the probabilities of nonradiative decay, k_d dissociation constant. For the quantitative description of pyrene excimerization in terms of diffusion-controlled approach, a vast majority of theoretical models has been proposed [54-60]. Among these, the most accepted is the model of random walk, developed by Montroll [58] and further extended by Galla [59]. The model postulates that lipid bilayer is the two-dimensional lattice on which the pyrene monomers form sublattice. The excited monomers locate in the centers of this sublattice and represent the "traps" for the nonexcited molecules. The average steps necessary for nonexcited monomer to fall in the trap and form the complex with an excited molecule, equals to:

$$\langle n_s \rangle = \frac{2}{\pi f_{pyr}} \ln \frac{2}{f_{pyr}} \tag{4}$$

where f_{pyr}

The parameter $\langle n_s \rangle$ relates to the frequency of collisions between excited and nonexcited monomers via the following equation:

$$\nu_{col} = \langle n_s \rangle^{-1} \cdot \nu_j \tag{5}$$

where ν_j is the frequency of diffusional steps.

From the other side, ν_{col}

$$\nu_{col} = \frac{E}{M \cdot \kappa} \cdot \frac{k_f}{k_f^{'}} \cdot \frac{1}{\tau_0^E} \tag{6}$$

where E and M define the excimer and monomer fluorescence intensities, respectively, κ is the proportionality coefficient between the ratio of the quantum yields and intensities of excimer and monomer fluorescence, τ_0^E is the excimer lifetime.

The eq. (6) may be rewritten if one suggests that $\kappa \approx 0.8$, $k_f / k_f^{'} \approx 0.1$, $\tau_0^E \approx 38$ ns [58, 59].

Combining the eqs. (4) – (6), one receives:

$$\nu_j = \frac{E}{M \cdot \kappa} \cdot \frac{k_f}{k_f^{'}} \cdot \frac{1}{\tau_0^E} \cdot \frac{2}{\pi f_{pyr}} \ln \frac{2}{f_{pyr}} \tag{7}$$

Pyrene diffusion coefficient may be calculated as:

$$D_{diff} = \frac{1}{4} \nu_j \cdot \lambda^2 \tag{8}$$

where $\lambda \approx 0.8$ nm denotes the average length of monomer jump.

In terms of the described model, pyrene is considered as the indicator of lipid bilayer free volume, which controls the frequency of collisions and pyrene diffusion coefficient. Bilayer free volume is a result of rotational isomers of lipid acyl chains (the so-called gauche-trans-gauche kinks) because of thermal undulations of hydrophobic tails. Kink formation results in lateral displacement of the chain and creation of the local free volume (Figure 3)

The diffusional step of pyrene is finished when the space, created because of molecule jump, will be closed due to the defect displacement of the neighboring acyl chain. Thus, the frequency of diffusional steps ν_j correlates with the frequency of the displacement of the defects along the lipid hydrophobic tail, ν_k

$$\nu_j = \frac{2d_k^2}{L^2} \nu_k \tag{9}$$

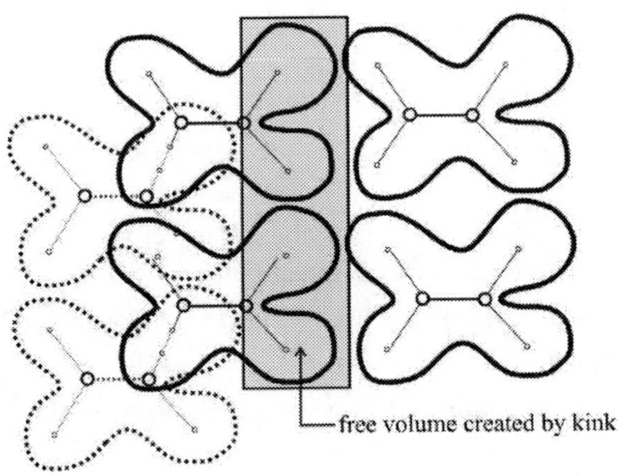

Figure 3. Illustration of lipid bilayer free volume formation in the acyl chain region. The view is perpendicular to the membrane plane. The contours represent the van der Waals surfaces of lipid molecules in the direction perpendicular to the chain axis. The left side depicts the lateral displacement of the chain due to kink formation. The new position of the chain fragment after the displacement is shown by dotted lines.

where $d_k = 0.13$ nm means the lateral displacement of hydrocarbon chain upon kink formation, $L = 0.7$ nm – is the length of pyrene molecule.

Figure 4 depicts the typical emission spectra of pyrene in PC and PC/Chol (30 mol%) model lipid membranes at the presence of fibrillar proteins.

Calculated set of parameters based on emission spectra, included the spectral (pyrene excimerization degree, E/M; intensity of the first-to-third vibronic bands, (I_I/I_{III}), and dynamical (ν_{col} ν_j, ν_k and D_{diff} (Tables 1-4).

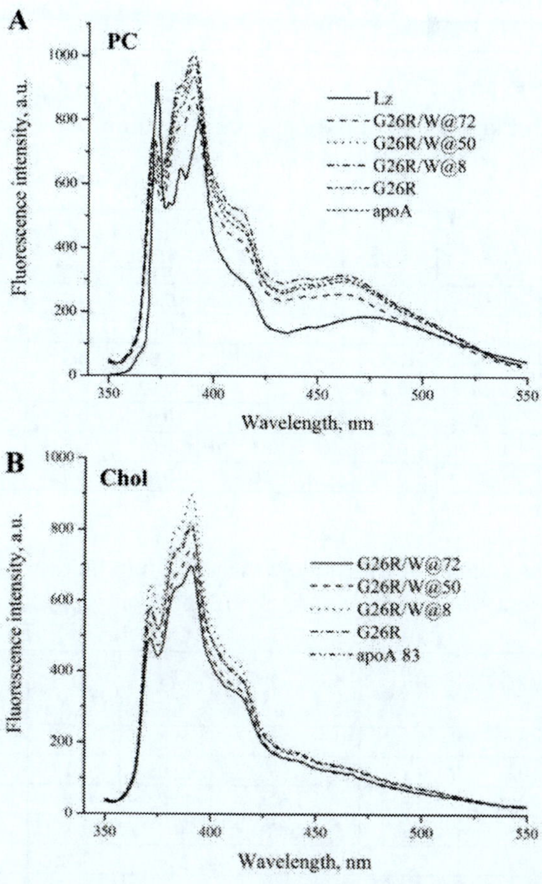

Figure 4. Pyrene emission spectra in PC (A) and Chol (B) lipid vesicles in the presence of aggregated Lz and apoA. Lipid concentration was 38.3 and 16 μM for Lz and apoA, respectively. The corresponding pyrene concentrations were 0.96 and 1 μM.

Table 1. Parameters of pyrene excimerization in the neat lipid systems without fibrillar proteins*

Lipid system	I_I/I_{III}	E/M	v_{col}, ns^{-1}, ×10^{-3}	v_j, ns^{-1}	v_k, ns^{-1}	D_{diff}, sm^2 s^{-1}, ×10^{-8}
PC	0.95	0.39	2.71	0.02	0.44	2.42
CL5	0.96	0.44	3.05	0.02	0.49	2.73
CL10	0.96	0.47	3.61	0.02	0.58	3.23
CL20	0.96	0.52	2.36	0.01	0.38	2.11
Chol30	0.94	0.34	2.42	0.01	0.39	2.17
CL5/Chol30	0.95	0.35	2.83	0.02	0.46	2.55
CL20/Chol30	0.95	0.41	2.71	0.02	0.44	2.42

*the error does not exceed 6%

Table 2. Parameters of pyrene excimerization in the presence of Lz amyloid fibrils*

Lipid system	I_I/I_{III}	E/M	v_{col}, ns^{-1}, ×10^{-3}	v_j, ns^{-1}	v_k, ns^{-1}	D_{diff}, sm^2 s^{-1}, ×10^{-8}
PC	0.94	0.37	2.56	0.01	0.41	2.31
CL5	0.72	0.28	1.94	0.01	0.32	1.74
CL10	0.69	0.33	2.29	0.01	0.39	2.13
CL20	0.64	0.39	2.71	0.02	0.44	2.42
Chol30	0.93	0.32	2.21	0.01	0.36	1.99
CL5/Chol30	0.85	0.27	1.87	0.01	0.31	1.68
CL20/Chol30	0.78	0.37	2.56	0.01	0.42	2.31

*the error does not exceed 6%

Table 3. Parameters of pyrene excimerization in PC liposomes in the presence of different variants of apoA amyloid fibrils*

ApoA variant	I_I/I_{III}	E/M	v_{col}, ns^{-1}, ×10^{-3}	v_j, ns^{-1}	v_k, ns^{-1}	D_{diff}, sm^2 s^{-1}, ×10^{-8}
W@72	0.81	0.29	0.97	0.05	0.42	0.89
W@50	0.83	0.29	0.97	0.05	0.43	0.89
W@8	0.82	0.31	1.02	0.06	0.42	0.91
G26R	0.81	0.31	1.02	0.05	0.44	0.91
1-83	0.81	0.32	1.04	0.06	0.42	0.95

*the error does not exceed 6%

The main conclusions which can be reached from the analysis of the obtained parameter set can be outlined as follows: i) both Lz and apoA exerted no statistically significant effect on pyrene parameters in non-charged lipid vesicles (PC and PC/Chol), ii) amyloid fibrils of Lz produced marked changes in pyrene excimerization in the negatively charged CL-containing lipid vesicles (CL5, CL20, CL5/Chol30 and CL20/Chol30), where the protein-lipid complexation resulted in the decrease of all estimated parameters, iii) the inclusion of cholesterol into the lipid bilayer suppressed the action of fibrillar protein on the membrane properties sensed by pyrene. Interestingly, pyrene fluorescence spectra in PC/Chol membranes were not characterized by clearly defined excimer peak (Figure 4, B). This implies that condensing effect of cholesterol [*] hinders the probe lateral diffusion, thereby reducing the probability of collisions between pyrene monomer and decreasing the rate of excimer formation.

Table 4. Parameters of pyrene excimerization in PC/Chol liposomes in the presence of different variants of amyloid fibrils*

ApoA variant	I_I/I_{III}	E/M	ν_{col}, ns^{-1}, ×10^{-3}	ν_j, ns^{-1}	ν_k, ns^{-1}	D_{diff}, sm^2 s^{-1}, ×10^{-8}
W@72	0.78	0.17	0.58	0.03	0.27	0.50
W@50	0.81	0.17	0.58	0.03	0.29	0.50
W@8	0.80	0.18	0.59	0.03	0.29	0.53
G26R	0.79	0.17	0.58	0.03	0.27	0.50
1-83	0.79	0.17	0.58	0.03	0.28	0.50

*the error does not exceed 6%

These findings suggest that the lysozyme, either in fibrillar or monomeric form, does not modify the free volume of the neutral vesicles, while the protein binding to the negatively charged bilayers is accompanied by the drop in the polarity of pyrene microenvironment (as judged from the decrease in I_I/I_{III} and the reduction of the membrane free volume (as concluded from the decrease in E/M, ν_{col} ν_j, ν_k and D_{diff}

that lipid bilayer composition controls the membrane-associating properties of amyloid fibrils and governs the extent to which fibrillar proteins alter the physicochemical properties of the lipid bilayer.

To replenish the above findings, another membrane probe DPH was used. The spectral behavior of this fluorophore is largely dependent on molecular organization and dynamics of bilayer hydrophobic region. The DPH fluorescence anisotropy characterizes the velocity of probe rotational diffusion and correlates with the changes in membrane microviscosity [61]. Notably, the measurements of DPH anisotropy reported in literature, allowed the authors to draw the ambiguous conclusions. For example, the group of Kremer et al. showed that aggregated β-amyloid peptide interact with the membrane nonpolar part and decrease the fluidity of unilamellar liposomes, the effect manifesting itself in the elevated values of DPH anisotropy [62]. Muller with co-workers revealed the analogous effect of Aβ-peptide on the membranes of human brain cells [63], however, the magnitude of the protein influence on membrane fluidity at the lipid/water interface was found to be smaller than that at level of acyl chains. In turn, according to the results of Ma, the aggregated Aβ-peptide affects predominantly the interfacial part of lipid bilayer with insignificant effect on its hydrophobic region [64]. The contradictions in the reported data may arise from the amyloid fibril heterogeneity and polymorphism. Recent studies showed that different destabilizing conditions lead to the various folding pathways of the same polypeptide chain. The result of these processes is the formation of polymorphic mature fibers with different conformational peculiarities [65].

Table 5 summarizes the results of DPH anisotropy measurements in different lipid vesicles in the presence and absence of Lz amyloid fibrils. As shown from the table, anisotropy values remain virtually unchanged at increasing concentrations of protein. These results indicate that fibrillar protein does not interact with the hydrocarbon core of the lipid bilayer and has no effect on the membrane microviscosity in the nonpolar region or the magnitude of this effect is not sufficient for DPH to record it.

Table 5. Parameters of pyrene excimerization in PC liposomes in the presence of different variants of apoA amyloid fibrils

Protein concentration, μM	PC	CL10	Chol30
0	0.080	0.081	0.141
0.4	0.082	0.081	0.142
0.8	0.081	0.082	0.141
1.2	0.082	0.081	0.143
3.1	0.081	0.081	0.142

Modifications of Lipid Bilayer Polar Region under the Influence of Amyloid Fibrils

Further experiments were directed towards the examination of the amyloid effect on the interfacial part of the lipid bilayer. To this end, the fluorescent probes Laurdan and MBA were used. MBA is characterized by poor solubility in water, but readily incorporates into the membrane, exhibiting high sensitivity to lipid bilayer hydration level [66]. However, the spectral behavior of this probe in membrane systems exhibit the complex pattern due to heterogeneity in MBA distribution resulting in two different binding sites for the probe in a membrane interior – in the vicinity of the phosphate groups and at the level of glycerol backbone. Such distribution of MBA populations is reflected in two-banded emission spectrum of the probe in the membrane-bound state [67].

In the context of amyloid fibril effect on the properties of a lipid bilayer, much more valuable information may be extracted not from the examination of MBA emission, but from the the measurements of fluorescence anisotropy. As it was mentioned in previous section, in general meaning, such parameters as anisotropy and polarization reflect the rate of rotational diffusion during the probe fluorescence lifetime, which, in turn, depends on solvent viscosity. As shown in Table 6, the complexation of lysozyme fibrillar species with model lipid membranes was accompanied by MBA anisotropy increase by approximately 30% in PC, Chol30 and CL10 bilayers at the lowest protein concentration. Increasing the concentration of amyloid fibrils up to 3.1 μM provoked the substantial rise in MBA anisotropy by ~350% for PC bilayers and ~220% for Chol30 and CL10 vesicles. These findings suggest that Lz amyloid fibrils induce the dehydration of membrane polar region and decrease in bilayer microviscosity.

Table 6. Relative increase in MBA anisotropy as a result of fibrillar lysozyme binding to the lipid membranes of different composition, %

Protein concentration, μM	PC	CL10	Chol30
0.4	26	32	28
0.8	61	75	72
1.2	82	113	84
3.1	350	224	222

Additional arguments in favor of this statement comes from the examination of spectral characteristics of another membrane probe, Laurdan. This dye is widely used in membrane biophysics due to its sensitivity to differences in lipid packing degree and hydration level. The unique extended structure of Laurdan provides the basis for energetically favorable probe redistribution in transversal direction within the membrane with the lauric acid tail residing in the acyl chain region of the lipid bilayer, and its fluorescing moiety localizing at the glycerol level of the phospholipid headgroups [68]. Depending on the extent of reorientation of solvent dipoles around the excited state of Laurdan, the probe emission spectrum is featured by two components, which are attributed to solvent-unrelaxed (shorter wavelength band) and solvent-relaxed (longer wavelength band) states [68-71]. Variations in membrane water content cause shifts in the Laurdan emission spectrum, which are quantified by calculating the generalized polarization (GP). The calculated values of GP in different amyloid-lipid systems are presented in Figure 5. While analyzing this figure, the first noticeable observation is that GP attains negative values in PC and CL-containing vesicles and have positive values in Chol30 lipid membranes indicating different location of Laurdan. For all types of liposomes, the association of fibrillar Lz was accompanied by an increase in GP, which can be interpreted in terms of bilayer dehydration and/or increase in lipid packing density.

A different mode of interactions was revealed for amyloid fibrils of different mutants of apoA. Accordingly, in PC/Chol bilayers the relative changes of GP did not exceed 4% which may indicate that sterol hampers the binding of the amyloid structures to the lipid bilayers. Interestingly, the preventing role of Chol in membrane modification by aggregated protein species was reported by a number of scientific groups [63, 72, 73]. Specifically, it was shown that elevating amounts of sterol inhibits the processes of liposome aggregation and fusion, induced by Aβ(1-40) peptide [72]. Furthermore, lipid membranes with high sterol content were found to have strong resistance to modifying action of prefibrillar aggregates [73].

In turn, the formation of amyloid-lipid complexes in PC membranes was followed by 25-70% decrease in GP values of Laurdan suggesting the increase in bilayer hydration due to the fibril binding. Furthermore, the observed GP decrease was revealed to correlate inversely with the degree of peptide fibrillization. Accordingly, the smallest changes in GP were induced by G26R/W@8 mutant which possess the highest aggregation potential.

Overall, the results presented in this contribution, showed that in the neutral PC and Chol30 liposomes the lysozyme fibrils do not produce any noticeable alterations in the pyrene parameters, but affect the Laurdan spectral behavior, suggesting that fibrillar protein lacks interactions with the uncharged membranes but its modifying effects are restricted to the interfacial bilayer region and do not propagate into the membrane core. This fact can be explained by the structural properties of the mature lysozyme fibrils.

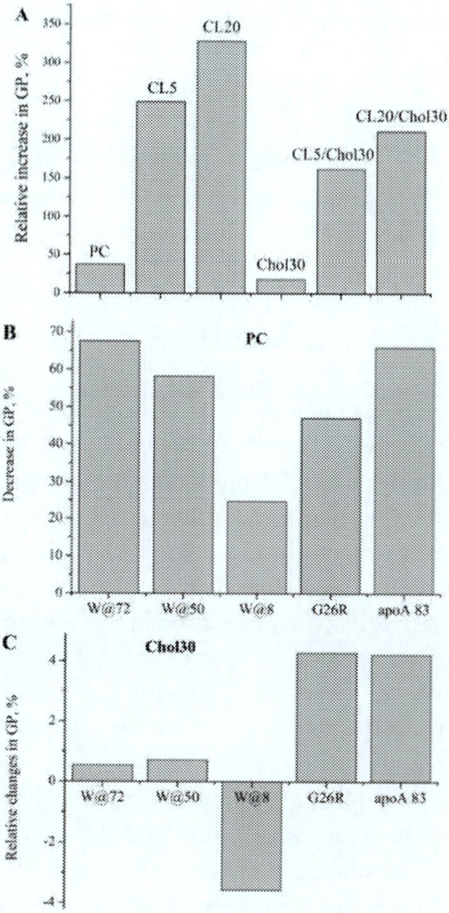

Figure 5. Effect of fibrillar Lz (A) and apoA variants (B, C) on generalized polarization of Laurdan in lipid membranes of different composition.

To penetrate into hydrocarbon core of the lipid bilayer, protein aggregates should have the exposed hydrophobic surfaces. It is noteworthy in this regard that protofibrils have been proposed to possess amphiphilic properties: the hydrophobic fibril surface associates efficiently with the membrane hydrocarbon core and the hydrophilic surface is responsible for the structural membrane permeabilizing defects (Kinnunen, 2009). A strong correlation among conformational, structural (exposure of hydrophobic patches and flexibility) and stability features of amyloid assemblies and their ability to permeabilize the cell membrane has been also reported. Highly flexible, less compact HypF-N oligomers with extended exposed hydrophobic surface area were found to be substantially more toxic [74]. Furthermore, it has been shown that structural and dynamic properties and, in particular, the presence of substantial non-core regions in lysozyme fibrils determinate their cytotoxicity. More specifically, lysozyme fibrils formed at physiological pH were less ordered in the cross-beta structure and generated a substantial cytotoxic effect, while the highly organized fibrils which were formed at acidic pH, had a negligible effect on cell viability [35, 75]. Likewise, a significance of the surface hydrophobicity of protein aggregates in amyloid-induced decrease of membrane fluidity was demonstrated for the aggregated Aβ peptide [62]. In view of these findings, it can be proposed that large extent of ordering and high stability, as well as a minor quantity of the exposed hydrophobic patches make it difficult for fibrillar lysozyme to accumulate on the membrane surface in a concentration high enough for producing the impact in the hydrocarbon region. In contrast, for negatively charged PC/CL membranes strong electrostatic interactions between the anionic phospholipid headgroups and positively charged amino acid side chains not only promote the lysozyme membrane anchoring, but also favor the protein bilayer embedment to the level of acyl chains and occupation of the membrane free volume. All these processes slow down the pyrene diffusion and decrease the frequency of the collision between its molecules, resulting thereby in the reduced excimerization extent.

In the meantime, the mode of association between amyloid fibrils of apoA and the model membranes seems to be more intricate. To explain the observed destabilization of a membrane polar part, it seems of interest to analyze different models of amyloid-lipid interactions proposed in the literature. One possible route of membrane disruption involves pore formation. It is proposed that β-sheet-rich peptides form toroidal or barrel-stave pores in the membranes. The end effect of this process is the loss of membrane integrity and eventually cell death [76]. Another way for the amyloid fibers to destabilize the lipid bilayer is detergent-like behavior. In this case, the fibrillar protein remove lipid molecules from the bilayer plane through the carpet mechanism, resulting in general destabilization and formation of lateral defects of big area [77]. Finally, it is proposed that amyloid fibrils may form extended domains inside the lipid bilayer, which also results in the global disintegration of the membrane structure. Apparent affinity of the amyloid fibers for the particular class of lipids is reported to underlie this process [53]. Whether these three models are mutually exclusive or they cooperate in triggering the membrane damage remains unclear. However, in the case of apoA amyloid fibrils, the pore-like pathway is unlikely, since it is expected that pore formation would result in perturbation of nonpolar region of the lipid bilayer, and this should be reflected in the changes in pyrene spectral behavior. The generation of amyloid or amyloid-lipid domains is also hardly probable because this mode of interactions is realized, apparently, during amyloid growth on the membrane surface, but in the present contribution we are dealing with the complexation of mature fibers with the lipid vesicles. Thus, we suppose that the most probable route of membrane destabilization induced by apoA amyloids, involves the detergent-like mechanism. Accordingly, the events underlying the loss of membrane integrity are assumed to go through the following sequence: amyloid adsorption onto bilayer surface → membrane invagination and thinning → bilayer defragmentation and loss of structural stability (Figure 6). Specific fibril-lipid interactions, involving intercalation of protein wedge-shaped domains, were assumed to result in lipid bilayer destabilization and eventual extraction of the membranous material [53]. Milanesi et al. identified such mode of fibril-liposome interactions as a specific membrane distortion by a peptide/protein self-assembly different from previously observed mechanisms of lipid bilayer disruption [53].

The proposed model of membrane destabilization by apoA amyloid fibers resembles the "carpet" mechanism of membranotropic activity of anti-microbial peptides. According to this mechanism, the peptides bind to the membrane in monomeric or oligomeric state and localize in such a way, that hydrophobic interface of amphipathic α-helices inserts into the bilayer while hydrophilic side is exposed to the solvent [78]. When the surface concentration of the adsorbed peptide reaches the threshold value, membrane destabilization occurs due to the alterations in a bilayer local curvature.

Alterations in Amyloid Conformation in a Membrane Environment

As the nest logical step, it seems of interest to analyze whether the formation of amyloid-lipid complexes is accompanied by perturbations in fiber conformation. To this end, we measured the protein intrinsic fluorescence and its quenching by a neutral polar quencher acrylamide. The hen egg white lysozyme contains six tryptophan residues located at positions 28, 62, 63, 108, 111 and 123, with Trp62 and Trp108 predominantly contributing to the overall emission spectrum of the protein [79]. The measurements of intrinsic fluorescence of Lz amyloid fibrils in the presence of the negatively charged liposomes (representative spectra are depicted in Figure 7) showed that increasing lipid concentrations gives rise to the ~5 nm blue shift of the amyloid emission maximum in CL5 and CL20 systems, suggesting the transfer of the protein fluorophores into the membrane environment with lower polarity. In contrast, in Chol-containing systems, the blue shift was only ~1 nm, indicating that cholesterol prevents the penetration of fibrillar lysozyme into the bilayer hydrophobic core, the observation being in accordance with the pyrene fluorescence measurements.

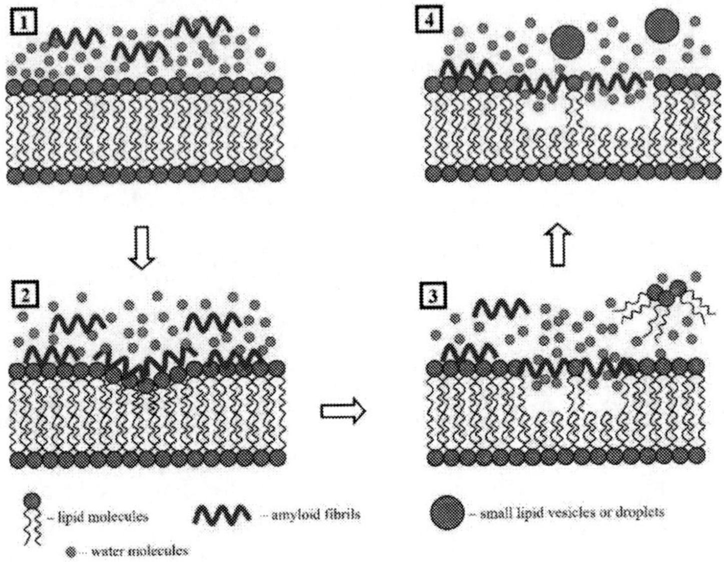

Figure 6. Schematic illustration of membrane destabilization induced by amyloid fibrils of apoA.

To obtain additional molecular level details of amyloid-lipid binding, the intrinsic lysozyme fluorescence was quenched by acrylamide. The emission maximum of lysozyme amyloids was found to undergo the blue shift upon increasing concentration of the quencher (data not shown). This indicates that individual Trp residues differ in their accessibility to the quencher, i.e., the solvent-exposed tryptophans are quenched more efficiently than the buried residues. This phenomenon is reflected in the dependence of the Stern-Volmer plots on the emission wavelength (data not shown).

The amyloid-lipid complexation led to the decrease in Stern-Volmer constants compared to the protein free in buffer solution (Table 7). This finding may be indicative of the reduced accessibility of Trp residues to the solvent in the lipid-bound state. Notably, comparison of the quenching data with those obtained for monomeric lysozyme showed that amyloid fibrils appeared to be less exposed to acrylamide quenching than the native protein assuming that stacking of the lysozyme molecules into fibrillar architecture shields the protein fluorophores from the solvent. Previously we proposed the hypothetical structural model of the lysozyme amyloid fibrils. Briefly, this model implies that the residues Gly54-Asn65 and Ile78-Thr89 constitute the cross-β fibril core with Trp62, one of the main protein emitters, and Trp63 locating almost in the fibril center [80]. This sterically hinders the collisions between tryptophans and acrylamide and weakens the overall quenching efficiency. Alternatively, one may suppose that close packing of monomer subunits within the protofilaments reduces the accessibility of the protein fluorophores to the quencher.

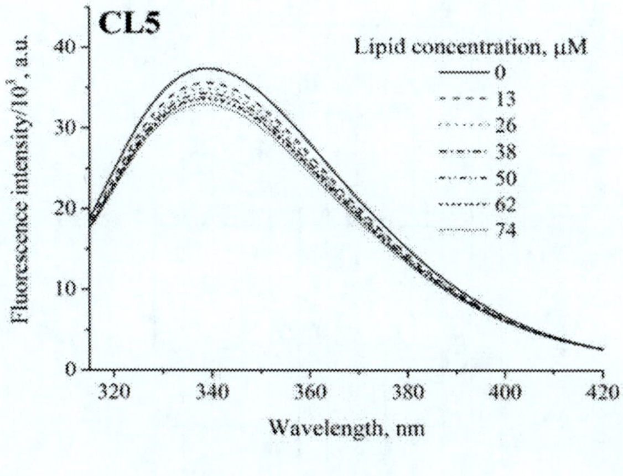

Figure 7. Fluorescence spectra of fibrillar lysozyme in the presence of CL5 model membranes. Protein concentration was 1.6 μM.

Furthermore, molecular dynamics simulations allowed us to suggest that the fragment embracing the residues Gly117-Leu129, characterized by labile unstructured or turn conformation within the fibril, anchors the fibrillized protein on the lipid matrix, the process which may results in the irreversible loss of bilayer integrity. The putative anchoring segment 117-129 is characterized by a high hydrophobicity favoring its embedment into a lipid bilayer, but at the same time it contains several polar residues capable of forming contacts with lipid polar groups. Based on the above rationales, it can be assumed that the fragment 117-129 resides at the lipid polar/nonpolar interface, while the fibril core is located on the membrane surface, as schematically illustrated in Figure 8.

Table 7. Comparison of Stern-Volmer constants for acrylamide quenching of lysozyme intrinsic fluorescence in monomeric (M) or fibrillar (F) state

Lipid membranes		Emission wavelength, nm					
		320	330	340	350	360	370
Buffer	F	4.68 ± 0.06	5.11 ± 0.06	5.33 ± 0.05	5.74 ± 0.07	5.92 ± 0.05	6.17 ± 0.06
	M	4.93 ± 0.08	5.19 ± 0.07	5.51 ± 0.07	5.83 ± 0.09	6.04 ± 0.05	6.38 ± 0.08
CL5	F	2.21 ± 0.03	2.43 ± 0.04	2.69 ± 0.03	2.81 ± 0.04	2.91 ± 0.03	2.98 ± 0.04
	M	2.89 ± 0.04	3.31 ± 0.05	3.73 ± 0.05	4.01 ± 0.06	4.17 ± 0.06	4.31 ± 0.07
CL5/Chol30	F	2.52 ± 0.04	2.81 ± 0.04	3.09 ± 0.05	3.24 ± 0.05	3.39 ± 0.05	3.51 ± 0.05
	M	2.56 ± 0.03	2.84 ± 0.04	3.19 ± 0.04	3.42 ± 0.05	3.55 ± 0.05	3.65 ± 0.06
CL20	F	1.96 ± 0.03	2.18 ± 0.03	2.42 ± 0.04	2.47 ± 0.04	2.51 ± 0.05	2.59 ± 0.03
	M	2.21 ± 0.02	2.36 ± 0.02	2.51 ± 0.02	2.56 ± 0.02	2.59 ± 0.03	2.61 ± 0.03
CL20/Chol30	F	1.17 ± 0.01	1.29 ± 0.01	1.38 ± 0.01	1.43 ± 0.02	1.47 ± 0.02	1.49 ± 0.02
	M	1.67 ± 0.05	1.91 ± 0.06	2.12 ± 0.06	2.25 ± 0.06	2.33 ± 0.06	2.42 ± 0.05

In such an orientation both the polar and hydrophobic membrane regions are affecred by amyloid fibrils as was discussed here in the previous sections.

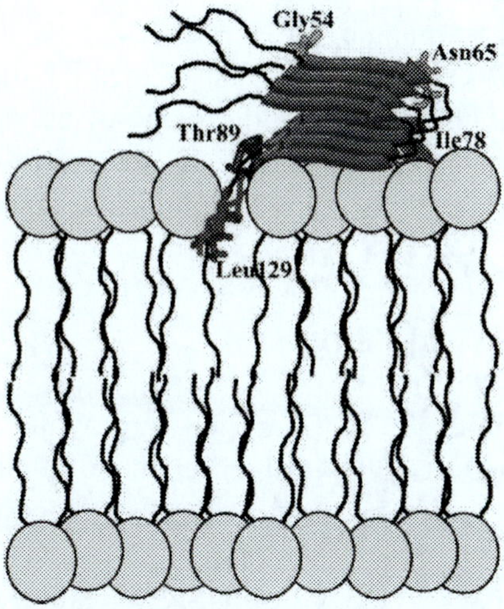

Figure 8. Hypothetical model for orientation of lysozyme amyloid fibrils on the lipid bilayer. Colored in red is the unstructured region of the lysozyme fibrils embracing the amino acids residues Gly117-Leu129. Blue arrows indicate β-strands.

Computer-Assisted Analysis of Amyloid-Lipid Complexation

It is becoming generally accepted that self-assembly of the proteins into the extended fibrillar structures displays a broad collection of morphologies, at the level of both protofibrillar and mature fibrillar species, with distinct molecular architecture being formed by the same amino acid sequence [81]. These polymorphs can coexist at given experimental conditions or can be controlled by changing the conditions of amyloid growth. Furthermore, a vast majority of theoretical and experimental studies suggest that different modes of intramolecular and side-chain interactions along the same polypeptide chain yield distinct cytotoxic performance [82]. In the light of these rationales, establishing the correlation between the structural peculiarities of amyloid aggregates and their lipid bilayer affinity represents one of the key factors in clarifying the membranotropic activity of amyloid fibrils. In this section we concentrated our efforts on the use of the methods of computer modeling for analyzing the interactions between different structural variants of amyloid fibrils and lipid membranes. Specifically, the main goals to be achieved in this section may be outlined as follows: i) using the computational framework of CreateFibril web resource to construct the polymorphic amyloid fibrils, differing in their twisting angle, ii) to evaluate the stability of constructed fibrils in water with AQUASOL tool and to identify the most energetically favorable polymorphs, iii) to predict the mode of amyloid association with lipid bilayer using the PPM web server, iv) to unravel the molecular-level details of amyloid-lipid binding using the molecular docking calculations.

Construction of Amyloid Fibrils

To model the amyloid fibrils from hen egg white lysozyme, Aβ and apolipoprotein A-I using the CreateFibril server, the following peptide sequences were analyzed:

1. K-peptide of lysozyme, GILQINSRW, residues 54-62 of wild-type protein (LzF1)
2. derivative of K-peptide, IFQINS (LzF2)
3. peptide (residues 24-50) from G26R-mutated apolipoprotein A-I, DSRRDYVSQFEGSALGKQLNLKLLDNW (apoA)
4. peptide from Aβ protein, LVFFAEDVGSNKGAIIGLMVGGVVIA, residues 17-42 of Aβ (Aβ).

The structure of LzF1 dimer, used as the input for CreateFibril, was generated from LzF1 monomer in the β-strand conformation with PatchDock and FireDock docking tools [21, 25, 26]. LzF2 dimer was obtained in similar way, but monomeric structure was taken from the Protein Data Bank (PDB code 4R0P). ApoA dimer was created as described in [32], while Aβ tetramer was taken from the Protein Data Bank (PDB code 2BEG). The choice of the peptides LzF1, apoA and Aβ was dictated by the fact that they represent the core of the amyloid fibrils from corresponding proteins. The peptide LzF2 was used to clarify the effect of amino acid sequence on the amyloid stability and fibril-lipid interactions.

The mathematical description of CreateFibril algorithms is based on the translational and rotational affine transformations which allow creating several identical copies of the fibril fragment and stacking them side-by-side, elongating thereby the fibrillar aggregate [21]. To construct the amyloid fibril model, the following parameters should be known: a) the number of protofilaments and their packing distance perpendicular to fibril axis, denoted as n and R, respectively; b) the direction of fibril axis; c) the fibril length (the number of monomers), r; d) the distance between β-strands along the fibril axis, d; e) the twisting angle of amyloid monomers along the fibril axis, θ. To create the amyloid structures, the parameters R and d were taken as 5 Å, n was 2, 4, 1 and 2 for LzF1, LzF2, Aβ and apoA, respectively. The parameter r was set as 4 for Aβ and 10 for the rest of the peptides. The twisting angle was varied from 0 to 40° with a 5° step. The differences in the input parameters are explained by the distinct structural features and amyloid-forming propensities of the examined peptides. Such kind of modeling yields nine fibrillar structures for every peptide, whose representative pictures are given in Table 8. As seen from this figure, the constructed fibers are characterized by repeated intertwined structure, enriched in β-sheets, which represents the hallmark of amyloid aggregates.

Table 8. The models of amyloid fibrils constructed using the CreateFibril server

Peptide	Side view	Orthogonal view
LzF1	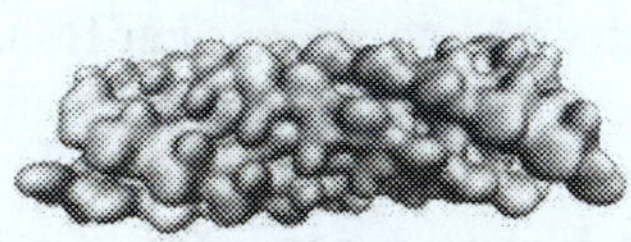	
LzF2	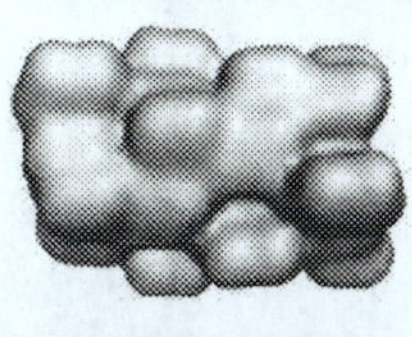	
Aβ	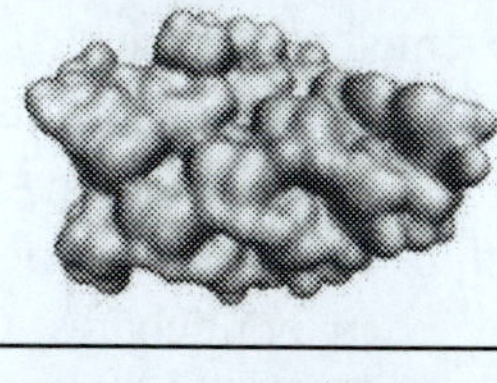	
apoA	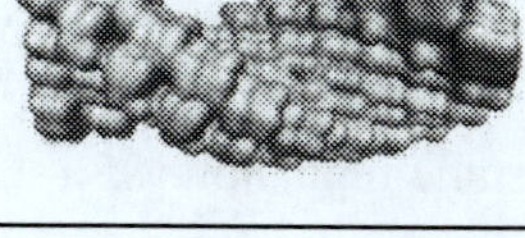	

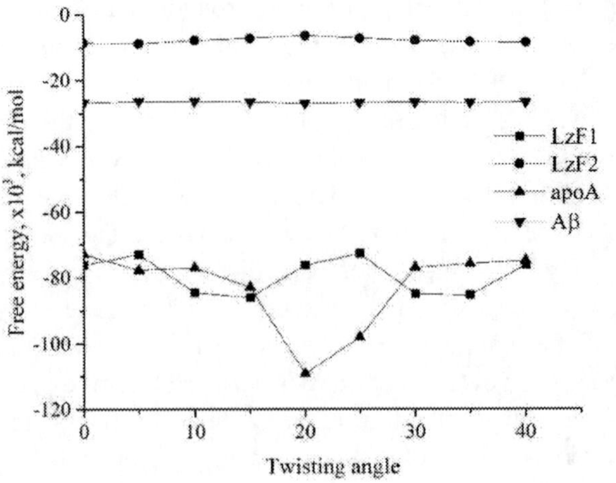

Figure 9. Amyloid free energy as a function of twisting angle calculated using AQUASOL server.

Next, we evaluated the stability of the constructed fibrillar assemblies. Accordingly, the free energies of the fibrils were calculated in water as a sum of Lennard-Jones, Coulomb and solvation energies using the AQUASOL web-server [83]. This program solves the dipolar Poisson-Boltzmann-Langevin (DPBL) equation for a given system. In terms of this formalism the solvent is considered as a collection of orientable dipoles with nonuniform concentration. This results in a nonlinear dependence of permittivity on the atom position and local electric field at this position. DPBL approach is supposed to be the most adequate in calculating the energetic profiles of biomolecules. The solution of DPBL equation yields the electrostatic potential at each position. The procedure of AQUASOL simulations includes the hydration of amyloid fibrils and their equilibration in water followed by calculations of the free energies. The recovered dependencies of the fibril energy on its twisting angle are presented in Figure 9. The analysis of these profiles showed that the most stable fibrillar aggregates have the twisting angles of 15, 5, 20 and 20° for LzF1, LzF2, Aβ and apoA, respectively.

In the present context it is worth of mentioning that molecular origin of the amyloid twist is still a matter of extensive debates. Originally, the main reason for the intertwining of the protofibrils is thought to be the intrinsic chirality of amino acid residues [84]. However, further progress in amyloid physics showed that twisting of fibrillar aggregates arises from the balance between fibril elasticity and electrostatic repulsion [85]:

$$U = \frac{1}{2} M\theta^2 + \frac{k_B T 2\pi \nu^2 Q e^{-\kappa D}}{\kappa \sin\theta} \tag{6}$$

where U is the total energy of the twisting, M is the torsional modulus of the fibrils, ν, κ Q denote the linear charge density of the fibril, the inverse Debye and the Bjerrum lengths, respectively, θ is the twist angle. The first term in this equation represents the elastic torsional energy, which opposes the twist, while the second term stands for screened electrostatic potential of the twisted fibrils, which promotes the twist. In a flat configuration the distance between the charges of the same sign may not be sufficient to compensate the energetic costs connected with the electrostatic repulsion. In turn, rotation around the long axis increases the distance between the charges, minimizing thereby the electrostatic repulsion. However, increased separation between the consecutive polypeptide segments creates the penalty in elastic energy favoring the flattered configuration. The final twist of the amyloid fibril is determined by the delicate balance between the above processes.

The widely accepted model of amyloid structure assumes that the stacked pleated β-sheets are twisted in such a way that a repeating unit of 24 β-strands constitutes a full helical turn around the fibril axis [86]. Such a configuration generates the twist of 15° between the neighboring β-strands. The twisting angles 15-20° recovered here for the most stable configurations of LzF1, Aβ and apoA fibrils are in a good harmony with the above model. The only exception is LzF2 fibrils, for which the most stable structure corresponds to the twisting angle 5°. This finding may be explained by the small length of the constructed fibrils insufficient to provide the high twist angle, since in this case the large elastic energy penalty is created. However, it is expected that propagation of the fibril would shift the twisting angle that characterizes the most stable structure, to the higher values.

Modeling of Amyloid Fibril Binding to the Lipid Membranes

In view of the crucial role of fibril-induced membrane disruption in amyloid cytotoxicity, the analysis of protein aggregate orientation within the membrane may be of particular importance for elucidating the molecular-level details of their destructive effect. The simulations were performed on the basis of online PPM server, which can predict the membrane positions of both peripheral and integral proteins [87]. Specifically, spatial arrangement of proteins in membranes is determined by minimizing their transfer energies from water to lipid bilayer. Additional output parameters include the depth of peptide protrusion into the nonpolar membrane region, tilt angle (the angle between the peptide axis and the bilayer normal) and the residues embedded into membrane hydrocarbon core. The calculation is based on the universal solvation model which accounts for hydrophobic, van der Waals, H-bonding and electrostatic solute-solvent interactions [88]. In this model lipid bilayer is represented as a fluid anisotropic medium characterized by dielectric constant (ε), solvatochromic dipolar parameter (π^*), and H-bonding acidity and basicity parameters (α and β). The nonpolar bilayer region is modeled as a planar slab of adjustable thickness whose interior resembles the decadiene. The interfacial polarity of the membrane is simulated based on EPR, X-ray and spin-labeling experimental measurements of water concentration in the membrane interior [89]. Within the framework of solvation model, the transfer energy of a solute molecule from aqueous phase to the arbitrary position in the membrane accounts for the dependence of the solvation parameters σ_i η on the atom position along the bilayer normal (z_i), and is given by:

$$\Delta G^{transfer}(d,\varphi,\tau)=\sum_{i=1}^{N}\sigma_i^{wat\to bil}(z_i)ASA_i+\sum_{i=1}^{M}\eta_j^{wat\to bil}(z_j)\mu_j+\sum_{k=1}^{L}\min\left\{\Delta E_k^{ion},\Delta E_k^{neutr}\right\} \tag{7}$$

where $\sigma_i(z_i)$ stands for the solvation parameter, depending on the atom type and describing the surface energy transfer per Å^2 of *i*-th atom from water to the point z_i on the membrane, ASA$_i$ is a solvent-accessible surface area of *i*-th atom, $\eta(z_j)$ defines the energy cost of the transferring the dipole moment of 1D from water to point z_j, μ_j – the dipole moment of group *j*, E_k^{ion} E_k^{neutr} represent the energies of ionized group *k* in ionized and neutral states, respectively, *N* is the number of atoms in the molecule, *M* is the number of group dipoles, *L* is the number of ionizable groups, d, φ and τ define the sift along the bilayer normal, rotational and tilt angle, respectively.

The individual dipolar contributions were assumed to linearly decrease to zero when ASA of the atom associated with the dipole moment becomes smaller than the hydration area occupied by a single molecule of water ($ASA_{water} \approx 14$ Å):

$$\Delta G_{dip,j} = \eta_j^{wat \to bil}\left(z_j\right)\mu_j, \text{ if } ASA_j > ASA_{water}$$

$$\Delta G_{dip,j} = \eta_j^{wat \to bil}\left(z_j\right)\mu_j\left(ASA_j / ASA_{water}\right), \text{ if } ASA_j \le ASA_{water}$$

Analogous approximation of Born energy for ionized groups yields:

$$\Delta E_k^{ion} = \frac{166 e_{Born}}{r_k}\left[F_{Abe}^{wat} - F_{Abe}\left(z_k\right)\right] f_k + \sum_{i=1}^{L_k} \sigma_{l,ion}^{wat \to bil}\left(z_l\right) ASA_l, \text{ if } ASA_k > ASA_{water} \tag{10}$$

$$\Delta E_k^{ion} = \frac{166 e_{Born}}{r_k}\left[F_{Abe}^{wat} - F_{Abe}\left(z_k\right)\right]\left(ASA_k / ASA_{water}\right) f_k + \sum_{i=1}^{L_k} \sigma_{l,ion}^{wat \to bil}\left(z_l\right) ASA_l, \text{ if } ASA_k \le ASA_{water} \tag{11}$$

where coefficient $f_k = 0.5$ if the group is involved in ion pairing, and $f_k = 1$ for the rest of cases. The analysis of the binding of constructed amyloid fibrils to the lipid membranes allowed us to determine the energetically preferable orientation of the aggregates relative to lipid-water interface, as well as the energetics of protein-lipid binding. Notably, we found it reasonable to analyze not only the most stable structures, but also those covering the entire range of the examined twisting angles.

Recent experimental data from single molecule atomic force microscopy showed that amyloid periodicity is a tunable parameter whose initial value is sensitive to the changes of experimental variables [23]. In particular, increasing salt concentration was reported to bring about continuous decrease in amyloid pitch and corresponding twisting angle, according to the relationship:

$$L_2 = L_1\left[\left(\frac{\kappa_1}{\kappa_2}\right)^{1/2} + \frac{\alpha_2}{4\pi}\left(\kappa_1 - \kappa_2\right) L_1^{3/2}\right]^{-2/3} \tag{12}$$

where L, κ are the fibril pitch and inverse Debye length, the subscript denotes salinity regime – 1 or 2 (i.e., ionic strength I_1 or I_2), α_2 is the parameter at ionic strength 2, depending on the intrinsic characteristics of the fibrils (charge density, geometry, elastic rigidity). Elevation of the ionic strength screens the electrostatic interactions within the fibrils and decreases the contribution of the second term in Eq. (6), causing the relaxation of fibril tilt and its untwisting. The weakened electrostatic interactions may alter the H-bonding distribution along the fibrillar aggregates and their mechanical stability, the parameters that also underlie the structural configuration of the amyloid.

Table 9. Structural and thermodynamic parameters characterizing membrane location and orientation of LzF1 and LzF2 fibrils

Twist angle	ΔG, kcal/mol		Depth*, Å		Tilt angle	
	LzF1	LzF2	LzF1	LzF2	LzF1	LzF2
0	-34	-18.8	7	3.9	89	90
5	-32	-18.4	5.1	4.4	90	87
10	-24	-16.7	4.4	4.7	98	86
15	-16	-16.1	4.2	7.9	87	80
20	-12.5	-12.4	2.5	5.8	89	82
25	-12.3	-11.5	2.1	5.2	90	88
30	-11	ND	1.7	ND	90	ND
35	-9.5	-9.8	1.5	5.2	89	87
40	-10.3	-8.2	1.7	5.3	90	90

*depth of penetration into the hydrocarbon core of lipid bilayer

Table 10. Structural and thermodynamic parameters characterizing membrane location and orientation of Aβ and apoA fibrils

Twist angle	ΔG, kcal/mol		Depth*, Å		Tilt angle	
	Aβ	apoA	Aβ	apoA	Aβ	apoA
0	-36.6	-26.8	18.4	2.0	90	88
5	-34	-23.4	19	2.1	85	88
10	-36.4	-18.8	17.8	3.6	85	87
15	-33.5	-14.9	16.4	4.3	84	83
20	-28.9	-11.9	16.8	4.1	89	79
25	-22.4	-9.9	17.4	4.4	59	81
30	-20	-8.5	19.2	5.3	47	74
35	-23.2	-8.3	23.2	2.0	14	87
40	-26.2	-8.9	23.8	1.4	7	85

The results of calculations performed with PPM are shown in Tables 9, 10, and the representative snapshots are given in Figure 10. As judged from the tables, the position of amyloid fibrils in the membrane and energetics of peptide-lipid binding is strongly determined by amino acid sequence and the twist of the polypeptide chain. The recovered values of the tilt angle indicate that preferable orientation of amyloid fibrils from LzF1, LzF2 and apoA peptides is parallel or slightly tilted with respect to the bilayer surface. In contrast, Aβ fibrils were found to reside parallel to membrane surface only up to twisting angle of *ca.* 20°, while further increase in θ was followed by the rise in tilt angle up to 7°, the value corresponding to almost perpendicular orientation to the bilayer surface.

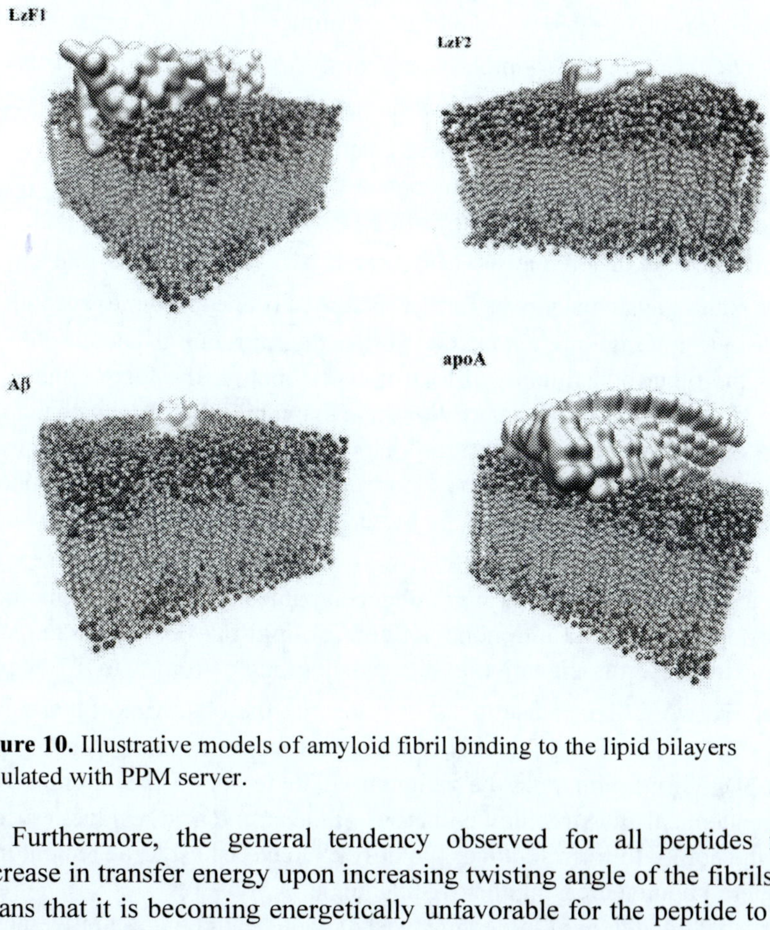

Figure 10. Illustrative models of amyloid fibril binding to the lipid bilayers simulated with PPM server.

Furthermore, the general tendency observed for all peptides is the decrease in transfer energy upon increasing twisting angle of the fibrils. This means that it is becoming energetically unfavorable for the peptide to insert into the lipid bilayer with increasing its periodicity and amyloid fibrils tend to locate at lipid/water interface or in the membrane polar region. This effect can be explained by at least two factors. The first one is that different amino acid residues participate in peptide-lipid binding depending on the fibril twist. According to White-Wimley hydrophobicity scale, the free energy of transfer of individual amino acid residues from water to the lipid phase varies substantially from residue to residue and is controlled by residue charge, chemical structure and its polarity [89]. If one take, for the sake of illustration, the peptide LzF1 and calculate roughly the sum of transfer energy contributions from each residue (see Figure 8C in [89] for the values of transfer energies for each amino acid) taking part in membrane penetration (ΔG^{sum}), it appears that at $\theta = 5°$, $\Delta G^{sum} = \Delta G^{I2} + \Delta G^{L3} + \Delta G^{W9} + \Delta G^{I10}$ $= -0.31 - 0.56 - 1.85 - 0.31 = -3.03$ kcal/ mol, while at $\theta = 40°$ $\Delta G^{sum} = \Delta G^{I10} = -0.31$ kcal/mol. Analogously, for apoA fibrils at $\theta = 5°$ $\Delta G^{sum} = \Delta G^{L38} = -0.56$ kcal/mol, while at $\theta = 40°$ $\Delta G^{sum} = \Delta G^{L38} + \Delta G^{D24}$ $= -0.56 - 1.23 = -1.59$ kcal/mol. These calculations show that variation even in one amino acid residue being in contact with nonpolar bilayer region may significantly change the free energy of transfer. Furthermore, one should bear in mind that the difference in ΔG^{sum} (and, as a consequence, the overall $\Delta G^{transfer}$) in our systems may be further increased if one takes into account the inter-residue interactions within the fibrils, their structural peculiarities, as well as the other constituents of total transfer energy, the factors that were neglected upon these rough calculations and that may substantially change the transfer energy of individual residues. However, evidently, the difference in the nature of amino acid residues, protruding the membrane hydrophobic region, is not the only reason for the observed increase in absolute values of $\Delta G^{transfer}$. Specifically, for all peptides we observed the rise in free energy at increasing twisting angle that was not accompanied by the change in amino acid residues embedded into nonpolar core of lipid bilayer. For example, in the case of LzF1 the elevation of the twisting angle from 5 to 15 degrees resulted in two-fold increase in $\Delta G^{transfer}$ despite the observation that amino acid residues penetrating the acyl chain region were the same in both cases (Table 9). This finding can be rationalized in terms of the alterations in physicochemical and structural properties of the amino acid residues (such as ASA, the ability to form H-bonds, polarity characteristics, arrangement in β-stacks, etc.) upon increasing the twisting angle.

Notably, modeling of increasing twist of peptide aggregates may resemble the maturation of amyloid from oligomeric nucleus. At the beginning of fibril growth phase, the protofilaments have a flat structure, and their intertwining is rare and random across the contour length. However, the magnitude of twisting angle and the number of twisting points increase with incubation time [90]. Furthermore, according to the prevailing concepts, amyloid intermediates (oligomeric aggregates with small twist) are thought to have much more pronounced impact on membrane structure compared to mature fibers with high twist. In view of this, the findings outlined here, may provide the possible explanation for this phenomenon, i.e., it is thermodynamically unfavorable for grown fiber to penetrate the membrane, since its twisted structure would require much more energy for membrane to adjust its contact region to the intertwined structure of the fibrils.

Molecular Docking Studies of Amyloid-Lipid Complexation

In the following, molecular docking calculations have been accomplished to unravel the molecular-level details of amyloid-lipid binding. Molecular docking has been implemented using the PatchDock server and refined using the FireDock server, followed by visualization of the docked complexes in the Discovery Studio. The prominent binding sites were also predicted through the DoGSiteScorer server.

Based on the predictions of PatchDock and FireDock servers, the binding modes of the ligands were ranked according to the global binding energy (GBE) of the amyloid-lipid complex, therefore the conformers with the lowest binding energy were analyzed for subsequent evaluations (Figures 11, 13, 15). Further analysis of the docked structures was conducted with respect to the following characteristics: i) lipid binding sites on/within amyloid fibril and interacting amino acid residues (Table 11), ii) types of interactions involved in amyloid-lipid complexation ensuring the most energetically stable conformation (Figures 12, 14, 16), iii) atomic contact energy (ACE) (Table 11).

Based on the results obtained, the following aspects can be recognized. The global energy of lipid binding to amyloid fibrils was shown to vary in the range from -69.25 to -7.26 kcal/mol. Interestingly, for all proteins under analysis, the values of GBE were found to decrease in the row Chol → CL → POPC, reflecting the most stable docking of POPC to the fibrillar proteins. Notably, analogous tendencies were revealed for atomic contact energies, i.e., the lowest ACE values were attributed to POPC, while the highest – to Chol. In the following, it seems reasonable to highlight the molecular level details of amyloid-lipid complexation.

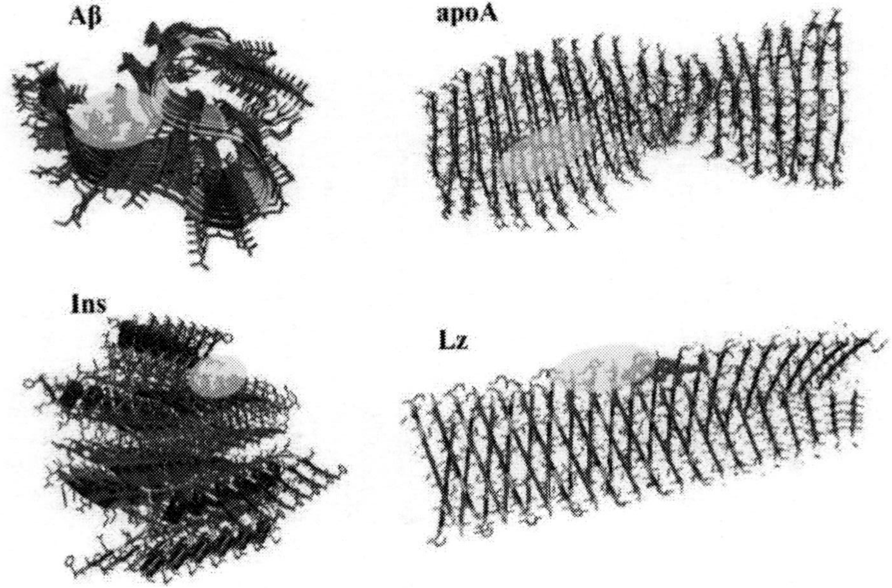

Figure 11. Cholesterol docked in the amyloid fibrils of Aβ, apoA, Lz and Ins amyloid fibrils. Light pink ovals represent the schematic illustration of the binding pocket for sterol predicted by DoGSiteScorer.

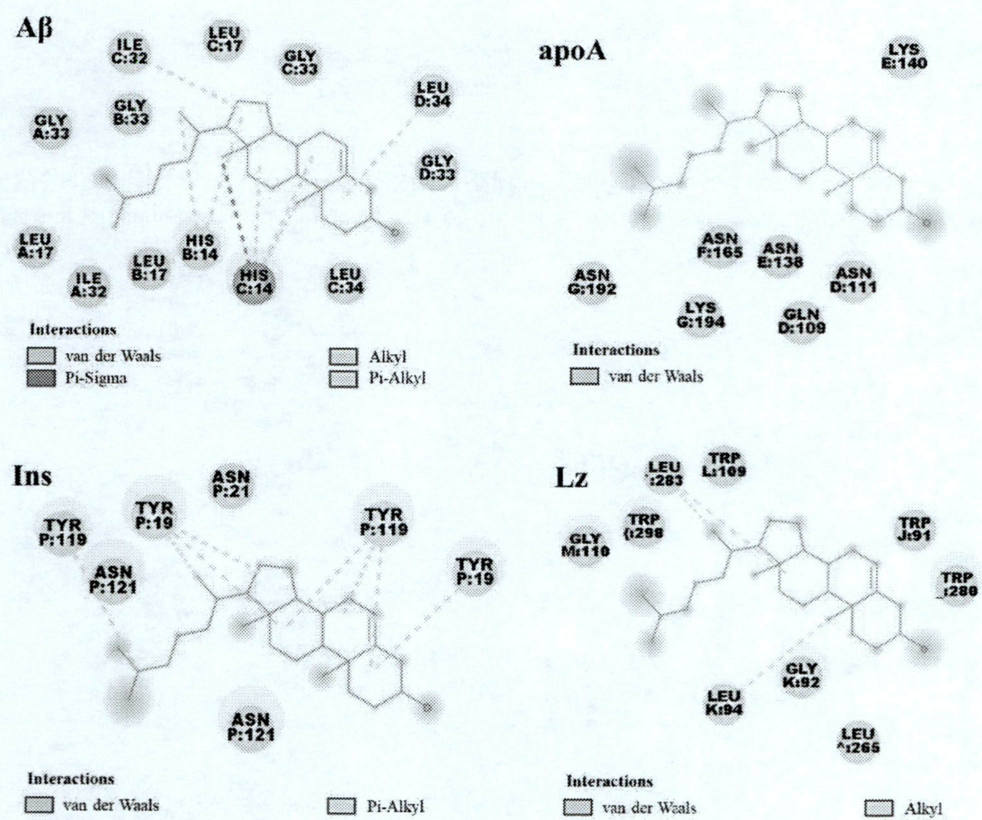

Figure 12. Binding interactions between Chol and various amyloid fibrils visualized by Discovery Studio.

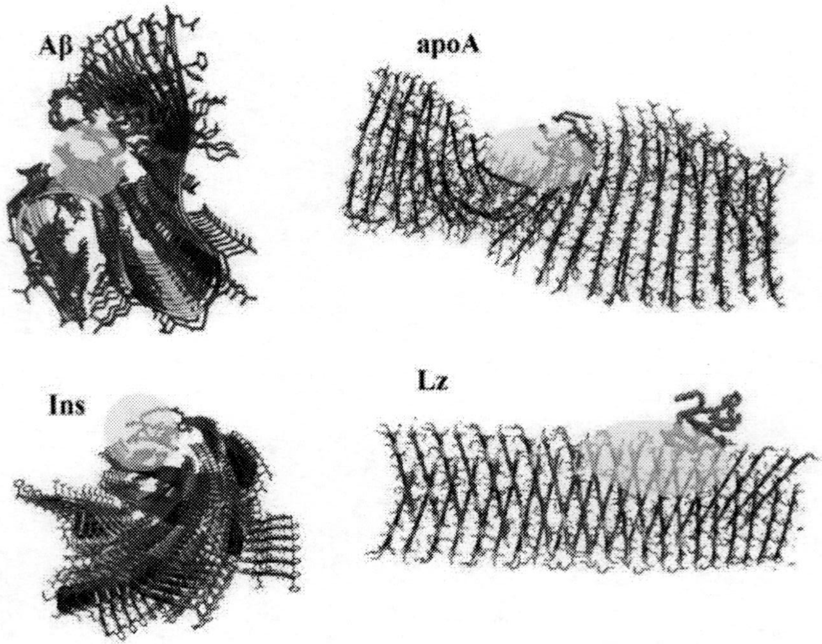

Figure 13. Docked pose for CL complexes with amyloid fibrils of Aβ, apoA, Lz and Ins amyloid fibrils. Light pink ovals represent the schematic illustration of the binding pocket for the lipid predicted by DoGSiteScorer.

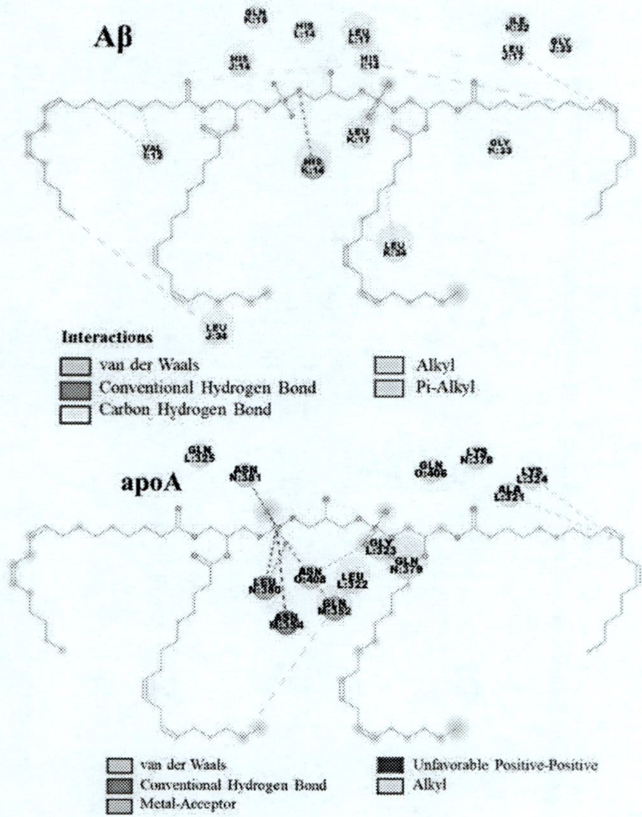

Figure 14. (Continued).

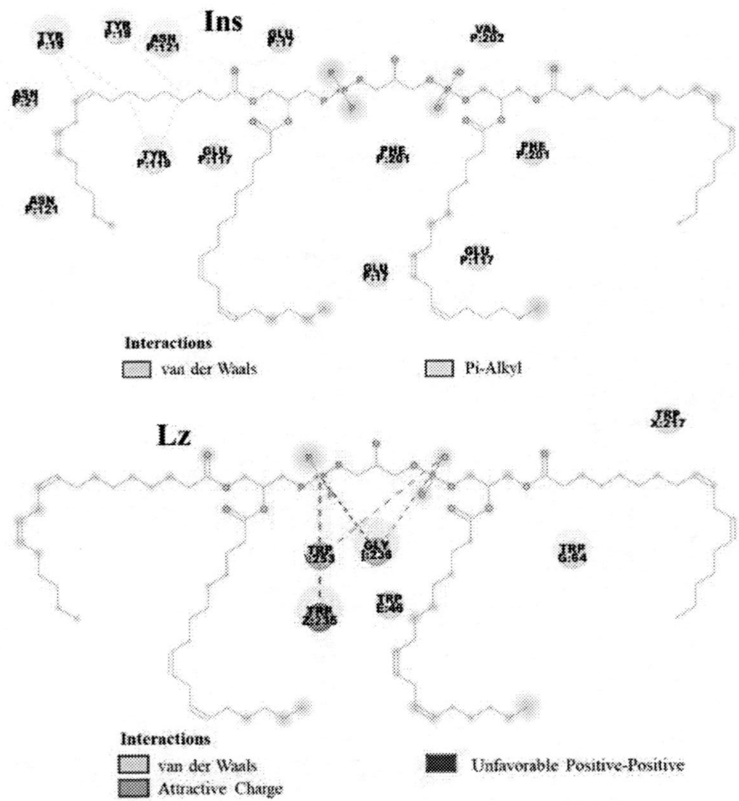

Figure 14. 2D plots of binding interactions between CL and various amyloid fibrils visualized by Discovery Studio.

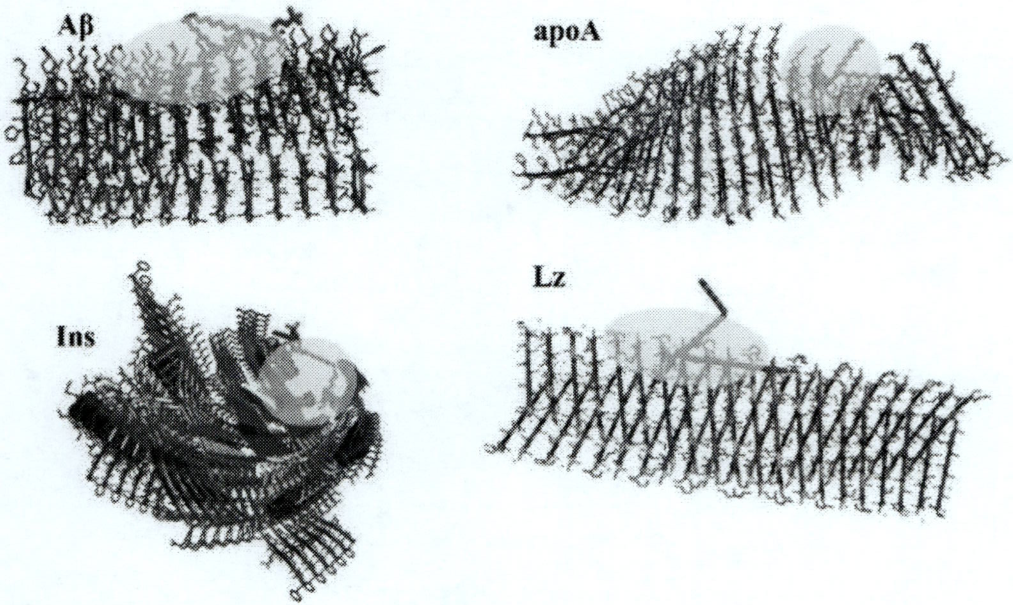

Figure 15. POPC docking with amyloid fibrils of Aβ, apoA, Lz and Ins amyloid fibrils. Light pink ovals represent the schematic illustration of the binding pocket for the lipid predicted by DoGSiteScorer.

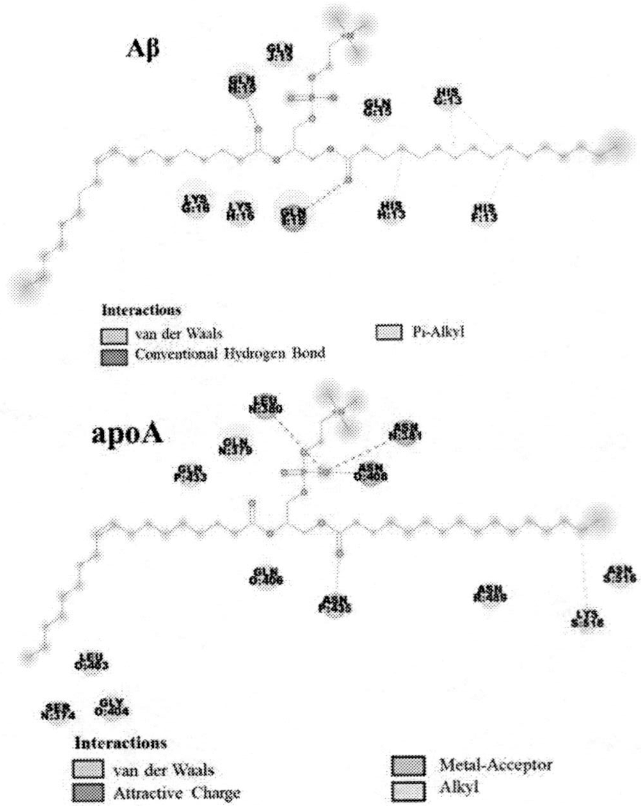

Figure 16. (Continued).

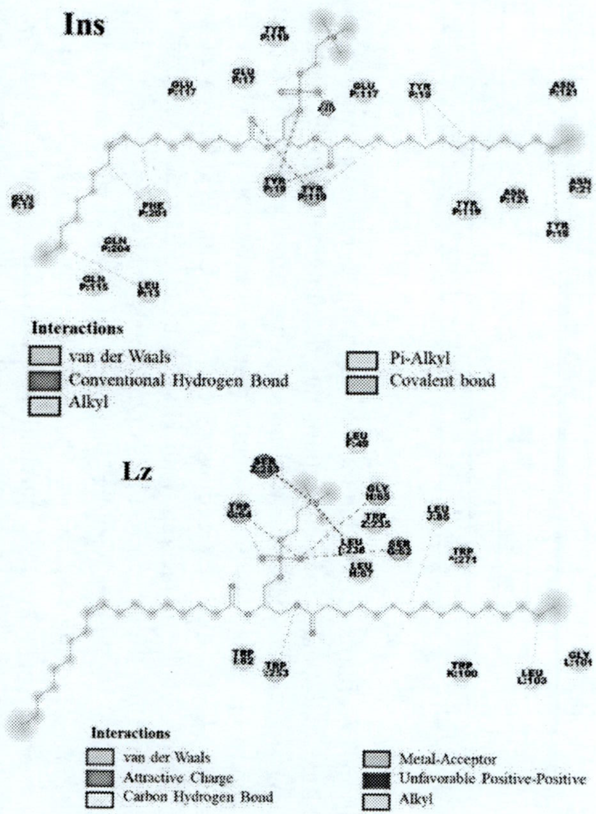

Figure 16. Binding interactions between POPC and various amyloid fibrils visualized by Discovery Studio.

Table 11. Parameters of molecular docking of amyloid fibrils with lipid molecules

Protein	Lipid	GBE, kcal/mol	ACE, kcal/mol	Amino acid residues participating in binding
Aβ	Chol	-33.19	-12.28	Gly, Leu, Ile, His
	CL	-44.39	-13.14	Gln, His, Leu, Ile, Gly, Val
	POPC	-63.18	-17.89	Lys, Gln, His
apoA-83	Chol	-7.26	-4.36	Asn, Lys, Gln
	CL	-13.84	-4.25	Gln, Asn, Ala, Lys
	POPC	-29.74	-4.32	Ser, Leu, Gly, Gln, Asn, Lys
Ins	Chol	-15.87	-1.89	Asn, Tyr
	CL	-28.24	-7.78	Asn, Glu, Tyr, Phe, Val
	POPC	-69.25	-10.34	Gln, Phe, Leu, Tyr, Glu, Asn
Lz	Chol	-29.11	-16.83	Gly, Trp, Leu
	CL	-32.12	-17.83	Trp, Gly
	POPC	-42.16	-22.38	Trp, Ser, Leu, Gly

Cholesterol

For all amyloid fibrils involved in the docking analysis cholesterol was found to exhibit some degree of penetration into the amyloid grooves with the shallowest location revealed for Lz fibers (Figure 10). Furthermore, as seen from Figure 11, all amyloid-cholesterol complexes are stabilized by van der Waals and alkyl/π-alkyl interactions. Specifically, Chol is involved in the strong alkyl/π-alkyl interactions with His, Leu, Ile amino acid residues of Aβ(1-42) amyloid fibrils, Tyr residues of Ins fibers and Leu amino acids in the case of fibrillar Lz. In turn, van der Waals protein-sterol contacts are ensured by Asn residues of apoA-83 and Ins fibers, and predominantly by Gly and Leu amino acids of Aβ(1-42) and Lz amyloid structures. Additionally, π-σ contacts between His residues of Aβ(1-42) and Chol strengthen the intercalation of sterol into the amyloid fibril.

Cardiolipin

As shown in Figure 12, the complexes of amyloid fibrils with cardiolipin are characterized mostly by superficial location of ligand on the fibrillar protein which can be attributed to the steric restrictions due to lipid bulky structure. No specific interactions were revealed for the association of Aβ(1-42) and Ins fibers with CL acyl tails, where amyloid-lipid assemblies were found to be stabilized predominantly by van der Waals and alkyl interactions (Figure 13). Furthermore, hydrogen bonds were established between His of Aβ(1-42) and lipid polar head group. In turn, the binding of fibrillar apoA-83 and Lz to anionic cardiolipin exhibited more complex pattern (Figure 13). Accordingly, in addition to van der Waals, H-bond and alkyl contacts, Asn residues of apoA-83 participate in the metal-acceptor and unfavorable positive-positive interactions with CL. The complexation of CL with Lz amyloid fibers is stabilized by a set of van der Waals, attractive charge, and unfavorable positive-positive contacts. Interestingly, Trp residues were found to be responsible for all these types of interactions.

POPC

Figure 14 depicts the docked poses of the complexes between amyloid fibrils and POPC. The amyloid-lipid assemblies are characterized by preferable surface location of POPC with some protrusion of its acyl chains into the protein grooves. Analogously to Chol and CL, the association of POPC with all types of amyloid fibers studied here was found to be stabilized by van der Waals and alkyl/π-alkyl interactions (Figure 15). Additionally, conventional and carbon hydrogen bonds were formed between Gln, Gly, Leu, Tyr amino acid residues and fibrillar Aβ(1-42), Ins and Lz (Figure 15). Furthermore, attractive charge and metal-acceptor contacts were revealed for apoA-83 and Lz amyloid fibrils, while covalent bond between Tyr residue of Ins and phosphate group of POPC was settled in the case of Ins fibers (Figure 15). Interestingly, unfavorable positive-positive interactions were found to contribute also into POPC complexation with fibrillar Lz (Figure 15). In the present context, it is noteworthy to mention that unfavorable interactions may appear due to several limitations of molecular docking protocol: i) unaccounted flexibility of protein backbone, ii) insufficient conformational sampling of the protein because of repeated fibrillar structure, iii) steric restrictions.

Overall, the analysis of molecular docking data showed that Asn, Gly, Glu, Leu and Trp are the most common amino acid residues in the formation of van der Waals amyloid-lipid contacts for all types of lipids and proteins studied in the present contribution. In turn, a predominant role of Leu, Tyr and Trp among other residues in providing the alkyl/π-alkyl interactions, was demonstrated.

Importantly, most of the interacting amino acid residues defined by the Discovery Studio based on the PatchDock/FireDock analysis, were found to be within the binding pockets predicted by the DoGSiteScorer server. The DoGSiteScorer is a grid-based method that uses a difference of Gaussian filter to detect potential binding pockets based on three-dimensional structure of the protein and ligand molecules. Specifically, the calculations of the amyloid fibril structures via DoGSiteScorer with respect to every lipid allowed us to identify from 2 to 8 binding pockets. However, for every amyloid-lipid complex the binding pocket with the highest score, calculated through the DoGSiteScorer, almost coincided with the binding site, revealed by docking. The structural properties of the binding pockets are shown in Table 12.

Table 12. Structural characteristics of binding pockets for lipids within the amyloid fibrils predicted by DoGSiteScorer

Protein	Lipid	Volume, Å^3	Surface, Å^2	Amino acid residues participating in binding
Aβ (1-42)	Chol	57.4	181.8	Ala, Asn, Asp, Gly, Ile, Leu, Phe, Ser
	CL	137.3	265.3	Gly, Ala, Leu, Ile, Val, Gln
	POPC	85.5	197.2	Asn, Leu, Ile, Lys, Phe, His
apoA-83	Chol	80.8	304.5	Gly, Ala, Asn, Ile, Lys, Gln
	CL	112.3	567.1	Ala, Gln, Lys, Ser, Asn
	POPC	99.3	434.6	Lys, Leu, Ser, Gly, Asn
Ins	Chol	75.1	115.2	Asn, Cys, Gln, Leu, Tyr, Val
	CL	104.6	347.9	Ile, Glu, Gln, Tyr, Val, Asn, Phe
	POPC	90.4	262.6	Phe, Val, Asn, Cys, Tyr, Glu, Leu
Lz	Chol	26.5	234.4	Ser, Trp, Arg, Gly, Leu, Ile, Asn
	CL	118.6	438.2	Arg, Gly, Ile, Trp, Ser
	POPC	97.2	311.1	Gly, Ile, Leu, Arg, Trp, Ser

Conclusion

Consistently growing number of reports suggests that detrimental effects of amyloidogenic proteins on membrane integrity represent the key mechanism underlying the amyloid cytotoxicity. However, despite the wealth of data collected up to date, the whole picture of toxic amyloid-lipid complexation is far from being emerged. Nevertheless, substantial evidence gives a strong support to the view that lipid-protein interactions may be involved *in vivo* in initiating and modulating the process of amyloid fibril formation. In this process lipids act as anti-chaperones facilitating protein transition into partially folded state and its subsequent assembly into protofibrillar and fibrillar structures. The membrane-related prerequisites for the nucleation of amyloid fibrils include: i) protein accumulation at lipid-water interface, commonly electrostatically-driven; ii) structural transition of polypeptide chain into aggregation-competent conformation enriched in the exposed areas complementary in terms of hydrophobicity, charge topography, hydrogen-bonding capability; iii) electrostatic attraction between the dipoles of antiparallel α-helixes; iv) protein self-association promoted by hydrophobic mismatch; v) aggregation-favoring arrangement of the solvent exposed and bilayer-buried portions of polypeptide chain. The changes in composition and physicochemical properties of lipid bilayer associated with aging or other physiological or pathological processes may increase membrane binding of specific proteins thereby triggering their amyloidogenesis. In this way, the proteins could be activated for proteosomal degradation or adverse amyloid deposition. It is tempting in this context to suppose that coupled evolution of protein sequences and their membrane environment was driven by negative selection against sequence changes that increase the tendency of a polypeptide chain to aggregate.

The main purpose of the present contribution was to recapitulate and summarized the results on amyloid-lipid interactions, obtained by our scientific group. Although this chapter is far from being exhaustive of integrated molecular mechanism lying behind the modulatory role of lipids in protein fibrillogenesis, we believe that our work contributes its piece of understanding on solving the amyloid puzzle. Further accumulation and integration of knowledge about this aspect of lipid-protein interactions may prove of great significance for development of new anti-amyloid therapeutic strategies.

References

[1] Diaz-Espinoza, R. (2021). Recent high-resolution structures of amyloids involved in neurodegenerative diseases. *Front. Aging Neurosci.*, 13, 1111-1154.

[2] Buell, K., Dobson C., and Knowles, T. (2014). The physical chemistry of the amyloid phenomenon: thermodynamics and kinetics of filamentous protein aggregation. *Essays Biochem.*, *56*, 11-39.

[3] Wentink, A., Krammer-Nussbaum, C., and Bukau, B. (2019). Modulation of amyloid states by molecular chaperones. *Cold Sprinng Harb. Perspect. Biol.*, *11*, a033969-a03994.

[4] Balasco, N., Diaferia, C., Morelli, G., Vitagliano, L., and Accardo, A. (2021). Amyloid-like aggregation in diseases and biomaterials: osmosis of structural information. *Front. Bioeng. Biotechnol.*, *9*, 1-22.

[5] Koike, H., and Katsuno, M. (2021). The ultrastructure of tissue damage by amyloid fibrils. *Molecules*, *26*, 4611-4621.

[6] Stefani, M. (2008). Protein folding and misfolding on surfaces. *Int. J. Mol. Sci.*, *9*, 2515-2542.

[7] Hebda, J., and Miranker, A. (2009). The interplay of catalysis and toxicity by amyloid intermediates on lipid bilayers, insights from type II diabetes. *Annu. Rev. Biophys.*, *38*, 125-152.

[8] Yeow, E. K. L., and Clayton, H. (2007). Enumeration of oligomerization states of membrane proteins in living cells by homo-FRET spectroscopy and microscopy, theory and application. *Biophys. J.*, *92*, 3098-3104.

[9] Trusova, V., and Gorbenko, G. (2019). Membrane interactions of fibrillar lysozyme, effect of lipid bilayer composition. *J. Mol. Liquids*., *274*, 338-344.

[10] Sciacca, M., La-Rosa, C., and Milardi, D. (2021). Amyloid-mediated mechanisms of membrane disruption. *Biophysica*., *1*, 137-156.

[11] Auer, S. (2015). Nucleation of polymorphic amyloid fibrils. *Biophys. J.*, *108*, 1176-1186.

[12] Gorbenko, G., and Kinnunen. P. (2006). The role of lipid-protein interactions in amyloid-type protein fibril formation. *Chem. Phys. Lipids*., *141*, 72-82.

[13] Gorbenko, G., Trusova, V., Girych, M., Adachi, E., Mizuguchi, C., and Saito, H. (2015) Interaction of lipid membranes with fibrillar protein aggregates. *Adv. Exp. Med. Biol.*, *855*, 135-155.

[14] Boeynaems, S., Alberti, S., Fawzi, N. L., Mittag, T., Polymenidou, M., Rousseau, F., et al. (2018). Protein phase separation, a new phase in cell biology. *Trends Cell Biol.*, *28*, 420-435.

[15] Knowles, T. P., Vendruscolo, M., and Dobson, C. M. (2014). The amyloid state and its association with protein misfolding diseases. *Nat. Rev. Mol. Cell Biol.*, *15*, 384-396.

[16] Vendruscolo, M. (2022). Lipid homeostasis and its links with protein misfolding diseases. *Front. Mol. Neurosci.*, *15*, 829291-829304.

[17] Sciacca, M., Tempra, C., Scollo, F., Milardi, D., and La Rosa, C. (2018). Amyloid growth and membrane damage, current themes and emerging perspectives from theory and experiments on Aβ and hIAPP. *Biochim. Biophys. Acta.*, *1860*, 1625-1638.

[18] Adachi, E., Nakajima, H., Mizuguchi, C., Dhanasekaran, P., Kawashima, H., Nagao, K., Akaji, K., Lund-Katz, S., Phillips, M. C., and Saito, H. (2013). Dual role of an N-terminal amyloidogenic mutation in apolipoprotein A-I, destabilization of helix bundle and enhancement of fibril formation. *J. Biol. Chem.*, *288*, 2848-2856.

[19] Groenning, M. (2010). Binding mode of Thioflavin T and other molecular probes in the context of amyloid fibrils – current status. *J. Chem. Biol.*, *3*, 1-18.

[20] Parasassi, T., Krasnowska, E., Bagatolli, L., and Gratton, E. (1998). Laurdan and Prodan as polarity-sensitive fluorescent membrane probes. *J. Fluoresc.*, *8*, 365-373.

[21] Schneidman-Duhovny, D., Inbar, Y., Nussimov, R., and Wolfson, H. (2005). PatchDock and SymmDock, servers for rigid and symmetric docking. *Nucl. Acid Res.*, *33*, W363-W367.

[22] Connolly, M. (1983). Solvent-accessible surfaces of proteins and nucleic acids. *Science.*, *221*, 709-713.

[23] Connolly, M. (1983). Analytical molecular surface calculation. *J. Appl. Crystallogr.*, *16*, 548-558.

[24] Zhang, C., Vasmatzis, G., Cornette, J., and De Lisi, C. (1997). Determination of atomic desolvation energies from the structures of crystallized proteins, *J. Mol. Biol.*, *267*, 707-726.

[25] Andrusier, N., Nussimov, R., and Wolfson, H. (2007). FireDock, fast interaction refinement in molecular docking. *Proteins.*, *69*, 139-159.

[26] Mashiach, E., Schneidman-Duhovny, D., Andrusier, N., Nussimov, R., and Wolfson, H. (2008). FireDock, a web-server for fast interaction refinement in molecular docking. *Nucl. Acid Res.*, *36*, W229-W232.

[27] Pepys, M., Hawkins, P., Booth, D., Vigushin, D., Tennent, G., Souter, A., and Totty, N., Nguyen, O., Blake, C., Terry, C., Feest, T., Zalin, A., and Hsuan, J. (1993). Human lysozyme gene mutations cause hereditary systemic amyloidosis. *Nature.*, *362*, 553-557.

[28] Ahn, M., Hagan, C., Bernardo-Gancedo, A., De Genst, E., Newby, F., Christodoulou, J., Dhulesia, A., Dumoulin, M., Robinson, C., Dobson, C., and Kumita, J. (2016). The significance of the location of mutations for the native-state dynamics of human lysozyme. *Biophys. J.*, *111*, 2358-2367.

[29] Obici, L., and Franceschini, L. (2006). Structure, function and amyloidogenic propensity of apolipoprotein A-I. *Amyloid.*, *13*, 191-205.

[30] Adachi, E., Kosaka, A., Tsuji, K., Mizuguchi, C., Kawashima, H., Shigenaga, A., Nagao, K., Akaji, K., Otaka, A., and Saito, H. (2004). The extreme N-terminal region of human apolipoprotein A-I has a strong propensity to form amyloid fibrils. *FEBS Lett.*, *588*, 389-394.

[31] Genschel, J., Haas, R., Propsting, M., and Schmidt, H. (1998). Apolipoprotein A-I induced amyloidosis *FEBS Lett.*, *430*, 145-149.

[32] Girych, M., Gorbenko, G., Trusova, V., Adachi, E., Mizuguchi, C., Nagao, K., Kawashima, H., Akaji, K., Lund-Katz, S., Phillips, M., and Saito, H. (2014). Interaction of Thioflavin T with amyloid fibrils of apolipoprotein A-I N-terminal fragment, resonance energy transfer study. *J. Struct. Biol.*, *185*, 116-124.

[33] Cao, A., Hu, D., and Lai, L. (2004). Formation of amyloid fibrils from fully reduced hen egg white lysozyme. *Protein Sci.*, *13*, 319-324.

[34] Frare, E., Mossuto, M., Polverino de Laureto, P., Dumoulin, M., Dobson, C., and Fontana, A. (2006). Identification of the core structure of lysozyme amyloid fibrils by proteolysis. *J. Mol. Biol.*, *361*, 551-561.

[35] Mossuto, M., Dhulesia, A., Devlin, G., Frare, E., Kumita, J., Polverino de Laureto, P., Dumoulin, M., Fontana, A., Dobson, C., and Salvatella, X. (2010). The non-core regions of human lysozyme amyloid fibrils influence cytotoxicity. *J. Mol. Biol.*, *402*, 783-796.

[36] Lara, C., Usov, I., and Mezzenga, R. (2011). Sub-persistence-length complex scaling behavior in lysozyme. *Phys. Rev. Lett.*, *107*, 238101.

[37] Stsiapura, V., Maskevich, A., Kuzmitsky, V., Uversky, V., Kuznetsova, I., and Turoverov, K. (2008). Thioflavin T as a molecular rotor, fluorescent properties of thioflavin T in solvents with different viscosity. *J. Phys. Chem. B.*, *112*, 15893-15902.

[38] Cecchini, P., de Franceschi, G., Frare, E., Fontana, A., and Polverino de Laureto, P. (2012). The role of tryptophan in protein fibrillogenesis, relevance of Trp7 and

Trp14 to the amyloidogenic properties of myoglobin. *Protein Eng. Des. Sel., 24*, 199-203.

[39] Adachi, E., Nakajima, H., Mizuguchi, C., Dhanasekaran, P., Kawashima, H., Nagao, K., Akaji, K., Lund-Katz, S., Phillips, M., and Saito, H. (2013). Dual role of an N-terminal amyloidogenic mutation in apolipoprotein A-I, destabilization of helix bundle and enhancement of fibril formation. *J. Biol. Chem. 288*, 2848-2856.

[40] Weldon, D. T., Rogers, S. D., Ghilardi, J. R., Finke, M. P., Cleary, J. P., O'Hare, E., Esler, W. P., Maggio, J. E., and Mantyh, P. (1998). Fibrillar β-amyloid induces microglial phagocytosis, expression of inducible nitric oxide synthase, and loss of a select population of neurons in the rat CNS *in vivo*. *J Neurosci., 18*, 2161-2173.

[41] Bucciantini, M., Rigacci, S., and Stefani, M. (2014). Amyloid aggregation, role of biological membranes and the aggregate-membrane system. *J. Phys. Chem. Lett., 5*, 517-527.

[42] Matsuzaki, K. (2011). Formation of toxic amyloid fibrils by amyloid beta-protein on ganglioside clusters. *Int. J. Alzheimers Dis., 956104*, 1-7.

[43] de Planque, R., Raussen, V., Contera, S., Rijkers, D., Liskamp, R., Ruysschaert, J., Ryan, J., Separovic, F., and Watts, A. (2007). β-sheet structured β-amyloid (1-40) perturbs phosphatidylcholine model membranes. *J. Mol. Biol., 368*, 982-987.

[44] Lee, C. C., Sun, Y., and Huang, H. (2012). How type II diabetes-related islet amyloid polypeptide damages lipid bilayers. *Biophys. J., 102*, 1059-1068.

[45] Verdier, Y., Zarandi, M., and Penke, B. (2004). Amyloid beta-peptide interactions with neuronal and glial cell plasma membrane, binding sites and implications for Alzheimer's disease. *J. Pept. Sci., 10*, 229-248.

[46] Sciacca, M. F., Brender, J., Lee, D., and Ramamoorthy, A. (2012). Phosphatidylethanolamine enhances amyloid-fiber dependent membrane fragmentation. *Biochemistry, 51*, 7676-7684.

[47] Butterfield, D. A., Castegna, A., Lauderback, C., and Drake, J. (2002). Evidence that amyloid beta-peptide-induced lipid peroxidation and its sequelae in Alzheimer's disease brain contribute to neuronal death. *Neurobiol. Aging., 23*, 655-664.

[48] Novitskaya, V., Bocharova, O., Bronstein, I., and Baskakov, I. (2006). Amyloid fibrils of mammalian prion protein are highly toxic to cultured cells and primary neurons. *J. Biol. Chem., 281*, 13828-13836.

[49] Gharibyan, A., Zamotin, V., Yanamandra, K., Moskaleva, O., Margulis, B., Kostanyan, I., and Morozova-Roche, L. (2007). Lysozyme amyloid oligomers and fibrils induce cellular death via different apoptotic/necrotic pathways. *J. Mol. Biol., 365*, 1337-1349.

[50] Huang, B., He, J., Ren, J., Yan, X., and Zheng, C. (2009). Cellular membrane disruption by amyloid fibrils involved inter molecular disulfide cross-linking. *Biochemistry, 48*, 5794-5800.

[51] Hirano, A., Yoshikawa, H., Matsushita, S., Yamada, Y., and Shiraki, K. (2012). Adsorption and disruption of lipid bilayers by nanoscale protein aggregates. *Langmuir, 28*, 3887-3895.

[52] Engel, M. F. M., Khemtémourian, L., Kleijer, C., Meeldijk, H., Jacobs, J., Verkeij, A., de Kruijff, B., Killian, A., and Höppener, J. (2008). Membrane damage by

human islet amyloid polypeptide thought fibril growth at the membrane. *Proc. Natl. Acad. Sci. U.S.A*, *105*, 6033-6038.

[53] Milanesi, L., Sheynis, T., and Xue, W. (2012). Direct three-dimensional visualization of membrane disruption by amyloid fibrils. *Proc. Natl. Acad. Sci. U.S.A.*, *109*, 20455-20460.

[54] Ioffe, V., and Gorbenko, G. (2005). Lysozyme effect on structural state of model membranes as revealed by pyrene excimerization. *Biohys. Chem.*, *114*, 199-204.

[55] Kinnunen, P., Koiv, A., Lehtonen, J., and Mustonen, P. (1994). Lipid dynamics peripheral interactions of proteins with membrane surfaces. *Chem. Phys. Lipids*, *73*, 181-207.

[56] Nakajima, A. (1971). Solvent effect on the vibrational structure of the fluorescence and absorption spectra of pyrene. *Bull. Chem. Soc. Japan*, *44*, 3272-3277.

[57] Barenholz, Y., Cohen, T., Haas, E., and Ottolenghi, M. (1996). Lateral organization of pyrene-labeled lipids in bilayers as determined from the deviation from equilibrium between pyrene monomers and excimers. *J. Biol. Chem.*, *271*, 3085-3090.

[58] Montroll, E. (1969). Random walks on lattices. *J. Math. Phys.*, *10*, 753-765.

[59] Galla, H., Hartmann, W., Theilen, U., and Sackmann, E. (1979). On two-dimensional random walk in lipid bilayers and fluid pathways in biomembranes. *J. Membr. Biol.*, *48*, 215-236.

[60] Blackwell, M., Gounaris, K., and Barber, J. (1986). Evidence that pyrene excimer formation in membranes is not diffusion-controlled. *Biochim. Biophys. Acta.*, *858*, 221-234.

[61] Osella, S., Paloncyova, M., Sahi, M., and Knippenberg, S. (2020). Influence of membrane phase on the optical properties of DPH. *Molecules*, *25*, 4262-4279.

[62] Kremer, J., Pallito, M., Sklansky, D., and Murphy, R. (2000). Correlation of β-amyloid aggregate size and hydrophobicity with decreased bilayer fluidity of model membranes. *Biochemistry*, *39*, 10309-10318.

[63] Muller, W., Kirsch, C., and Eckert, G. (2001). Membrane-disordering effects of β-amyloid peptides. *Biochem. Soc. Trans.*, *29*, 617-624.

[64] Ma, X., Sha, Y., and Lin, K. (2000). The effect of fibrillar Aβ1-40 on membrane fluidity and permeability. *Protein Pept. Lett.*, *9*, 173-178.

[65] Fandrich, M., Nystrom, S., Nilsson, K., Bockmann, A., LeVine, H., and Hammarstrom, P. (2018) Amyloid fibril polymorphism – a challenge for molecular imaging and therapy. *J. Inetrn. Med.*, *283*, 218-237.

[66] Dobretsov, G., Petrov, V., Mishiev, V., Klebanov, V., and Vladimirov, Yu. (1977). 4-dimethylaminochalcone and 3-methoxybenzanthrone as fluorescent probes to study biomembranes. I. Spectral characteristics. *Stud. Biophys.*, *65*, 91-98.

[67] Dobretsov, G., Petrov, V., and Vladimirov, Yu. (1997). 4-dimethylaminochalcone and 3-methoxybenzanthrone as fluorescent probes to study biomembranes. II. Sensitivity of 4-dimethylaminochalcone to water molecules in the surface layer of a membrane. *Stud. Biophys.*, *65*, 91-98.

[68] Parasassi, T., de Stasio, G., Ravagnan, G., Rusch, R., and Gratton, E. (1991). Quantitation of lipid phases in phospholipid vesicles by the generalized polarization of Laurdan fluorescence. *Biophys. J.*, *60*, 179-189.

[69] Parasassi, T., Stefano, M., Loiero, M., Ravagnan, G., and Gratton, E. (1994). Influence of cholesterol on phospholipid bilayer phase domains as detected by Laurdan fluorescence. *Biophys J.*, *66*, 120-132.

[70] Harris, F., Best, K., and Bell, J. (2002). Use of Laurdan fluorescence intensity and polarization to distinguish between changes in membrane fluidity and phospholipid order. *Biochim. Biophys. Acta*, *1565*, 123-128.

[71] Scheinpflug, K., Krylova, O., and Strahl, H. (2017). Measurement of cell membrane fluidity by Laurdan GP, fluorescence spectroscopy and microscopy. *Methods Mol. Biol.*, *1520*, 159-174.

[72] Sponne, I., Fifre, A., Koziel, V., Oster, T., Olivier, J., and Pillot, T. (2004). Membrane cholesterol interferes with neuronal apoptosis induced by soluble oligomers but not fibrils of amyloid-β peptide. *FASEB J.*, *838*, 836-838.

[73] Cecchi, C., Baglioni, S., and Fiorillo, C. (2005). Insights into the molecular basis of the differing susceptibility of varying cell types to the toxicity of amyloid aggregates. *Cell Sci.*, *118*, 3459-3470.

[74] Campioni, S., Mannini, B., Zampagni, M., Pensalfini, A., Parrini, C., Evangilisti, E., Relini, A., Stefani, M., Dobson, C., Cecchi, C., and Chiti, F. (2010). The causative link between the structure of aberrant protein oligomers and their ability to cause cell dysfunction. *Nat. Chem. Biol.*, *6*, 140-147.

[75] Aubrey, L., Blakeman, B., Lutter, L., Serpell, C., Tuite, M., Serpell, L., and Xue, W. (2020). Quantification of amyloid fibril polymorphism by nano-morphometry reveals the individuality of filament assembly. *Commun. Chem.*, *3*, 1-10.

[76] Di Scala, C., Yahi, N., Boutemeur, S., Flores, A., Rodriguez, L., Chahinian, H., and Fantini, J. (2016). Common molecular mechanism of amyloid pore formation by Alzheimer's β-amyloid peptide and α-synuclein. *Scientific Reports*, *6*, 28781-28790.

[77] Bode, D., Freeley, M., Nield, J., Palma, M., and Viles, J. (2019). Amyloid-β oligomers have a profound detergent-like effect on lipid membrane bilayers, imaged by atomic force and electron microscopy. *J. Biol. Chem.*, *294*, 7566-7572.

[78] Dean, R., O'Brien, L., Thwaite, J., Fox, M., Atkins, H., and Ulaeto, D. (2010). A carpet-based mechanism for direct antimicrobial peptide activity against vaccinia membranes. *Peptides*, *31*, 1966-1972.

[79] Gorbenko, G., Ioffe, V., and Kinnunen, P. (2007). Binding of lysozyme to phospholipid bilayers, evidence for protein aggregation upon membrane association. *Biophys. J.*, *93*, 140-153.

[80] Trusova, V., and Gorbenko, G. (2017). Molecular dynamics simulations of lysozyme-lipid systems, probing the early steps of protein aggregation. *J. Biomol. Struct. Dyn.*, *36*, 2249-2260.

[81] Usov, I., Adamcik, J., and Mezzenga, R. (2013). Polymorphism in bovine serum albumin fibrils, morphology and statistical analysis. *Faraday Discuss.*, *166*, 151-162.

[82] Marshall, K., Marchante, R., Xue, W., and Serpell, L. (2014). The relationship between amyloid structure and cytotoxicity. *Prion*, *8*, 192-196.

[83] Koehl, P., and Delarue, M. (2010). AQUASOL, an efficient solver for the dipolar Poisson-Boltzmann-Langevin equation. *J. Chem. Phys.*, *132*, 064101-064117.

[84] Dzwolak, W., and Pecul, M. (2005). Chiral bias of amyloid fibrils revealed by the twisted conformation of Thioflavin T, an induced circular dichroism/DFT study. *FEBS Lett.*, *579*, 6601-6603.

[85] Adamcik, J., and Mezzenga, R. (2011). Adjustable twisting periodic pitch of amyloid fibrils. *Soft Matter*, *7*, 5437-5443.

[86] Sunde, M., Serpell, L., Bartlam, M., Fraser, P., Pepys, M., and Blake, C. (1997). Common core structure of amyloid fibrils by synchrotron X-ray diffraction. *J. Mol. Biol.*, *273*, 729-739.

[87] Lomize, M., Pogozheva, I., Joo, H., Mosberg, H., and Lomize, A. (2012). OPM database and PPM web server, resources for positioning of proteins in membranes. *Nucl. Acids Res.*, *40*, D370-D376.

[88] Lomize, A., Pogozheva, I., Joo, H., and Mosberg, H. (2011). Anisotropic solvent model of the lipid bilayer. 2. Energetics of insertion of small molecules, peptides, and proteins in membranes. *J. Chem. Inf. Model.*, *51*, 930-946.

[89] White, S., and Whimley, W. (1999). Membrane protein folding and stability, physical principles. *Annu. Rev. Biophys. Biomol. Struct.*, *28*, 319-365.

[90] Adamcik, J., and Mezzenga, R. (2012). Protein fibrils from a polymer physics perspective. *Macromolecules*, *45*, 1137-1150.

Chapter 4

Updating Curvature of Graphitic Materials: Molecular Modeling Studies on the Spiral Structure of Graphene and Eumelanin Models

Juan C. Stockert[1,2,*], MD, PhD
and Marcelo N. Felix-Pozzi[3]
[1]Institute of Oncology "Angel H. Roffo", Research Area, Faculty of Medicine, University of Buenos Aires, Argentina
[2]Integrative Center of Biology and Applied Chemistry, University Bernardo O'Higgins, Santiago, Chile
[3]Department of Micro and Nanotechnology, Institute of Nanosciences and Nanotechnology, Constituyentes Atomic Center, National Commission of Atomic Energy, Buenos Aires, Argentina

Abstract

In the last few years, molecular modeling has become a relevant tool for research and teaching purposes in chemistry and molecular biology, allowing the design and development of new molecular structures and materials. In the present chapter, the focus of molecular modeling methodology is addressed towards the mechanistic origin of spiral organization of graphitic materials. A recently proposed catechol-porphycene (CPo) model for eumelanin, and both "arm-chair" (A) and "zig-zag" (Z) single layer graphene spirals were chosen as case studies to be analyzed by molecular modeling after MM+, PM3, and extended Hückel geometry optimization using HyperChem v7 software. The poly-CPo eumelanin model corresponds to a highly conjugated graphite-like planar structure based on the polymerization of 5,6-dihydroxy-indole monomers, using all the reactive carbon radicals. Formation of ether

* Corresponding Author's Email: jcstockert@institutoroffo.uba.ar.

In: Advances in Biology. Volume 2
Editor: Charles D. Grant
ISBN: 979-8-88697-422-5

bridges between catechol groups of the polymer induces a constrained curvature due to the newly formed furan rings, which in turn results in a spiral organization of poly-CPoe. Interestingly, published molecular modeling studies on graphene properties show a spiral structure, which generates improved opto-electronic properties. However, it is not obvious why a graphene layer should be curved lacking any chemical modification of the original planar polymer. Regarding oxidation, it is tempting to speculate that the possible attack of graphene edges by oxidant agents (e.g., singlet oxygen), could introduce hydroxyls along the reactive A and Z edges, which after ether bridging would result in curved and then spiral graphenes just as it is revealed by molecular modeling. Likewise, in addition to furan, cyclopenta-dienyl and pyrrole rings, oxo and peroxide bridges along graphene edges do result in a spiral structure. In all cases, oligomer stacking generates fused LUMOs at low energies. Stacked poly-CPoe and graphene spirals show compact and very close HOMO (valence band, VB) and LUMO (conduction band, CB) levels, fused LUMOs and strongly reduced energy gap (Eg). It is worth to note the clear similarity between the fused LUMOs of a spiral graphene and the poly-CPoe eumelanin model, which is related to the striking semiconductor characteristics of these amorphous graphitic compounds. Likewise, molecular modeling methods allow the design of new bioinspired materials (e.g., planar and spiral poly-catechol-imidacene) and explain the massive fusion of longitudinal and transversal LUMOs continuum from stacked, intercalated and bilayered graphitic spirals yielding compact and close VB and CB.

Keywords: eumelanin, graphitic materials, graphene oxidation, polymer curvature, spiral graphene

Introduction

Recently, an ever-growing literature on molecular modeling methods and applications has been published, mainly regarding chemical, biochemical and biomedical research and teaching. Common biological substrates are macromolecules such as proteins, nucleic acids and other biopolymers, as well as their specific interaction with drugs and small molecules (Braga et al., 2004; Meng and Kaxiras, 2008; Peng et al., 2010; Zhu and Li, 2013; Stockert, 2014; d'Ischia et al., 2014; Norouzi and Fakhrabadi, 2019; Pelemiš et al., 2021; Istifly, 2022).

Graphitic materials are a varied collection of polymers with fascinating chemical, physical and biological properties (Novoselov et al., 2004;

AZoNano, 2006; Zhu et al., 2010; Park and Aluru, 2011; Li et al., 2017; Liu et al., 2018; Hauser et al., 2020; Gaur et al., 2021; Stockert et al., 2022), which lead to formulate the following questions about their molecular structures: (1) which properties do the graphite, graphene layers, fullerenes, carbon dots, poly-dopamine and eumelanin share? (2) which is the common structural pattern of these graphite-like materials? (3) why do a curved and spiral structure appear in these polymers? In response to (1) and (2), broad-band absorption spectrum (spanning the ultraviolet, visible, and near infrared regions), semi-conductivity and even superconductivity, stacking interactions between aromatic sheets separated by about 3.4 Å, and photothermal effects are shared by all these materials. The answer to (3) accounts for the present chapter using a molecular modeling approach to illustrate the morphology, chemical origin and molecular orbitals of spiral graphitic structures.

The fascinating spiral shape, both Archimedean and logarithmic type, among others, are extensively represented in Physics, Chemistry and Biology. Examples of spirals are from the path of ionizing radiations in the Wilson's chamber, Belousov-Zhabotinski's chemical reactions, pinecones, sunflowers and snail shells, to tropical hurricanes and spiral galaxies. Likewise, the Archimedean spiral shape can be also found in some graphitic molecular structures.

Regarding the chemistry of eumelanin (a highly relevant biological graphitic material), the precursor monomers 5,6-dihydroxy-indole (DHI) and indole-quinone (IQ), are presented in Figure 1, A. Previous precursors are L-tyrosine, L-DOPA, and dopamine (DA), which can easily form poly-DA (PDA) by oxidative auto-polymerization. X-ray crystallography and electron microscopic studies indicate that synthetic PDA-melanin and tyrosine-melanin are essentially like natural eumelanin in their local atomic arrangements and ultrastructure (Bridelli et al., 1990; Cheng et al., 1994; Watt et al., 2009; Chen et al., 2013). Although PDA-melanin does not contain the 2-carboxyl group, its properties correspond to a true eumelanin (d'Ischia et al., 2015), and it has similar applications (Hauser et al., 2020).

Possible models of eumelanin polymers are a linear flexible poly 4-7 DHI (Liebscher et al., 2013; d'Ischia et al., 2015), and a planar cyclic tetramer of 2-7 IQ (Kaxiras et al., 2006; Meng and Kaxiras, 2008). In contrast with flexible chain models for eumelanin, a planar and rigid "porphyrin-like graphitic structure" had been suggested earlier (Olivieri and Nicolaus, 1999; Nicolaus, 2005) and revisited more recently (Stockert, 2021). In this poly-catechol-porphycene (poly-CPo) model (Figure 1, B), the building block is a catechol derivative of the porphycene ring (Po), namely a structural isomer of the porphyrin ring.

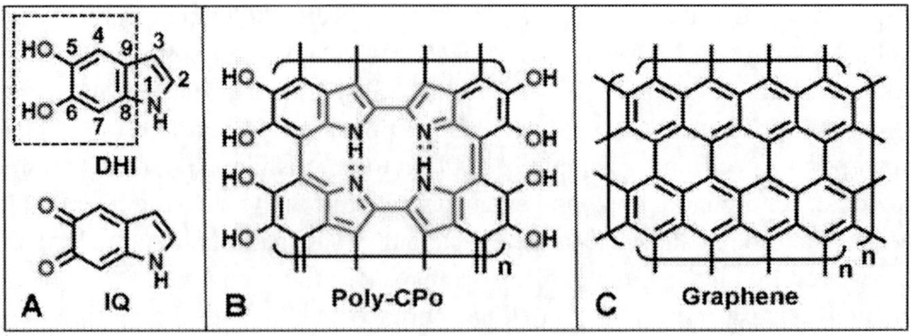

Figure 1. A: Chemical structure of eumelanin precursors: dihydroxy-indole (DHI) and indole-quinone (IQ), showing formal double bonds, atom numbering, reactive atoms (C2, C3, C4, C7) available for polymerization, and the catechol group (dashed rectangle). B: DHI tetramer, forming the eumelanin unit model, catechol-porphycene (CPo), with the porphycene ring (violet), and hydrogen bonds (NH:N, blue points indicate donor electron pairs). C: part of a graphene sheet showing the hexagonal skeleton of sp^2 carbon atoms and conjugated bonds. Polymers of B and C are indicated by n.

It is known that C2, C3, C4 and C7 atoms are the most reactive sites in the DHI ring (Nicolaus, 1997), and they allow easy polymerization by formation of intermediate radicals (Zhou et al., 2019) according the proposed synthesis mechanism of poly-CPo, involving all the reactive carbon atoms of DHI units. Molecular modeling methods have been recently applied to illustrate the synthesis of eumelanin models (Blázquez-Castro and Stockert, 2021; Stockert and Blázquez-Castro, 2022).

On the other hand, graphene is a bidimensional (2D) graphitic monolayer formed by sp^2 C atoms, which has attracted much attention from Materials scientists, chemists and physicists. A portion of the planar pristine graphene sheet is shown in Figure 1, C. In our case, regarding the possible curvature of graphitic materials, the underlying concept to support the spiral organization of chemically modified graphene is the already suggested spiral structure of the poly-CPoe eumelanin model caused by formation of ether bridges and furan rings (Stockert and Blázquez-Castro, 2022; Stockert et al., 2022). Due to the increasing interest in controllably tailoring the electronic structure of graphene, efforts are now devoted to its chemical functionalization by adsorbing or reacting atoms, molecules and functional groups on its surface or at their edges (Boukhvalov et al., 2008; Shenoy et al., 2008; Zhue and Li, 2013).

Methodological Aspects

Molecular modeling (MM) methods are commonly applied in biochemical research (Stockert et al., 2005; Stockert et al., 2014; Pelemiš et al., 2021) and teaching (Stockert, 1994; Stockert and Blázquez-Castro, 2017) and for illustrating and analyzing molecular structures and processes (Braga et al., 2004; Del Castillo et al., 2010; Zhu and Li, 2013; Stockert, 2014; Stockert, 2020), but also as a theoretical tool to correct inaccurate or misleading chemical assumptions, as well as to compare and predict properties, structure and binding modes of new compounds (Raghavan et al., 1990; Stockert et al., 2008; Peng et al., 2010; Istifly, 2022).

Typical small molecules subjected to MM are drugs, bioactive agents, dyes and fluorochromes, but also macromolecules and their interactions with substrates are often studied by MM. Although rather simple software (e.g., Dtmm87, ACD-ChemSketch 10) has been early applied for MM studies (Stockert, 1994; Stockert and Abasolo, 2011), advanced software (Chem3D Ultra 8, HyperChem) and MM methods such as AM1, PM3, and density functional theory (DFT) are routinely used for calculation of several parameters and molecular orbitals (MOs).

To substantiate molecular structures and processes, MM was here performed following previous work (Stockert et al., 2022). Chemical structures with formal double bonds were drawn with ChemDraw Ultra v12.0 software, whereas geometry optimization (GO) and energy minimization (EM) were performed with the HyperChem v7 software, using Molecular Mechanics Force Field (MM+), PM3 Polak-Riviere conjugate gradient converged at 1, 0.1, or 0.01 kcal/(Å mol), or extended-Hückel method (Galvão and Caldas, 1990) for calculation of MOs, which were represented as Gouraud shaded 3D isosurfaces. Along illustrative examples and case studies in this chapter, the used values for MM parameters are mentioned in the legend of figures.

Small aromatic molecules (e.g., colorants, drugs) have a rather simple electronic structure (Figure 2, A), whereas larger polymeric molecules are formed by many aromatic units (Figure 2, B). In these two cases, the distribution of energy levels is different. The ground singlet (S) state and the first photo-excited state of small aromatic molecules after energy absorption correspond to the S^0 and S^1 levels, respectively (Figure 2, C), which are also represented by different MOs. The highest-occupied (HOMO), and lowest-unoccupied (LUMO) states represent the energy levels of the ground state and first excited state of the molecule, respectively (Stockert and Blázquez-Castro, 2017). In HOMO-d and LUMO+d, d is a value from 0 to the maximum energy level, and the HOMO-LUMO separation corresponds to the forbidden Fermi's energy gap (Eg). When electronic energy levels are highly compact, HOMO and LUMO states form continuum bands, which correspond to the well-known "semiconductor-like" valence band (VB) and conduction band (CB), respectively (Fig, 2, D). The orbital phases are denoted by colors or signs of π-electron orbital lobes. Fused lobes with the same color or sign are in-phase, and those with isolated (unfused) lobes with different colors or signs are out-of-phase.

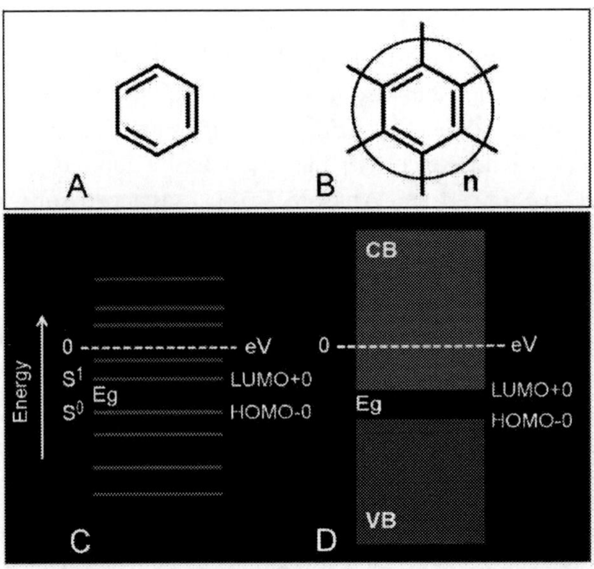

Figure 2. A: The benzene ring as a prototype of small aromatic chromophores (e.g., colorants and drugs with well-structured absorption spectra). B: An aromatic polymerized unit in large chromophores (e.g., graphitic materials with broad-band absorption spectra). C: Electronic energy levels in ground (blue lines) and excited states (red lines) of an aromatic compound like (A). D: Electronic energy levels (HOMO and LUMO, blue and red, respectively) organized as valence (VB) and conduction (CB) bands in a graphitic material like (B). Both energy graphics show the energy gap (Eg), isolated energy levels (S^0 and S^1), as well as HOMO-0 and LUMO+0 states.

Light absorption curves from mammalian, invertebrate (cuttlefish), and synthetic melanins have the same spectral characteristics, showing a monotonic broad-band absorption spectrum with exponential decay, as occurs in all graphitic materials (Plaetzer et al., 2009; Micillo et al., 2016; Blázquez-Castro and Stockert, 2021). The broad-band photonic absorption of eumelanin reminds more of an inorganic semiconductor material with close VB and CB and a small energy gap (about 0.5-1.5 eV) (Nicolaus, 1997; Mavridi-Printezi et al., 2020), than to an organic chromophore with structured absorption peaks typically associated to transitions from bonding π to antibonding π^* orbitals. Thus, eumelanins and black graphitic materials such as synthetic eumelanins, pyrolyzed polydopamine (PDA) films, graphite, graphene, and PDA-based carbon spheres (Liu et al., 2014) are supramolecular structures and share the broad-band absorption that characterizes amorphous semiconductors.

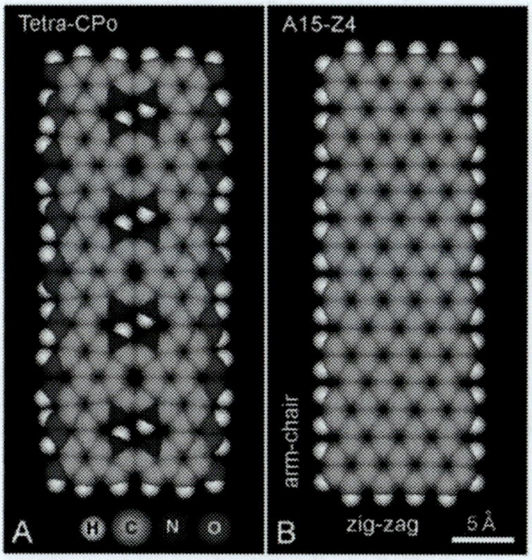

Figure 3. Frontal view of atomic volume models of A: tetra-CPo, and B: the graphene segment A15-Z4, in which 7 and 4 represent the number of benzene units along arm-chair (A) and zig-zag (Z) edges, respectively. GO by MM+ converged at 1 kcal/(Å mol). Observe the planar structure of both oligomers, and the occurrence of catechol groups in tetra-CPo. Excepting the chain ends, no H atoms can be removed from CH groups to further polymerize the CPo chain. Hydrogen bond (NH:N) distance is 2.8 Å. Elements are shown in different colors, which also correspond to the color code of later figures.

Following MM methods, both eumelanin CPo and graphene models clearly show a planar structure (Figure 3), which has been previously described for a CPo model (Stockert, 2021). In the case of tetra-CPo, the LUMO+0 state shows an extended longitudinal π*-electron conjugation (as well as fused MOs in stacked bilayers), whereas the HOMO-0 state has a rather transversal conjugation (Stockert et al., 2022). The energy levels of HOMO and LUMO states appear as compact VB and CB, respectively.

Curvature of Graphitic Materials

Regarding the poly-CPo model of eumelanin (Figure 3, A), dehydration and formation of ether (e) bridges and additional furan rings lead to a local chain curvature (Figure 4), a feature that was early observed by Olivieri and Nicolaus (1999) and Nicolaus (2005) using MM methods. Thus, the formation of a fused pentagonal furan ring results in a constrained molecular curvature of CPoe units, which in turn, can generate circular or spiral organizations of oligomers (Stockert and Blázquez-Castro, 2022). It is somehow surprising that only a small change, such as the addition of ether bridges, can largely modify the geometry of CPo oligomers and polymers to form curved chains. Considerable longitudinal and mild transversal curvatures are apparent in this model (Figure 4).

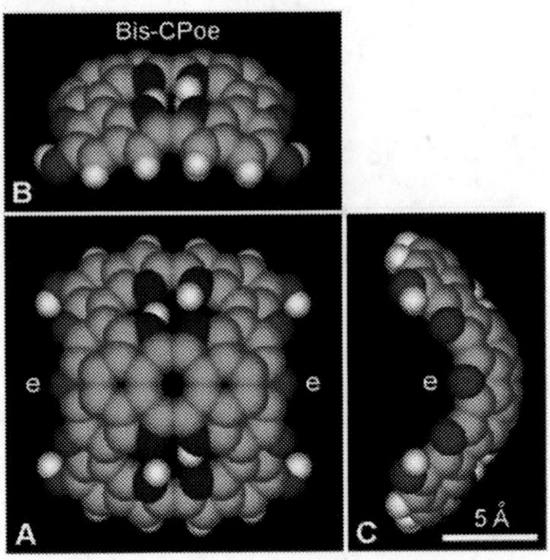

Figure 4. Orthogonal atomic volume views of the bis-CPoe model with 3 ether bridges (e). A: Frontal view. B and C: Side views observed by 90° tilting of image A to top (B) and side (C). GO by MM+ converged at E = 0.1 kcal/(Å mol). Distances of NH:N and OH:O hydrogen-bonds are 2.9 Å and 3.0 Å, respectively.

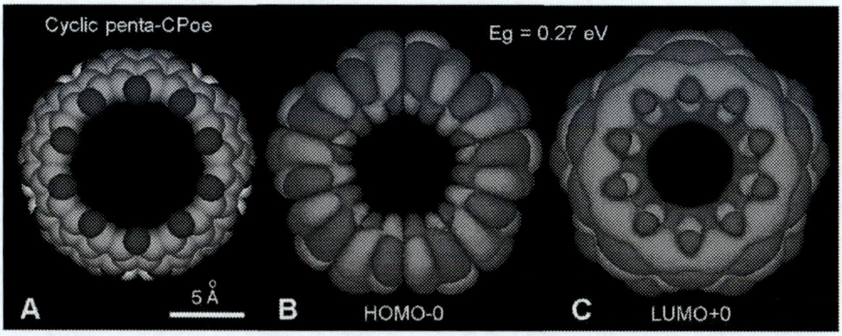

Figure 5. A: Atomic volume model of a fullerene-like cyclic penta-CPoe. EM by MM+ converged at 0.1 kcal/(Å mol). All O atoms form ether bridges. B, C: HOMO-0 (-10.26 eV) and LUMO+0 (-9.99 eV), showing radial and annular MOs patterns, respectively. Extended-Hückel, Gouraud 3D isosurface, orbital contour value (1/orbital radius): 0.0003, Eg = 0.27 eV). (Reproduccd from the Open Access Ref. Stockert and Blázquez-Castro, 2022).

Considering either a mild or a marked curvature of bis- and oligo-CPoe, cyclic or spiral arrangements of chains would be a logical consequence of these chemical changes. Thus, an open fullerene-like, tetra-CPoe with 16 indole units has been previously modeled for DHI-melanin (Olivieri and Nicolaus, 1999; Nicolaus, 2005). In keeping with this, anterior works showed that a cyclic penta-CPoe with 20 indole units is a geometrically improved ring structure showing clear radial and annular HOMO-0 and LUMO+0 patterns, respectively (Figure 5).

It is important to emphasize that from the morphological point of view, transmission electron microscopy (TEM) studies of both natural and synthetic eumelanins show stacked planar electron-dense sheets, which is a characteristic feature of the graphitic organization. Likewise, graphite-like soft or marked wavy layers, as well as concentric onion-like structures, are often seen in natural and synthetic eumelanin samples observed by TEM, in all cases with ~3.4 Å-interlayer separation (Figure 6, A, B). This is also confirmed by X-ray diffraction studies (Thathachari and Blois, 1969; Bridelli et al., 1990; Cheng et al., 1994; Littrell et al., 2003; Watt et al., 2009; Chen et al., 2013; Schroeder et al., 2015; Wasir and Kundi, 2016).

On the other hand, it is worth to note that both planar and curved graphitic structures easily bind to numerous inorganic cations, small molecules, and aromatic colorants, which remain trapped between interlayer spaces forming intercalation complexes. Examples of eumelanin-binding drugs and dyes have been recently reviewed, as well as possible binding processes using MM methods (Stockert et al., 2022).

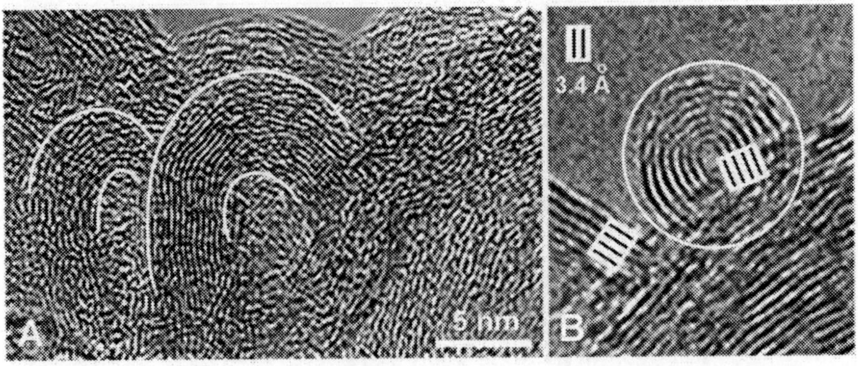

Figure 6. A: TEM image of synthetic poly-dopamine (PDA) eumelanin, showing the planar and wavy structure of the polymer layers, as well as a rather concentric graphitic organization (curved white lines). B: TEM image of PDA at higher magnification. Observe both the planar and curved (onion-like, white line) stacked structure, and the interlayer spacing of 3.4 Å between aromatic planes of the polymer (Adapted from Chen et al. (2013), ACS Nano 7: 1524-1532, with copyright permission).

When studied by MM methods, the possible formation and growth of a spiral growing structure for an oligo-CPoe is illustrated in Figure 7, A, B. The spiral model shows 1.5 turns and is continued by a short planar CPo segment. From a mechanistic point of view, a planar chain would first begin to add ether bridges to curve it, and curvatures would then increase resulting in the spiral rolling model. Different curvature degrees depend on the amount of ether bridges. When one ether group appears for every O atom, the O_{ether}/O_{total} ratio (ether index, EI) is 1/1 = 1. With 1/2 and 1/5 ratios, EI = 0.5 and 0.2, respectively. The curvature is greater with EI values between 0.5 and 1, whereas it is reduced or absent with EI of 0.2 or 0. The corresponding structures with different EIs are represented in Figure 7, C, facilitating the sequential scrolling process. It is worth to note that for CPoe oligomers, Eg values reduce as the amount of ether bridges increases.

At present, the attention of chemists has turned towards curved graphitic networks such as spherical fullerenes, carbon nanotubes, and onion-like graphitic spheres (Ugarte, 1992). To this respect, fullerenes are closed-cage sp^2 carbon spheres with great theoretical and practical importance. In addition to multilayered planar graphite regions, TEM studies revealed that carbon black nano- and micro-particles have onion-like structures formed by concentrically arranged graphene sheets (Iijima, 1980).

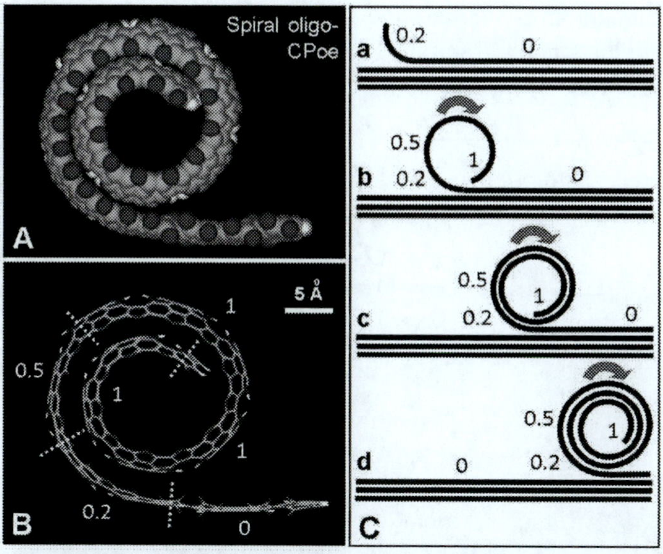

Figure 7. Spiral model of the proposed organization of curved regions of eumelanin. A: Lateral atomic volume view of the spiral oligo-CPoe model with 1.5 turns followed by a short planar bis-CPo segment. B: Wire model of (A) showing different number of ether bridges (EI = 0, 0.2, 0.5, and 1). C: Schematic model of the rolling process (a–d) of a planar oligo-CPo region into a curved stacked oligo-CPoe spiral. Numbers are EI values. Arrows indicate the direction of the rolling up process (Reproduccd from the Open Access Ref. Stockert and Blázquez-Castro, 2022).

Regarding nucleation and growth of C60 fullerene (Kroto and McKay, 1988), a curved sheet is first formed due to energy minimization (Ugarte, 1992; Kroto, 1990), and then spiral growth to form a multilayered fullerene continues by wrapping of a hexagonal network with occasional pentagonal rings to optimize curvature (Kroto and McKay, 1988). Thus, a partially closed nautilus-like sphere first appears, but as edge bypass occurs the closure is no longer possible, so the curved hemisphere grows to form spiral nano- and micro-particles (Kroto, 1990).

It is known that curved carbon structures have onion-like patterns formed by concentrically arranged graphene sheets. Examples are single- and multi-walled carbon nanotubes (Karousis et al., 2010), closed cage fullerenes (Iijima, 1980; Ugarte, 1992), carbon spheres (Diudea et al., 2014), nautilus-like fullerene spirals (Kroto and McKay, 1988), Möbius carbon nanobelts

(Segawa et al., 2022), and graphene nano-scrolls structured as Archimedean spirals (Xie et al., 2009; Liu et al., 2018) (Figure 8). The occurrence of spiral fullerenes and graphenes supports the spiral organization of natural and synthetic eumelanins (Stockert and Blázquez-Castro, 2022). To this respect, the similarity between the spiral CPoe-eumelanin model, graphene spirals and the scrolling process is astonishing (compare Figure 7 and Figure 8).

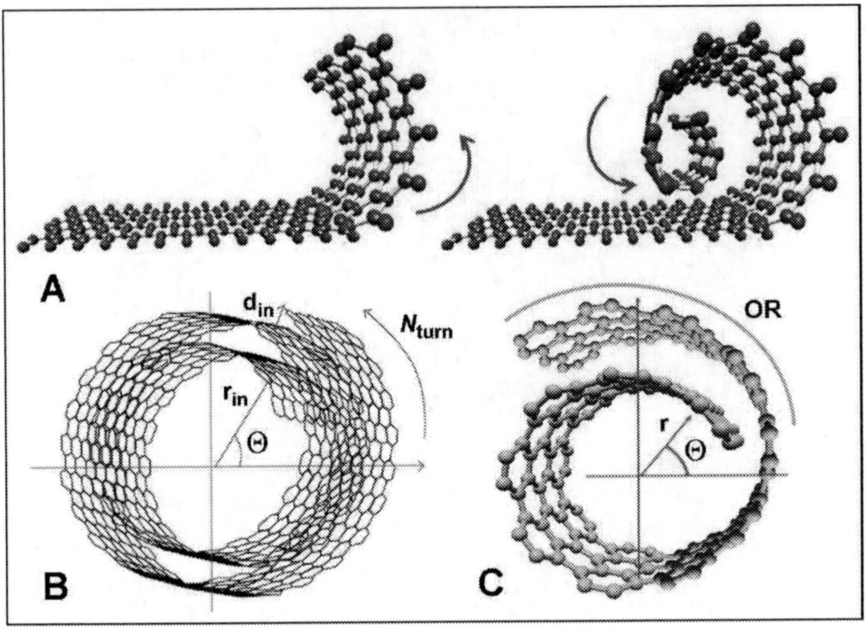

Figure 8. Typical graphene nanoscroll models showing Archimedean spiral geometries. A: Hedayat's model for the formation of a Z-graphene nano-scroll. Red arrows indicate the rolling process (Reproduced from the Open Access Ref. Hedayat et al., 2017). B: Khaledian's model of an A-graphene nano-scroll (Reproduced from the Open Access Ref. Khaledian et al., 2014). C: Ahmadi's model of a spiral A-graphene (Reproduced from the Open Access Ref. Ahmadi et al., 2019). B and C models show coordinate axis (red lines), innermost radius (r and r_{in}), interlayer space (d_{in}), rolling angle (Θ), and overlapping (stacking) region between layers (OR). No oxidation appears in these models.

Carbon or graphene nanoscrolls (CNSs, GNSs) are very interesting structures that comprise both graphene and nanotube properties. They were initially observed by Bacon in 1960 while reporting on graphite whiskers formed by concentric tubes, each tube being in the form of scrolls or rolled-up sheets of graphite layers (Bacon, 1960). In the last few years, numerous papers have been published about the fabrication, physical features, and electronic properties of graphene spirals (Xie et al., 2009; Zhu and Li, 2013; Hamzah et al., 2013; Perim et al., 2014; Khaledian et al., 2014; Hamzah et al., 2015; Khaledian et al., 2016; Hedayat et al., 2017; Ahmadi et al., 2019).

Accordingly, these carbon nanomaterials are an emerging family of spirally wrapped graphene sheets with a tubule-like organization resembling that of multi-walled carbon nanotubes (see Figure 8). Due to its unique topological structure, GNSs share the remarkable mechanical, electronic properties and thermal conductivity exhibited by carbon nanotubes and graphene. For multiwall carbon nanotubes, only the outermost shell (which is in direct contact with the metal electrodes) sustains current at low voltage, whereas for GNSs, the current extends to the whole scrolled graphene layer (Xie et al., 2009). Preparative strategies for synthesis of GNSs include arc-discharge, chemical vapor deposition (CVD), freeze-drying, cold quenching, functional groups or nanoparticles modification, mechanical ball milling, ultrasound-assisted and other methods (Hamzah et al., 2015).

At present, the most efficient methods for synthesis of spiral graphenes are (1) graphite intercalation with potassium, followed by dispersion in ethanol and then sonication, and (2) mechanical exfoliation of graphite using the scotch tape method, deposition of graphene on SiO_2 and then treatment by isopropyl alcohol/water followed by evaporation (Viculis et al., 2003; Perim et al., 2013; Perim et al., 2014). Other monoatomic planar materials are also suitable for fabrication of nano-scrolls. Thus, hexagonal boron nitride (hBN) has demonstrated to form spiral structures (hBNNSs) similarly to GNSs (Perim and Galvao, 2009; Perim et al., 2013; Li et al., 2013).

GNSs provide interlayer spaces that can be intercalated with dopants forming inclusion compounds, and accordingly, the spiral diameter can be expanded to increase the volume of the intercalant (Perim et al., 2014). Thus, the potential of GNSs for capture and storage of gases such as H_2, CO_2 and CH_4 is of increasing interest (Mpourmpakis et al., 2007; Peng et al., 2010). Likewise, MM methods reveal that GNSs serve as tunable water channels (Shi et al., 2010), with possible applications in cell biology.

A completely rolled-up graphene spiral is energetically more favorable than the pristine graphene in a planar form, because the reduction in the van der Waals (vdW) interaction energy among carbon layers in the spiral overbalances the increase in strain energy due to graphene bending (Braga et al., 2004; Zhu and Li, 2013). Formation of a spiral graphene, however, requires a sufficiently large driving force to overcome the energy barrier for the initial curling and rolling of the planar graphene before its continuous spiral growth. Once the scrolling can proceed into the planar graphene region, further scrolling involves wrapping of the graphene around a GNS of increasing radius, and the energy barrier for further scrolling is ever decreasing. Chemical methods to overcoming such an initial energy barrier to form GNSs include hydrogenation and oxidative reactions.

It must be noted that specific chemical causes for explaining the curvature of graphene have been rather overlooked. However, Zhu and Li (2013) have reported that hydrogenation of a C atom in planar graphene produces a sp^3 carbon-hydrogen (C-H) bond, which gives rise to a local out-of-plane distortion and bending around that C atom. Molecular dynamics simulations show the formation of curved structures by suitable single-side hydrogenation on the graphene surface, followed by spontaneous scrolling of the planar graphene sheet, which remains stable at room temperature.

A striking feature of this mechanism is that only a few hydrogenation points are needed to initiate the spiral structure, which is followed by a curved overlapping of the remaining planar (or wavy) graphene segments, because carbon nanoscrolls have a lower energy than the pristine planar graphene (Braga et al., 2004). Since hydrogenation is a reversible process, the chemically adsorbed hydrogen can be evaporated at an elevated temperature (Zhu and Li, 2013). Thus, the resulting graphene spiral could be further treated to be hydrogen free while maintaining its scrolled form, as the lower energy favors a stacked spiral structure. It is known that the geometric structure of spiral graphene and overlapping region strongly affect its semiconducting electronic properties. Constant increasing the length of the overlapping region decreases the Eg continuously (Hamzah et al., 2015), a feature already indicated for planar CPo and spiral CPoe polymers (Stockert and Blázquez-Castro, 2022).

Chemical Origin of Graphitic Curvatures

It is obvious that chemical modifications of polymers provoke substantial structural changes, and oxidative processes are often the cause of severe molecular alterations. To this respect, relevant examples of photo-oxidation of several dyes are well known. The chemistry of dye bleaching and fluorescence fading, induced by the attack of singlet oxygen ($^1O_2$) on double bonds of stilbene, triphenylmethane, indigoid and Bodipy dyes has been described (Kuramoto and Kitao, 1982; Stockert and Blázquez-Castro, 2017; Stockert, 2020).

Oxidative reactions and other chemical modifications are expected to affect the geometry of planar monolayer polymers, and this just occurs in the case of some eumelanin models and graphene. The molecular curvature due to the formation of ether bridges in oligo- or poly-CPoe chains has been already mentioned in this chapter, and MM features may also be applied in the case of graphene sheets. Graphene edges are more reactive than the planar surface (Shenoy et al., 2008; Syama and Mohanan, 2019; Centi and Perathoner, 2022), and then oxidation processes should mostly occur along these regions. It is well known that because of oxidative reactions of graphene, hydroxyl, carboxyl, carbonyl, peroxide, and epoxide groups are generated giving hydrophilic graphene oxides (Marcano et al., 2010; Zhu et al., 2010; Nezakati et al., 2014; Wazir and Kundi, 2016; Abid et al., 2018; Banerjee, 1918; Bianco et al., 2022). Most oxidative changes are represented in the He-Klinowski-Forster-Lerf model of graphene oxide structure (He et al., 1998; Zhu et al., 2010). Oxidative changes frequently induced by $^1O_2$ are illustrated in Figure 9.

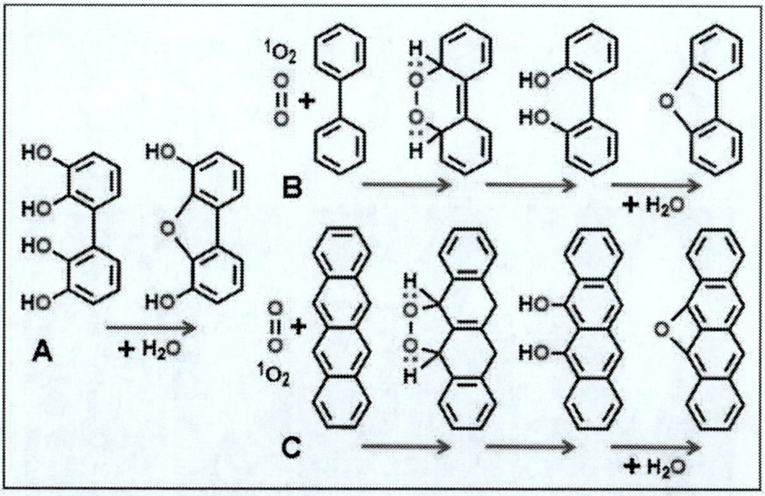

Figure 9. Possible oxidation processes through singlet oxygen ($^1O_2$) attack, with formation of peroxides (C-O-O-C), and then ether bridges (C-O-C) from hydroxyls (arrows). A: In the case of catechol from a CPo model, the polymer does not require further oxidation but only dehydration. B and C: Oxidation of graphene along its arm-chair and zig-zag edges, respectively. Two parallel blue points indicate donor electron pairs for H bonds.

Structural consequences of some oxidative modifications inducing a clear molecular curvature like that of bis-CPoe are illustrated in Figure 10, which shows the structural changes related to the formation of ether bridges, carbonyls, and epoxides along the A and Z edges of graphene. In carboxylated A and Z edges, closing a pentagonal and tetragonal ring, respectively, by a carbonyl group can be easily achieved.

Illustrative examples of similar structural consequences from other chemical changes (formation of oxo (peroxide), cyclopentadienyl and pyrrole rings along the A and Z edges of graphene, as well as hydrogenation) inducing significant curvatures of graphene sheets are shown in Figure 11. However, the curvature by hydrogenation has the disadvantage of generating saturated sp^3 C atoms, which decreases aromatic conjugation and could modify electronic properties, whereas conjugation is retained in curved graphene sheets by oxidation. Formation of pentagonal pyrrole ring is also possible (Bianco et al., 2022), and results in graphene curvature (Figure 11).

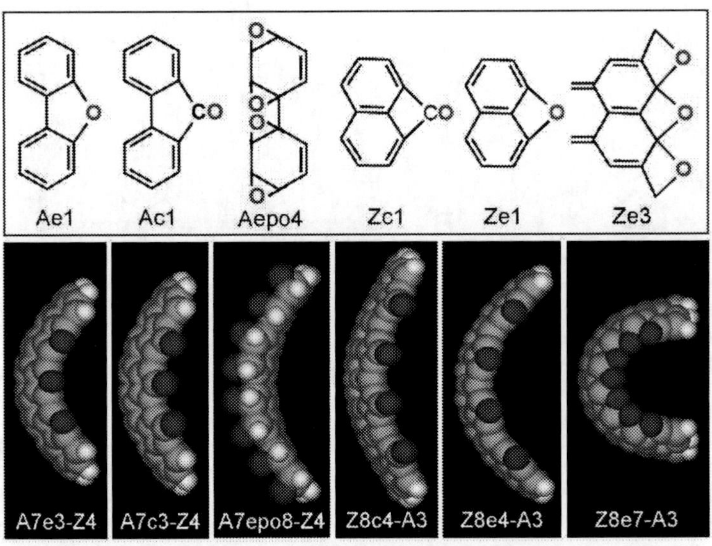

Figure 10. Segments of graphene showing edges with different chemical modifications along the A and Z direction, which include formation of ether (e), carbonyl (c), and epoxide (epo) groups, followed by the number of these groups. Top: Chemical structures of modified edges. Bottom: Atomic models of the corresponding curved structures shown above. The number after A and Z directions represents the amount of adjacent benzene rings in the oligomer.

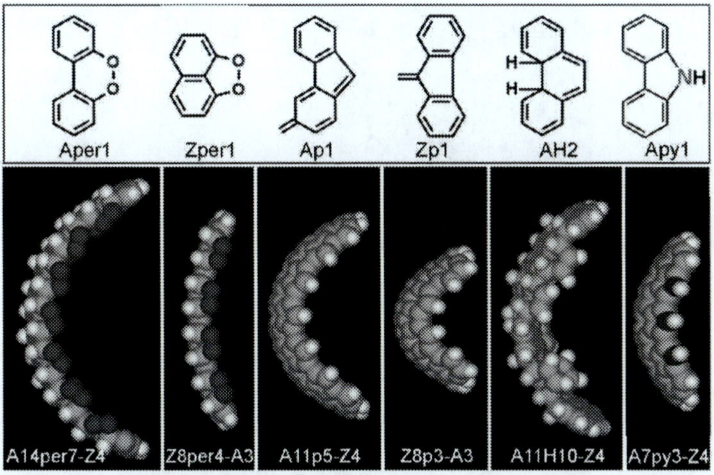

Figure 11. Segments of graphene showing edges with other chemical modifications along the A and Z direction, which include peroxide (per), cyclopentadienyl (p), and pyrrole (py) groups, as well as hydrogenation (H), followed by the number of these groups. Top: chemical structures of modified edges. Bottom: Atomic models of the corresponding curved structures. The number after A and Z directions is the amount of adjacent benzene rings in the oligomer.

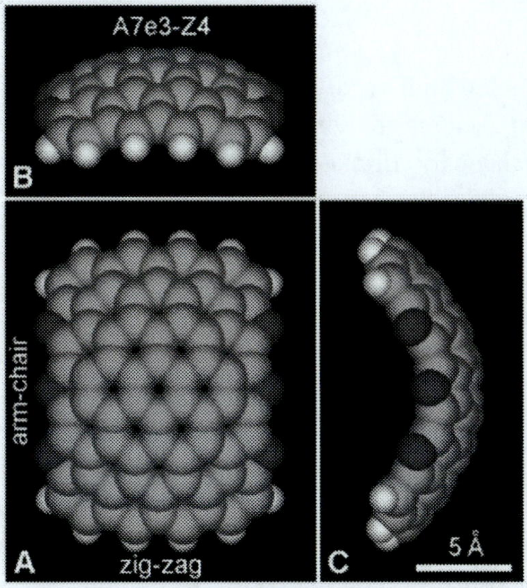

Figure 12. Orthogonal views of the graphene segment A7e3-Z4 model with 3 ether bridges. A: Frontal view. B and C: side views observed by 90° tilting of image A to top (B) and side (C) position. GO by MM+ converged at E = 0.1 kcal/(Å mol).

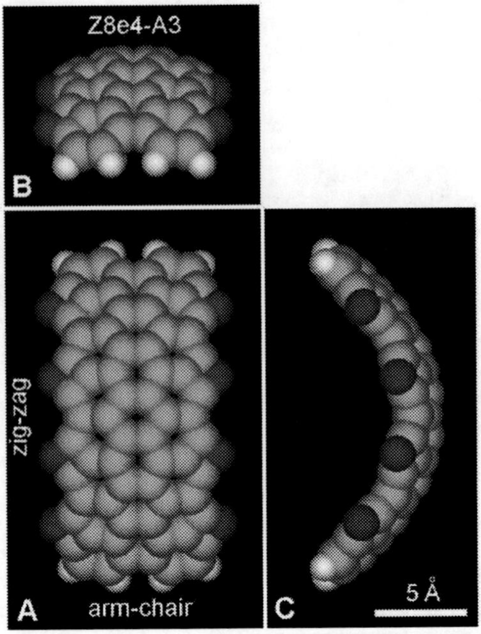

Figure 13. Orthogonal views of the graphene segment Z8e4-A3 model with 4 ether bridges. A: Frontal view. B and C: side views observed by 90° tilting of image A to top (B) and side (C) position. GO by MM+ converged at E = 0.1 kcal/(Å mol).

It must be noted that regarding graphene spirals, curvatures not only affecting the A or Z edges are possible, but also a curvature along both edges can occur simultaneously (Braga et al., 2004; Zhu and Li, 2013). Examples of ether-induced graphene curvatures are illustrated in Figures 12 and 13.

Spiral Structure of Eumelanin CPoe Model and Graphene

The presence of ether groups in bis- and oligo-CPoe models (see Figures 4 and 5), induces curvature and facilitates extended linear LUMOs. In hydrophobic face-to-face stacked aromatic compounds, vdW forces and π–π
take place resulting in either unfused or fused MOs. Overlapping of LUMOs from stacked structures depends on the energy level of the excited state. Figure 14 illustrates stacked bis-CPoe and A7e3-Z4 dimers, showing the similarly fused LUMO patterns. This feature indicates that strong π–π
these graphitic materials occur not only along the linear conjugated polymers

but also between their stacked aromatic units (Park and Aluru, 2011; Stockert and Blázquez-Castro, 2022).

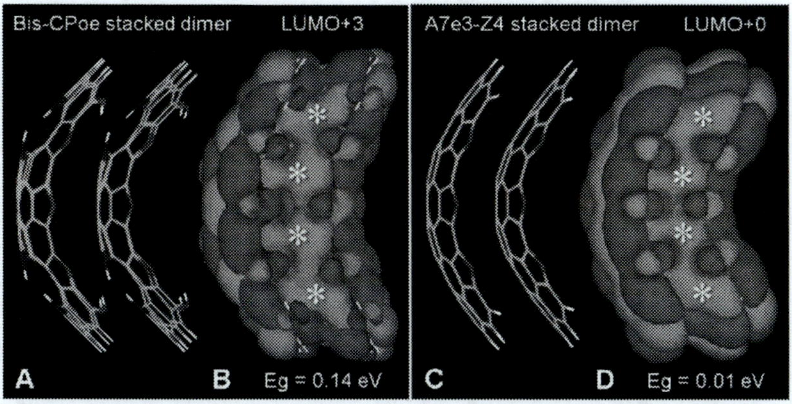

Figure 14. Lateral views of the bis-CPoe stacked dimer (A, B), and the stacked A7e3-Z4 dimer segment of graphene (C, D), each with 3 ether bridges, represented as a wire structure (A, C) and its corresponding LUMOs (B, D), showing a similar pattern. In both cases, EM was performed by MM+ converged at E = 0.1 kcal/(Å mol), followed by extended-Hückel method. LUMOs are shown with orbital contour value: 0.0005, and asterisks indicate the fused LUMOs pattern.

Recently, the new category of quasi-one dimensional graphene nanoscroll (GNS) has captivated the researchers because of its exceptional electronic properties (large stability and carrier mobility), as opposed to their counterpart configurations like nanoribbons, nanotubes, and bilayer graphenes. The formation of a perfect Archimedes spiral is dominated by two major energetic contributions, (a) the mechanic energy increase caused by bending the layer (decreasing stability), and (b) the free energy decrease from vdW interactions between overlapping layers (increasing stability).

A spiral organization of stacked CPoe sheets has been proposed to explain the formation of protoparticles of eumelanin and synthetic PDA-melanin (Stockert and Blázquez-Castro, 2022). Interestingly, comparative pictures between spiral structures from a graphitic eumelanin model and graphene show striking similarities. A CPoe dodecamer forming two spiral turns is illustrated in Figure 15, A. The LUMO+1 state (Figure 15, B) clearly shows a spiral pattern with fused MO lobes, as well as the in-block distribution of HOMO and LUMO energy levels. The structure, LUMO+1 state, and energy levels of the spiral graphene segment A53e26-Z4 can be seen in Figure 15, C and D, and the corresponding LUMO similarities are apparent.

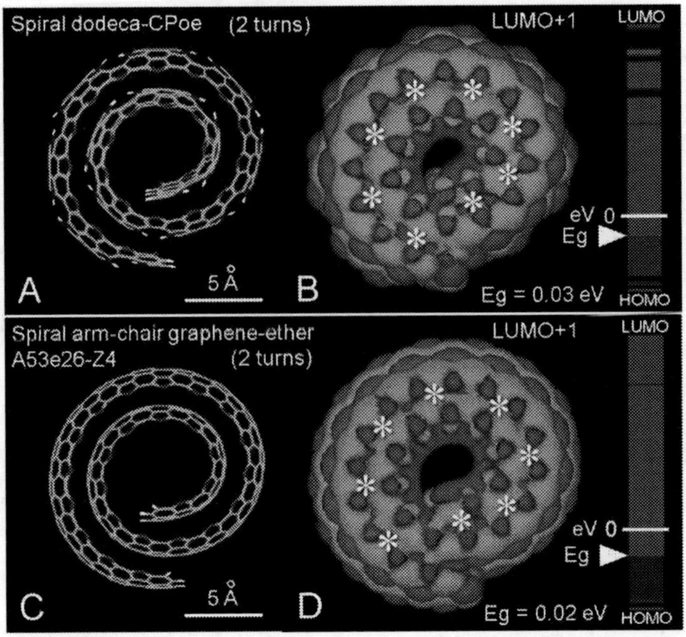

Figure 15. A: Lateral wire model of dodeca-CPoe. GO by MM+ converged at E = 0.1 kcal/(Å mol), followed by extended-Hückel (calculated:1984 orbitals). B: LUMO+1, contour value: 0.0001, Eg = 0.03 eV. HOMO-0: -9.88 eV, LUMO+0: -9.85 eV. Stacking distance 3.6 A (Panel A and B, reproduced from the Open Access Ref. Stockert and Blázquez-Castro, 2022). C: Lateral wire model of the arm-chair A53e26-Z4 segment. MM+ converged at E = 0.1 kcal/(Å mol), followed by extended-Hückel (calculated: 2164 orbitals). D: LUMO+1, contour value: 0.0001, Eg = 0.02 eV. HOMO-0: -10.50 eV, LUMO+0: -10.48 eV. In both cases, the spiral fused LUMOs (asterisks) are shown. On the right-hand side, gallery of energy levels showing the massive occurrence of LUMO (red) and HOMO (blue) states, which represent the CB and VB, respectively. As the HyperChem software shows HOMO and LUMO states with the same colors of the π lobes from pz MOs, the former (equivalent to VB and CB) are represented in blue and red, respectively.

Thus, it is most exciting to consider that such spiral organization also occurs in eumelanin protomolecules. Regarding the nucleation and growth of spiral graphenes, it is tempting to assume that a similar mechanism takes place to form spiral CPoe sheets, resulting in a supramolecular organization of spiral protomolecules formed by aggregation of small and then large globular structures (Stockert and Blázquez-Castro, 2022).

Considering multilayer spiral graphenes, important stacking and vdW forces appear between graphene sheets (Park and Aluru, 2011), and it is expected that different MOs patterns can occur depending on the energy level and visualization of the orbital contour. The structure of the multilayer spiral graphene A116e53-Z4 with 53 ether bridges recorded using different LUMOs and orbital contours is shown in the panels of Figure 16. It is interesting to note that three types of low-energy LUMO lobes occur in this structure: (B) unfused segments, (C) overlapped LUMO surface, and (D) linear fused alternating spirals. In this molecular spiral, all the LUMO states form a compact and non-fragmented CB with a very small Eg (0.03 eV).

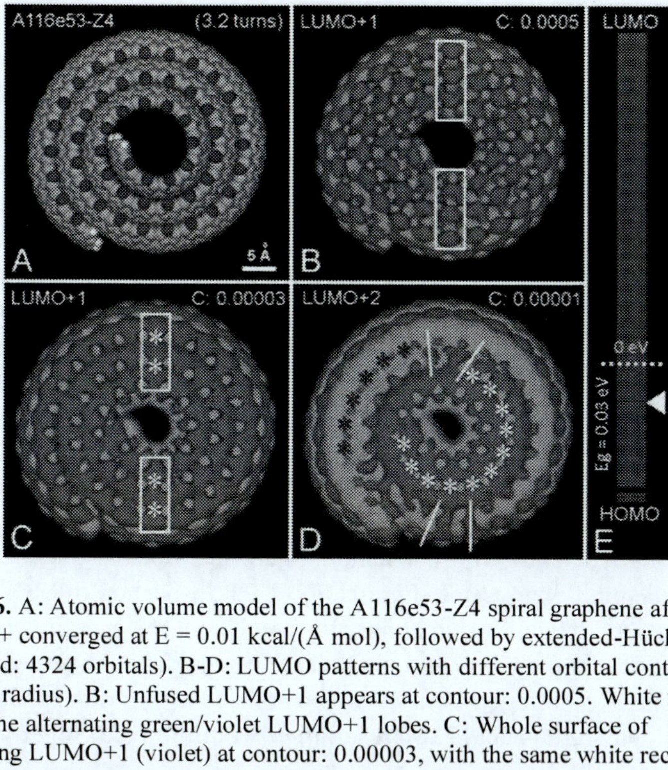

Figure 16. A: Atomic volume model of the A116e53-Z4 spiral graphene after GO with MM+ converged at E = 0.01 kcal/(Å mol), followed by extended-Hückel (calculated: 4324 orbitals). B-D: LUMO patterns with different orbital contour value (1/orbital radius). B: Unfused LUMO+1 appears at contour: 0.0005. White rectangles indicate the alternating green/violet LUMO+1 lobes. C: Whole surface of overlapping LUMO+1 (violet) at contour: 0.00003, with the same white rectangles as in B but showing fused LUMO+1 (asterisks) on contiguous turns. D: Two spiral lobes (white and black asterisks) of fused LUMO+2, and two transition radial lobes between spirals (white lines) appear at contour: 0.00001. E: Energy levels of A116e53-Z4, showing the energy gap (Eg = 0.03 eV, white triangle) from HOMO-0 (-10.50 eV) to LUMO+0 (-10.47 eV), and massive occurrence of LUMO (red) and HOMO (blue) states, corresponding to CB and VB, respectively.

A rationale for the occurrence of overlapping and fusion of lobes is that in small model molecules, MOs from near but non-bound sp^2 C atoms often fuse at suitable energy levels and orbital contour values. This occurs in structures such as C1=C2-O-C3=C4, in which a direct MO can span the non-bound C2 and C3 atoms, keeping outside the central O. It is worth noting here that when compact CB and VB appear and no isolated energy levels are seen, specific HOMO or LUMO states only represent particular cases in a 3D continuum of long and massively fused MOs.

Molecular Modeling Perspectives and Conclusion

At present, graphitic nanomaterials such as carbon quantum dots, nanotubes, fullerenes, graphenes and GNSs are receiving a great deal of attention due to their unique physical, chemical and biological properties. And so, the creation of functional polymers for electronics, nanotechnology, optics, and biomedical applications is growing permanently. Several of these compounds are based on biopolymers and therefore they are "bioinspired" materials. An illustrative example can be put forward here to describe the potential of MM approaches for the design and development of new graphitic materials.

The 3,6,13,16-tetraazaporphycene skeleton, termed "imidacene" (Sargent et al., 2003) is a porphycene analogue that could be used to form a polymeric chain like the poly-CPo model of eumelanin. Using a suitable monomer such as benzimidazole oxidized to catechol-imidazole (5,6-dihydroxy-benzimidazole), the catechol-imidacene (CIm) ring can be formed (Figure 17, A). Like other eumelanin precursors (DOPA, DA), spontaneous oxidative auto-polymerization of CIm units should generate a planar oligo-CIm and a dehydrated spiral oligo-CIme (Figure 17, B, C). Inspection of MOs from the later compound reveals a spiral pattern of the LUMO+1 energy level (Figure 17, D), as occurs for CPoe and graphene-ether polymers. Dye intercalation and complexes of nitrogen atoms from the imidacene ring with suitable metals (e.g., Cu^{2+}, Ni^{2+}, Zn^{2+}) could also result in new or improved properties of CIm polymers.

A special mention of interactions between graphitic materials with colorants should be noted here: an intercalating dye model being copper-phthalocyanine (CuPc) macrocycle. A tetra-sulfonated (CuPcS4) dye deposited on multiwall carbon nanotubes has been used as an electronic conductor in photovoltaics (Hatton et al., 2007). Likewise, the cationic dye Alcian blue pyridine variant (ABPV) was also applied in studies on PDA-melanin. A new composite was formed by layer-by-layer film deposition of PDA-melanin and ABPV, which displayed an electrical conductivity five orders of magnitude higher than that of pure PDA-melanin films (Chen et al., 2013).

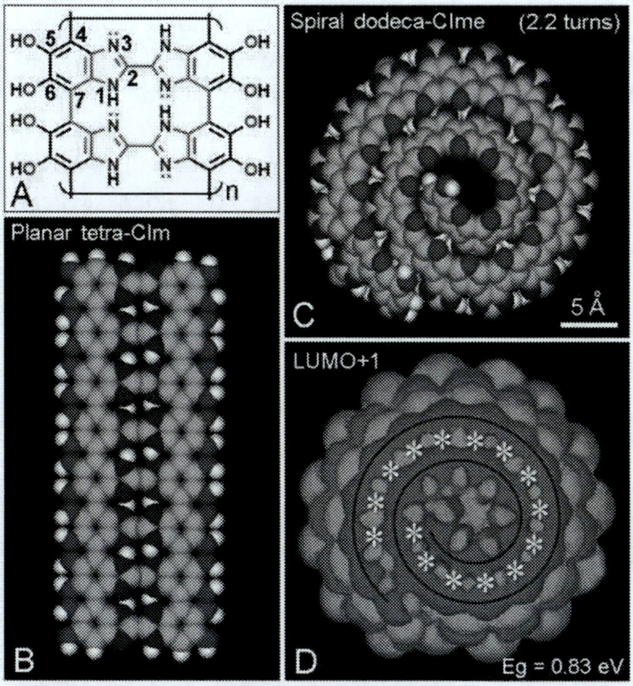

Figure 17. The imidacene (Im) analogue of porphycene (Sargent et al., 2003), as a new starting material to form planar and spiral catechol-imidacene (CIm) derivatives. A: Chemical structure of the CIm unit, showing the imidacene ring (violet) with formal double bonds, atom numbering, and H bonds (blue dots). B: Frontal view of a planar atomic volume model of tetra-CIm. C: Lateral view of a spiral dodeca-CIme model. GO with MM+ converged at E = 0.1 kcal/(Å mol). D: LUMO+1 of dodeca-CIme, extended-Hückel (calculated: 1984 orbitals), contour value: 0.0001, Eg = 0.83 eV (HOMO-0: -10.68 eV, LUMO+0: -9.85 eV), showing the spiral structure (black line) and fused LUMO+1 lobes (asterisks).

Intercalation of colorants within eumelanin has been proposed using MM methods (Stockert and Blázquez-Castro, 2022). In the intercalative binding mode, the planar ligands slip between aromatic units of eumelanin layers and remain trapped as an inclusion complex or "graphitic sandwich," allowing both linear and stacked fused LUMOs (Stockert et al., 2022). Thus, massive π^*-π^* electron interactions are found in dye-intercalated CPo and CPoe eumelanin models.

Interestingly, a clear 610-nm metachromatic peak appears after addition of successive CuPc-PDA layers (Chen et al., 2013). This is an intriguing finding, because dye intercalation involves binding of isolated monomers, canonically leading to orthochromasia, and not to metachromasia. This unexpected metachromatic reaction can be explained because of the massive occurrence of overlapping LUMOs in the dye-graphitic layers, resulting in π^*-π^* electron coupling between all intercalated dye chromophores, which represents a highly aggregated state with metachromatic reaction.

Colorants from acridine, thiazine, xanthene, azine, oxazine, phthalocyanine, porphyrin and porphycene groups are all well suited to intercalate between layers of graphitic materials. Coupling LUMO patterns just occur in aggregated colorants commonly found in tissue histochemistry, leading to a strong metachromatic shift (Stockert and Blázquez-Castro, 2017). Natural and synthetic eumelanins are also capable of binding small molecules with high dielectric constants (dimethyl sulfoxide, formamide, diethylamine, etc.). The high affinity of eumelanin for organic compounds allows the establishment of charge-transfer complexes between the host pigment (electron acceptor) and the included compound (electron donor), leading to a strongly increased conductivity (Nicolaus, 1997). Interesting perspectives could be imagined regarding new intercalation complexes formed between macrocyclic dyes and graphitic or planar compounds.

It is obvious that in addition to the orbital radius and energy levels within VB and CB, the molecular geometry and topology can influence the form and distribution of MOs in graphitic structures. Molecular rings and spirals are illustrative examples of these energy-geometry relations, which in the case of penta-CPoe, dodeca-CPoe, graphene and dodeca-CIme generate circular and spiral LUMOs (see Figures 5, 15, 16 and 17). It can be noted that simple monolayer spirals do not allow the continuous fusion of LUMO states, but the presence of an intercalated layer or a graphitic bilayer induce the formation of continuous fused LUMO spiral modes. A schematic diagram of fused LUMOs as coupled states from intercalated and bilayer structures displayed at high orbital radius is shown in Figure 18.

Some of the present results using MM methods could be useful in giving a better understanding of the molecular geometry of spiral graphitic materials and their possible improved properties. It should be also pointed out here that the misleading name of "helical graphenes" has been assigned to the spiral organization of graphene scrolls (Norouzi and Fakhrabadi, 2019). Overlooked geometrical differences between helical (helicene-like) and spiral graphene are evident, and thus the true name for helical aromatic polymers should correspond to helicenes (Gingras, 2013).

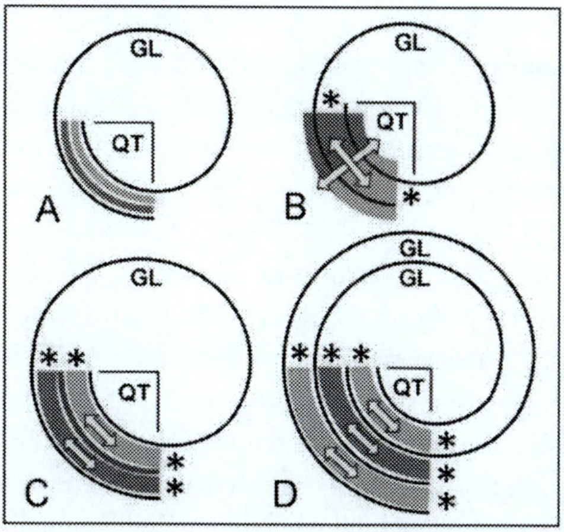

Figure 18. Schematic patterns of graphitic LUMO spirals at low energies for quarter-turn (QT) segments of one or two graphitic layers (GL). A: Single GL spiral with unfused LUMO lobes at low orbital radius. B: Single GL spiral with surface-overlapped (transversal and longitudinal fused) LUMO lobes (in front and behind of all turns) at high orbital contour radius. C: Intercalated GL spiral, with intercalating layer (red line) that leads to continuous fused LUMO lobes between turns. D: Two GL spirals, also allowing fused LUMO lobes between turns. Asterisks indicate fused LUMOs, and yellow arrows show their longitudinal and transversal (B) direction.

Recently reviewed potential applications of classic and new graphene and bioinspired graphitic materials have been proposed for use in various specialized areas such as graphene-based electronics, solid-state memories, multilayer *moiré* graphenes, composite materials and batteries, super-capacitors and oscillators, molecular gas sensors and biosensors, photocatalytic and photovoltaic materials, hydrogen storage and energy conversion, solar and fuel cells, etc. (Zhu et al., 2010; Liu et al., 2014; Hamzah et al., 2015; Liu et al., 2018; Abid et al., 2018; Sharma and Das, 2019; Park et al., 2022).

Relevant biomedical applications of graphitic materials (carbon micro- and nanoparticles, natural and synthetic melanins, etc.) are their use in regenerative tissue engineering (Eivazzadeh-Keihanet al., 2019), and in photodynamic, photothermal, sono-electric, radio- and microwave-sensitized therapy of cancer (Nezakati et al., 2014; Albert and Hsu, 2016; Youssef et al., 2017; Li et al., 2017; Banerjee, 2018; Syama and Mohanan, 2019; Hauser et al., 2020; Mavidi-Printezi et al., 2020; Ross et al., 2020; Gaur et al., 2021). In the case of photothermal therapy, the simultaneous decay of excited states from the CB continuum after photo-excitation results in a massive and immediate conversion of electronic to vibrational energy due to the photon-electron-phonon coupling in graphitic materials (Corry et al., 1976; Kono et al., 1979; Crippa et al., 1991; Migliaccio et al., 2018; Sharma and Das, 2019; Stockert and Blázquez-Castro, 2022), which produces a highly efficient lethal photothermal effect on tumoral cells (Blázquez-Castro et al, 2018; Colombo et al., 2019).

At present, structural, spectral and electronic properties of natural and synthetic graphitic compounds, and particularly spiral polymers, represent a wide and important crossroad in the successful research field of materials sciences. Accordingly, possible developments of exciting next-generation molecular electronic compounds, quantum devices, hybrid nanostructures, and biotechnological platforms can be envisaged in the scenario of the future design and new applications of the amazing spiral graphitic molecules.

Acknowledgments

We thank M. I. Abasolo, J. L. Bella, M. M. Blanco, A. Blázquez-Castro, A. Casas, L. L. Colombo, O. Gabay, J. Herkovits, R. W. Horobin, S. A. Romero, G. R. Solarz, and A. Stockert, for valuable collaboration, stimulating discussions, and critical reading of the manuscript.

Disclosure Statement

The authors declare no conflict of interest.

References

Abid, Sehrawat P, Islam, Mishra P, Ahmad S. (2018). Reduced graphene oxide (rGO) based wideband optical sensor and the role of temperature, defect states and quantum efficiency. *Sci Rep* 8: 3537. Doi: 10.1038/s41598-018-21686-2.

Ahmadi MT, Ahmadi R, Nguyen TK. (2019). Graphene nanoscroll geometry effect on transistor performance. *J Electr Mater* 49: 544-550. Doi: 10.1007/s11664-019-07801-7.

Albert K, Hsu HY. (2016). Carbon-based materials for photo-triggered theranostic applications. *Molecules* 21: 1585. Doi: 10.3390/molecules21111585.

AZoNano. (2006). Carbon Nanotubes and Buckytubes - The complete what, when, how, why and applications. https://www.azonano.com/article.aspx?ArticleID=1597 (accessed on 3 July 2022).

Bacon R. (1960). Growth, structure, and properties of graphite whiskers. *J Appl Phys* 31(2): 283-290. Doi: 10.1063/1.1735559.

Banerjee AN. (2018). Graphene and its derivatives as biomedical materials: future prospects and challenges. *Interface Focus* 8: 20170056. Doi: 10.1098/rsfs.2017.0056.

Bianco GV, Sacchetti A, Grande M, D'Orazio A, Milella A, Bruno G. (2022). Effective hole conductivity in nitrogen-doped CVD-graphene by singlet oxygen treatment under photoactivation conditions. *Sci Rep* 12: 8703. Doi: 10.1038/s41598-022-12696-2.

Blázquez-Castro A, Colombo LL, Vanzulli SI, Stockert JC. (2018). NIR pointer laser for in vivo photothermal therapy of murine LM3 tumor using intratumoral China ink as a photothermal agent. *Lasers Med Sci* 33: 1307-1315. Doi: 10.1007/s10103-018-2483-z.

Blázquez-Castro A, Stockert JC. (2021). Biomedical overview of melanin. 1. Updating melanin biology and chemistry, physico-chemical properties, melanoma tumors, and photothermal therapy. *Biocell* 45: 849-862. Doi: 10.32604/biocell.2021.015900.

Boukhvalov, W, Katsnelson MI, Lichtenstein AI. (2008). Hydrogen on graphene: Electronic structure, total energy, structural distortions and magnetism from first-principles calculateons. *Phys Rev B* 77: 035427. Doi: 10.1103/PhysRevB.77.035427.

Braga SF, Coluci VR, Legoas SB, Giro R, Galvao DS, Baughman RH. (2004). Structure and dynamics of carbon nanoscrolls. *Nano Lett* 4: 881-884. Doi: 10.1021/nl0497272.

Bridelli MG, Crippa PR, Ugozzoli F. (1990). X-ray diffraction studies on melanins in lyophilized melanosomes. *Pigm Cell Res* 3: 187-191. Doi: 10.1111/j.1600-0749.1990.tb00288.x.

Centi G, Perathoner S. (2022). Revealing the role of edges in the electrocatalytic synthesis of H_2O_2 over metal-free nanocarbon. *Chem Catalysis* 2: 1245-1256. Doi: 10.1016/j.checat.2022.05.017.

Chen CT, Ball V, de Almeida Gracio JJ, Singh MK, Toniazzo V, Ruch D, Buehler MJ. (2013). Self-assembly of tetramers of 5,6–dihydroxyindole explains the primary physical properties of eumelanin: Experiment, simulation, and design. *ACS Nano* 7: 1524-1532. Doi: 10.1021/nn305305d.

Cheng J, Moss SC, Eisner M. (1994). X-ray characterization of melanins-II. *Pigm Cell Res* 7: 269-273. Doi: 10.1111/j.1600-0749.1994.tb00061.x.

Colombo LL, Vanzulli SI, Blázquez-Castro A, Sanchez Terrero C, Stockert JC. (2019). Photothermal effect by 808-nm laser irradiation of melanin: A proof-of-concept study of photothermal therapy using B16-F10 melanotic melanoma growing in BALB/c mice. *Biomed Opt Express* 10(6): 2932-2941. Doi: 10.1364/BOE.10.002932.

Corry PM, McGinness JE, Armour E. (1976). Semiconductor properties of melanins related to preferential killing of melanoma cells. *Pigm Cell* 2: 321-326.

Crippa PR, Martini F, Viappiani C. (1991). Direct evidence of electron-phonon interaction in melanins. *J Photochem Photobiol B: Biol* 11: 371-375. Doi: 10.1016/1011-1344(91)80042-G.

Del Castillo P, Horobin RW, Blázquez-Castro A, Stockert JC. (2010). Binding of cationic dyes to DNA: distinguishing intercalation and groove binding mechanisms using simple experimental and numerical models. *Biotech Histochem* 85: 247-256. Doi: 10.3109/10520290903149620.

d'Ischia M, Napolitano A, Ball V, Chen CT, Buehler MJ. (2014). Polydopamine and eumelanin: From structure–property relationships to a unified tailoring strategy. *Acc Chem Res* 47(12): 3541-3550. Doi: 10.1021/ar500273y.

d'Ischia M, Wakamatsu K, Cicoira F, di Mauro E, Garcia-Borron JC, Commo S, Galván I, Ghanem G, Kenzo K, Meredith P, Pezzella A, Santato C, Sarna T, Simon JD, Zecca L, Zucca FA, Napolitano A, Ito S. (2015). Melanins and melanogenesis: From pigment cells to human health and technological applications. *Pig Cell Melan Res* 28: 520-544. Doi: 10.1111/pcmr.12393.

Diudea MV, Bende A, Nagy CL. (2014). Carbon multi-shell cages. *Phys Chem Chem Phys* 16: 5260-5269. Doi: 10.1039/c3cp55309d.

Eivazzadeh-Keihan R, Maleki A, de la Guardia M, Bani MS, Chenab KK, Pashazadeh-Panahi P, Baradaran B, Mokhtarzadeh A, Hamblin MR. (2019). Carbon based nanomaterials for tissue engineering of bone: Building new bone on small black scaffolds: A review. *J Adv Res*18: 185-201. Doi: 10.1016/j.jare.2019.03.011.

Galvão DS, Caldas MJ. (1990). Theoretical investigation of model polymers for eumelanins. I. Finite and infinite polymers. *J Chem Phys* 92: 2630-2636. Doi: 10.1063/1.457957.

Gaur M, Misra C, Bihari A, Swaroop S, Maolmhuaidh FO, Bechelany M, Barhoum A. (2021). Biomedical applications of carbon nanomaterials: Fullerenes, quantum dots, nanotubes, nanofibers, and graphene. *Materials* 14: 5978. Doi: 10.3390/ma14205978.

Gingras M. (2013). One hundred years of helicene chemistry. Part 3: applications and properties of carbohelicenes. *Chem Soc Rev* 42: 1051. Doi: 10.1039/c2cs35134j.

Hamzah MAN, Johari Z, Hamid FKA, Ahmadi MT, Ismail R. (2013). Geometry effect on graphene nanoscrolls band gap. *J Comput Theoret Nanosci* 10: 581-586. Doi: 10.1166/jctn.2013.2737.

Hamzah A, Ahmadi MT, Ismail R. (2015). Analytical study of electronic structure in Archimedean type-spiral zig-zag graphene nanoscroll. *Curr Nanosci* 11(1): 87-94. Doi: 10.2174/1573413710999140918201702.

Hatton RA, Blanchard NP, Stalojan V, Miller AJ, Silva SRP. (2007). Nanostructured copper phthalocyanine-sensitized multiwall carbon nanotubes films. *Langmuir* 23: 6424-6430. Doi: 10.1021/la070156d.

Hauser D, Septiadi D, Turner J, Petri-Fink A, Rothen-Rutishauser B. (2020). From bioinspired glue to medicine: Polydopamine as a biomedical material. *Materials* 13: 1730. DOI 10.3390/ma13071730.

He H, Klinowski J, Forster M, Lerf A. A new structural model for graphite oxide. (1998). *Chem Phys Lett* 287: 53-56.

Hedayat SN, Ahmadi MT, Sedghi H, Goudarzi H. (2017). Quantum transmission modelling in graphene nano scrolls (GNSs) of double barrier. *Int J New Tech Res* 3: 12-19. ISSN:2454-4116.

Iijima S. (1980). Direct observation of the tetrahedral bonding in graphitized carbon black by high resolution electron microscopy. *J Crystal Growth* 50: 675-683. Doi: 10.1016/0022-0248(80)90013-5.

Istifli ES. (2022). Preliminary in silico studies of the interactions of certain genotoxic azo dyes with different double-stranded DNA conformations. *Colorants* 1(2): 236-255. Doi: 10.3390/colorants1020015.

Karousis N, Tagmatarchis N, Tasis D. (2010). Current progress on the chemical modification of carbon nanotubes. *Chem Rev* 110(9): 5366-5397. Doi: 10.1021/cr100018g.

Kaxiras E, Tsolakidis A, Zonios G, Meng S. (2006). Structural model of eumelanin. *Phys Rev Lett* 97: 218102. Doi: 10.1103/PhysRevLett.97.218102.

Khaledian M, Ismail R, Akbari E. (2016). Band structures of graphene nanoscrolls and their dispersion relation near the Fermi point. *RSC Adv* 6(45): 38753-38760. Doi: 10.1039/C5RA27789B.

Khaledian M, Ismail R, Saeidmanesh M, Ahmadi MT, E. Akbari E. (2014). Carrier statistics and quantum capacitance models of graphene nanoscroll. *J Nanomater* Doi: 10.1155/2014/762143.

Kono R, Yamaoka T, Yoshizaki H, McGinness JE. (1979). Anomalous and dispersion of sound waves in diethylamine melanin. *J Appl Phys* 50: 1236-1244. Doi: 10.1063/1.326143.

Kroto HW, McKay K. (1988). The formation of quasi-icosahedral spiral shell carbon particles. *Nature* 331: 328-331. Doi: 10.1038/331328a0.

Kroto HW. (1990). Fullerene cage clusters. The key to the structure of solid carbon. *J Chem Soc Faraday Trans* 86: 2465-2468. Doi: 10.1039/FT9908602465.

Kuramoto N, Kitao T. (1982). Contribution of singlet oxygen to the photofading of some dyes. *J Soc Dyers Colour* 98: 334-340. Doi: 10.1111/j.1478-4408.1982.tb03604.x.

Li X, Hao X, Zhao M, Wu Y, Yang J, Tian Y, Qian G. (2013). Exfoliation of hexagonal boron nitride by molten hydroxides. *Adv Mater* 25(15): 2200-2204. Doi: 10.1002/adma.201204031.

Li Q, Hong L, Li H, Liu C. (2017). Graphene oxide-fullerene C60 (GO-C60) hybrid for photodynamic and photothermal therapy triggered by near-infrared light. *Biosens Bioelectron* 89: 477-482.

Liebscher J, Mrówczynski R, Scheidt HA, Filip C, Hadade ND, Turcu R, Bende A, Beck S. (2013). Structure of polydopamine: A never-ending story? *Langmuir* 29: 10539-10548. Doi: 10.1021/la4020288.

Littrell KC, Gallas JM, Zajac GW, Thiyagarajan P. (2003). Structural studies of bleached melanin by synchrotron small-angle X-ray scattering. *Photochem Photobiol* 77(2): 115-120.

Liu Y, Ai K, Lu L. (2014). Polydopamine and its derivative materials: Synthesis and promising applications in energy, environmental, and biomedical fields. *Chem Rev* 114(9): 5057-5115. Doi: 10.1021/cr400407a.

Liu H, Le T, Zhang L, Xu M. (2018). Carbon nanoscrolls: synthesis and applications. *J Mater Sci: Mater Electron* 29: 18891-18904. Doi: 10.1007/s10854-018-0074-1.

Marcano DC, Kosynkin DV, Berlin JM, Sinitskii A, Sun Z, Slesarev A, Alemany LB, Lu W, Tour JM. (2010). Improved synthesis of graphene oxide. *ACS Nano* 4(8): 4806-4814. Doi: 10.1021/nn1006368.

Mavidi-Printezi A, Guernelli M, Menichetti A, Montalti M. (2020). Bio-applications of multifunctional melanin nanoparticles: From nanomedicine to nanocosmetics. *Nanomaterials* 10: 2276. Doi: 10.3390/nano10112276.

Meng S, Kaxiras E. (2008). Theoretical models of eumelanin protomolecules and their optical properties. *Biophys J* 94: 2095-2105. Doi: 10.1529/biophysj.107.121087.

Micillo R, Panzella L, Koike K, Monfrecola G, Napolitano A, d'Ischia M. (2016). Fifty shades of black and red or how carboxyl groups fine tune eumelanin and pheomelanin properties. *Int J Mol Sci* 17: 746. Doi: 10.3390/ijms17050746.

Migliaccio A, Gryszel M, Đerek V, Pezzella A, Głowacki ED. (2018). Aqueous photo(electro)catalysis with eumelanin thin films. *Mater Horiz* 5: 984-990. Doi: 10.1039/C8MH00715B.

Mpourmpakis G, Tylianakis E, Froudakis GE. (2007). Carbon nanoscrolls: a promising material for hydrogen storage. *Nano Lett* 7: 1893-1897.

Nezakati T, Cousins BG, Seifalian AM. (2014). Toxicology of chemically modified graphene-based materials for medical application. *Arch Toxicol* 88(11): 1987-2012. Doi: 10.1007/s00204-014-1361-0.

Nicolaus RA. (1997). Coloured organic semiconductors: Melanins. *Rend Accad Sci Fis Matem Napoli* 64: 325-360.

Nicolaus RA. (2005). *Perspectives in Melanin Chemistry*. Available online: http://www.tightrope.it/nicolaus/link%2023.htm (accessed on 7 December 2018).

Norouzi S, Fakhrabadi MMS. (2019). Nanomechanical properties of single- and double-layer graphene spirals: a molecular dynamics simulation. *Appl Phys A* 125(5): 321. Doi: 10.1007/s00339-019-2623-8.

Novoselov KS, Geim AK, Morozov SV, Jiang D, Zhang Y, Dubonos SV, Grigorieva IV, Firsov AA. (2004). Electric field effect in atomically thin carbon films. *Science* 306(5696): 666-669. Doi: 10.1126/science.1102896.

Olivieri M, Nicolaus RA. (1999). Sulla DHI-melanina. *Rend Accad Sci Fis Matem Napoli* 66: 85-96.

Park JH, Aluru NR. (2011). Self-assembly of graphenes. *Surface Sci* 605: 1616-1620. Doi: 10.1016/j.susc.2011.02.011.

Park JM, Cao Y, Xia LQ, Sun S, Watanabe K, Taniguchi T, Jarillo-Herrero P. (2022). Robust superconductivity in magic-angle multilayer graphene family. *Nature Mater* Doi: 10.1038/s41563-022-01287-1.

Pelemiš S, Armaković SJ, Tomić BT, Armaković S. (2021). Computational study of amiloride – a WADA banned molecule. *J Eng Process Manag* 13(1): 12-15. Doi: 10.7251/JEPM2101012P.

Peng X, Zhou J, Wang W, Cao D. (2010). Computer simulation for storage of methane and capture of carbon dioxide in carbon nanoscrolls by expansion of interlayer spacing. *Carbon* 48: 3760-3768. Doi: 10.1016/j.carbon.2010.06.038.

Perim E, Galvao DS. (2009). The structure and dynamics of boron nitride nanoscrolls. *Nanotechnol* 20: 335702. Doi: 10.1088/0957-4484/20/33/335702.

Perim E, Paupitz R, Galvo DS. (2013), Controlled route to the fabrication of carbon and boron nitride nanoscrolls: A molecular dynamics investigation. *J Appl Phys* 113 (5): 113, 054306. Doi: 10.1063/1.4790304.

Perim E, Machado LD, Galvao DS. (2014). A brief review on syntheses, structures, and applications of nanoscrolls. *Front Mater* 1: 31. Doi: 10.3389/fmats.2014.00031.

Plaetzer K, Krammer B, Berlanda J, Berr F, Kiesslich T. (2009). Photophysics and photochemistry of photodynamic therapy: Fundamental aspects. *Lasers Med Sci* 24: 259-268. Doi: 10.1007/s10103-008-0539-1.

Raghavan PR, Zane PA, Tripp SL. (1990). Calculation of drug-melanin binding energy using molecular modeling. *Experientia* 46: 77-80. Doi: 10.1007/BF01955422.

Ross S, Wu RS, Wei SC, Ross GM, Chang HT. (2020). The analytical and biomedical applications of carbon dots and their future theranostic potential: A review. *J Food Drug Anal* 28(4): 12. Doi: 10.38212/2224-6614.1154.

Sargent AL, Hawkins IC, Allen WE, Liu H, Sessler JL, Fowler CJ. (2003). Global versus local aromaticity in porphyrinoid macrocycles: Experimental and theoretical study of "imidacene," an imidazole containing analogue of porphycene. *Chem Eur J* 9: 3065-3072. Doi: 10.1002/chem.200204694.

Schroeder RL, Double KL, Gerber JP. (2015). Using sepia melanin as a PD model to describe the binding characteristics of neuromelanin—A critical review. *J Chem Neuroanat* 64-65: 20-32. Doi: 10.1016/j.jchemneu.2015.02.001.

Segawa Y, Watanabe T, Yamanoue K, Kuwayama M, Watanabe K, Pirillo J, Hijikata Y, Itami K. (2022). Synthesis of a Möbius carbon nanobelt. *Nature Synth* Doi: 10.1038/s44160-022-00075-8.

Sharma A, Das J. (2019). Small molecules derived carbon dots: Synthesis and applications in sensing, catalysis, imaging, and biomedicine. *J Nanobiotechnol* 17: 92. Doi: 10.1186/s12951-019-0525-8.

Shenoy VB, Reddy C, Ramasubramaniam A, Zhang YW. (2008). Edge-stress-induced warping of graphene sheets and nanoribbons. *Phys Rev Lett* 101: 245501. Doi: 10.1103/PhysRevLett.101.245501.

Shi X, Cheng Y, Pugno NM, Gao H. (2010). Tunable water channels with carbon nanoscrolls. *Small* 6(6): 739-744. Doi: 10.1002/smll.200902286.

Stockert JC. (1994). Stereoscopy of computer drawn molecular structures. *Biochem Education* 22(1): 23-25.

Stockert JC, Abasolo MI. (2011). Inaccurate chemical structure of dyes and fluorochromes found in the literature can be problematic for teaching and research. *Biotech Histochem* 86: 52-60. Doi: 10.3109/10520295.2010.489428.

Stockert JC, Blázquez-Castro A. (2017). *Fluorescence Microscopy in Life Sciences*. Bentham Science Publishers. Sharjah, U.A.E. Doi: 10.2174/97816810851801170101.

Stockert JC, Blázquez A, Galaz S, Juarranz A. (2008). A mechanism for the fluorogenic reaction of amino groups with fluorescamine and 2 methoxy 2,4-diphenyl 3(2H)-furanone. *Acta Histochem* 110: 333-340. Doi: 10.1016/j.acthis.2007.10.011.

Stockert JC, Pinna-Senn E, Bella JL, Lisanti JA. (2005). DNA-binding fluorochromes: correlation between C-banding of mouse metaphase chromosomes and hydrogen bonding to adenine-thymine base pairs. *Acta Histochem* 106: 413-420. Doi: 10.1016/j.acthis.2004.09.001.

Stockert JC. (2014). Predictive binding geometry of ligands to DNA minor groove: isohelicity and hydrogen bonding pattern. *Meth Mol Biol* 1094: 1-12. Doi: 10.1007/978-1-62703-706-8_1.

Stockert JC, Blázquez-Castro A, Horobin, RW. (2014). Identifying different types of chromatin using Giemsa staining. *Meth Mol Biol* 1094: 25-38. Doi: 10.1007/978-1-62703-706-8_3.

Stockert JC. (2020). Lipid peroxidation assay using BODIPY-phenylbutadiene probes: a methodological overview. *Meth Mol Biol* 2202: 199-214. Doi: 10.1007/978-1-0716-0896-8_16.

Stockert JC. (2021). Melanin and melanoma: Updating molecular structure and photothermal therapy. *InVet* 23(1): 1-15. http://www.fvet.uba.ar/archivos/publicaciones/invet/vol23-1-2021/art-4-vol23-1-2021.pdf (accessed on 15 February 2022).

Stockert JC, Blázquez-Castro A. (2022). Biomedical overview of melanin. 2. Updating molecular modeling, synthesis mechanism, and supramolecular properties regarding melanoma therapy. *Biocell* 46(6): 1391-1415. Doi: 10.32604/biocell.2022.019493.

Stockert JC, Espada J, Blázquez-Castro A. (2022). Melanin-binding colorants: Updating molecular modeling, staining and labeling mechanisms, and biomedical perspectives. *Colorants* 1: 91-120. Doi: 10.3390/colorants1010007.

Syama S, Mohanan PV. (2019). Comprehensive application of graphene: Emphasis on biomedical concerns. *Nano-Micro Lett* 11: 6. Doi: 10.1007/s40820-019-0237-5.

Thathachari YT, Blois MS. (1969). Physical studies on melanins. II. X-ray diffraction. *Biophys J* 9: 7-89.

Ugarte D. (1992). Curling and closure of graphitic networks under electron-beam irradiation. *Nature* 359: 707-709. Doi: 10.1038/359707a0.

Viculis LM, Mack JJ, Kaner RB. (2003). A chemical route to carbon nanoscrolls. *Science* 299(5611): 1361-1361. Doi: 10.1126/science.1078842.

Watt AAR, Bothma JP, Meredith P. (2009). The supramolecular structure of melanin. *Soft Matter* 5: 3754-3760. Doi: 10.1039/B902507C.

Wazir AH, Kundi IW. (2016). Synthesis of graphene nano sheets by the rapid reduction of electrochemically exfoliated graphene oxide induced by microwaves. *J Chem Soc Pak* 38(1): 11-16.

Xie X, Ju L, Feng X, Sun Y, Zhou R, Liu K, Fan S, Li Q, Jiang K. (2009). Controlled fabrication of high-quality carbon nanoscrolls from monolayer graphene. *Nano Letters* 9(7): 2565-2570. Doi: 10.1021/nl900677y.

Youssef Z, Vanderesse R, Colombeau L, Baros F, Roques-Carmes T, Frochot C, Wahab H, Toufaily J, Hamieh T, Acherar S, Mohd Gazzali AM. (2017). The application of titanium dioxide, zinc oxide, fullerene, and graphene nanoparticles in photodynamic therapy. *Cancer Nano* 8: 6. Doi: 10.1186/s12645-017-0032-2.

Zhou X, McCallum NC, Hu Z, Cao W, Gnanasekaran K, Feng Y, Stoddart JF, Wang Z, Gianneschi NC. (2019). Artificial allomelanin nanoparticles. *ACS Nano* 13: 10980-10990. Doi: 10.1021/acsnano.9b02160.

Zhu Y, Murali S, Cai W, Li X, Suk, JW, Potts JR, Rouff RS. (2010). Graphene and graphene oxide: synthesis, properties, and applications. *Adv Mater* 22: 3906-3924. Doi: 10.1002/adma.201001068.

Zhu S, Li T. (2013). Hydrogenation enabled scrolling of graphene. *J Phys D: Appl Phys* 46: 075301. Doi: 10.1088/0022-3727/46/7/075301.

Biographical Sketches

Marcelo Nicolás Felix-Pozzi

Affiliation: Micro and Nanotechnology Department, Institute of Nanosciences and Nanotechnology, National Atomic Energy Commission of Argentina (CNEA).

Education:
2022: Bsc. Degree in Physics, Department of Physics, Faculty of Exacts and Natural Sciences, University of Buenos Aires (UBA).

Business Address: Constituyentes Atomic Center, CNEA, Av. General Paz 1499, Pdo. de San Martin, Buenos Aires, Argentina.

Research and Professional Experience:
2007 – 2010 Fellowship from State Society for Applied Researches (INVAP S.E.).
2014 – 2017 Fellowship form CNEA.

Professional Appointments:
2018 - present: Permanent position at CNEA as Technician and Research Member
2022: Assistant Professor of Physics and Introduction to Biophysics, UBA
2022: Assistant Professor of Artificial Intelligence, Chair of Data Sciences, Digital House Coding School, Buenos Aires (private institution)

Juan Carlos Stockert

Affiliation: Institute of Oncology "Angel H. Roffo," Research Area, Faculty of Medicine, University of Buenos Aires (UBA), Argentina.

Education:
1964: Medical Doctor, Faculty of Medicine, UBA, Argentina
1972: PhD Degree in Medicine, UBA
1979: PhD Degree in Medicine and Surgery, Autonomous University of Madrid (UAM), Spain

Business Address: Institute of Oncology "Angel H. Roffo," Research Area, Faculty of Medicine, University of Buenos Aires (UBA), Avda. San Martin 5481, Buenos Aires C1417DTB, Argentina

Research and Professional Experience:
1965-1968: Fellowship from LALCEC (Argentinean League Against Cancer), Buenos Aires.
1968-1971: Fellowship from the International Agency for research on Cancer (WHO), Madrid, Spain.
1971-1973: Associate Professor of Histology, Faculty of Medicine, UBA.
1973-1976: Fellowship from Alexander-von-Humboldt Stiftung. Tübingen, Federal Republic of Germany.
1976-1986: Full Professor and Head of the Department of Cell Biology, UAM.
1987-2010: Scientific Member of the High Council of Scientific Research, Madrid, Spain.

Professional Appointments:
Honorary Member of the Research Staff, Institute of Oncology "Angel H. Roffo," Research Area, Faculty of Medicine, UBA

Honors:
- PhD *Honoris causa*, National University of Rio Cuarto, Cordoba, Argentina, May 19, 2008
- PhD *Honoris causa*, University of Buenos Aires, Argentina, September 17, 2009

Publications from the Last 3 Years:
2020
Carrasco, E., Stockert, J. C., Juarranz, A., Blázquez-Castro, A.: Plasmonic hot-electron reactive oxygen species generation: Fundamentals for redox biology. *Front. Chem. Chem. Biol.* 8: 591325, 2020. doi: 10.3389/fchem.2020.591325 .

Stockert, J. C., Carou, M. C., Casas, A. G., García Vior, M. C., Ezquerra Riega S. D., Blanco, M. M., Espada, J., Blázquez-Castro, A., Horobin, R. W., Lombardo, D. M.: Fluorescent redox-dependent labeling of lipid droplets in cultured cells by reduced phenazine methosulfate. *Heliyon* 6: e04182, 2020. doi: 10.1016/j.heliyon.2020.e04182.

Stockert, J.C.: Lipid peroxidation assay using BODIPY-phenylbutadiene probes: a methodological overview. *Meth. Mol. Biol.* 2202: 199-214, 2020. doi: 10.1007/978-1-0716-0896-8_16.

2021

Blázquez-Castro, A., Stockert, J. C.: Biomedical overview of melanin. 1. Updating melanin biology and chemistry, physico-chemical properties, melanoma tumors, and photothermal therapy. *BIOCELL* 45: 849-862, 2021. doi: 10.32604/biocell.2021. 015900.

Horobin, R. W., Stockert, J. C., Zhang, H.: Reactive dyes for living cells — Applications, artefacts, and some comparisons with textile dyeing. *Color. Technol.* 2021. doi: 10.1111/cote.12577.

Horobin, R. W., Stockert, J. C.: Using an integrated QSAR model to check whether small-molecule xenobiotics will accumulate in biomembranes, with particular reference to fluorescent imaging probes. In: Methods in Pharmacology and Toxicology. Quantitative analysis of cellular drug transport, disposition, and delivery, Chapter 8 (G. R. Rosania and G. M. Thurber, eds.). 2021, pp. 163-177. Springer *Nature*. doi: 10.1007/978-1-0716-1250-7_8.

Stockert, J. C., Herkovits, J.: Melanin adhesivity for possible trapping of SARS-CoV-2 on chin straps: a proof-of-concept assay using model nanoparticles. *Acta Sci. Microbiol.* 4: 2-5, 2021. doi: 10.31080/ASMI.2020.04.0759.

Stockert, J.C.: Melanin and melanoma: Updating molecular structure and photothermal therapy. *InVet* 23(1): 1-15, 2021.
http://www.fvet.uba.ar/archivos/publicaciones/invet/vol23-1-2021/art-4-vol23-1-2021.pdf.

2022

Stockert, J. C., Blázquez-Castro, A.: Biomedical overview of melanin. 2. Updating molecular modeling, synthesis mechanism, and supramolecular properties regarding melanoma therapy. *BIOCELL* 46(6): 1391-1415, 2022. doi: 10.32604/biocell.2022. 019493.

Stockert, J. C., Espada, J., Blázquez-Castro, A.: Melanin-binding colorants: Updating molecular modeling, staining and labeling mechanisms, and biomedical perspectives. *Colorants* 1: 91-120, 2022. doi: 10.3390/colorants1010007.

Stockert, J. C., Durantini, E. N., Gonzalez Lopez, E. J., Durantini, J. E., Villanueva, A.; Horobin, R. W. Fluorescence labeling of mitochondria in living cells by the cationic photosensitizer ZnTM2,3PyPz, and the possible role of pseudobase formation and redox processes in facilitating dye uptake. *Biotech. Histochem*. 2022. doi: 10.1080/10520295.2022.2090603.

Chapter 5

Generation of Antigen Specific CD8+ T Cells by B Cells and Dendritic Cells

Mansi Ahuja, Tanmay K. Dalai and Sarat K. Dalai*
Institute of Science, Nirma University, Ahmedabad, India

Abstract

The role of $CD8^+$ T cells in protecting the host from the intracellular pathogens and cancer is well established. APCs (antigen presenting cells) present the antigens to $CD8^+$ T cells in the form of MHC-I/peptide complex. Following the antigen specific stimulation, $CD8^+$ T cells proliferate and differentiate; the cytotoxic T cells generated kill the target cells (pathogen infected cells or cancer cells), thus clear the infection or regress the tumor. The initial interaction between the antigen bearing APCs and $CD8^+$ T cells occurs in the peripheral lymphoid organs. Dendritic cells are the most efficient APCs and known to initiate $CD8^+$ T cell response. However, emerging evidence suggests the role of other APCs such as B cells in activation of $CD8^+$ T cells through a mechanism known as antigen cross-presentation. Since $CD8^+$ T cells play a significant role during antiviral immune response, various efforts are being made to design vaccines that can generate specific antiviral $CD8^+$ T cells. The present chapter discusses the generation of $CD8^+$ T cells during a natural infection or vaccination. It also highlights the mechanisms behind $CD8^+$ T cell generation and provides insights to design better vaccines.

* Corresponding Author's Email: sarat.dalai@nirmauni.ac.in.

In: Advances in Biology. Volume 2
Editor: Charles D. Grant
ISBN: 979-8-88697-422-5

Introduction

Host defence comprises of innate and adaptive immune systems. First line of defence i.e., the innate immune system responds quickly following an encounter with a pathogen [1]. Whereas, adaptive immunity takes longer time to develop but is more specific in nature. Both innate and adaptive arms of immune system work closely together to destroy the pathogen [1].

Innate immune system consists of physical barriers, chemical barriers and cellular responses. Physical barriers such as epithelial layer of the skin prevent pathogens from entering [2]. Chemical barriers include soluble substances which have antimicrobial activity or acidic pH. For example, lysozyme an enzyme found in saliva, tears functions as an antibacterial agent [2]. Pathogens can breach these chemical or physical barriers and hence cellular responses are required. Phagocytic cells such as macrophages, neutrophils, monocytes can rapidly engulf and destroy the invading pathogen [2, 3]. In addition, various soluble and cell surface receptors known as pattern recognition receptors(PRRs) recognize molecular patterns on the surface of microbes known as pathogen associated molecular patterns(PAMPs) and induces secretion of antimicrobials, cytokines and chemokines [2, 3]. Cytokines and chemokines activate and recruit other cells involved in innate and adaptive immune response [2].

Pathogens possess many features that allow them to evade innate immune responses and that's why innate immune response are not sufficient to resolve the infection fully. In order to eradicate the pathogen completely, antigen specific response generated by the adaptive immune system is required.

B and T lymphocytes are the key players of adaptive immunity [4]. Humoral and cell mediated immunity are the two major mechanisms of adaptive immunity.[4]. Antibodies secreted by B cells mediate the humoral immune response T cells are the key players of cell mediated immune response. B cells are capable of recognizing antigens directly but T cells can only recognize antigens coupled to MHC molecule on the surface of antigen presenting cells (APCs) [4]. Some naïve T cells commit to $CD4^+$ helper T cell lineage and some to $CD8^+$ cytotoxic T cell lineage [4]. $CD4^+$ T cells recognize peptide bound to MHC II and differentiate into helper cells. $CD8^+$T cells recognize peptide bound to MHC I and differentiate into cytotoxic cells [4].

This chapter discusses the classical MHC class I and MHC class II antigen presentation pathways. Further, it discusses how APCs activate $CD8^+$ T cell through cross-presentation. Lastly, it describes strategies to design better vaccines.

The Classical Preparation of Endogenous Antigens-MHC Class I Pathway

The MHC class I pathway is essential in alerting the immune system against the intracellular pathogens such as viruses. MHC I molecule is expressed on all nucleated cells. They present peptides derived from intracellular proteins, these can be misfolded, abnormal or denatured proteins of the cell itself or viral proteins derived from a virus infected cell [4]. Antigen specific cytotoxic $CD8^{+}$T cells recognize the peptide-MHCI complex on the surface of cells and destroy the infected cells.

Immunoproteasome-a cytosolic proteolytic machinery found in professional APCs and cells of the infected tissues degrade the endogenous antigens into small peptides [4]. Immunoproteasome is distinct from the traditional proteasome found in all cells; it has different catalytic subunits which increases its efficiency to generate peptides that are able to bind to MHC class I molecule [4]. Once the peptides are generated in the cytosol, they are transported to RER (rough endoplasmic reticulum) through TAP (transporter associated with antigen processing). TAP is a heterodimer that spans the membrane of RER, it is composed of TAP1 and TAP2 proteins. Both TAP1 and TAP2 have ATP binding domain, transport of peptides from cytosol to ER through TAP require ATP hydrolysis [4]. Inside ER an enzyme endoplasmic reticulum amino peptidase (ERAP) trims the peptide to a length of 9 amino acids, which is an optimal length for binding with MHC class I [4]. MHC I is synthesized on ribosome on RER. Different chaperons are required for the assembly of MHC class I molecule [4]. The first molecular chaperon in this process is calnexin which along with Erp57 associates with free class I alpha chain to promote its folding [4]. In the next step beta 2 micro-globulin associates with alpha chain which allows the release of calnexin and the next two chaperons calreticulin and tapsin associate with MHC class I molecule [4]. Role of tapsin is to bring TAP in close proximity of MHC I molecule [4]. As a result, MHC I molecule acquires the peptide from the transporter and exhibits increased stability [4]. Finally, MHC class I molecule dissociates form the chaperons and now it progresses towards cell surface through Golgi complex [4]. Further $CD8^{+}$ T cell bind to this peptide-MHCI complex through its T cell receptor (TCR). Signals generated through TCR along with other signals result in proliferation and differentiation of $CD8^{+}$ T cell into effector cells that kill infected cells.

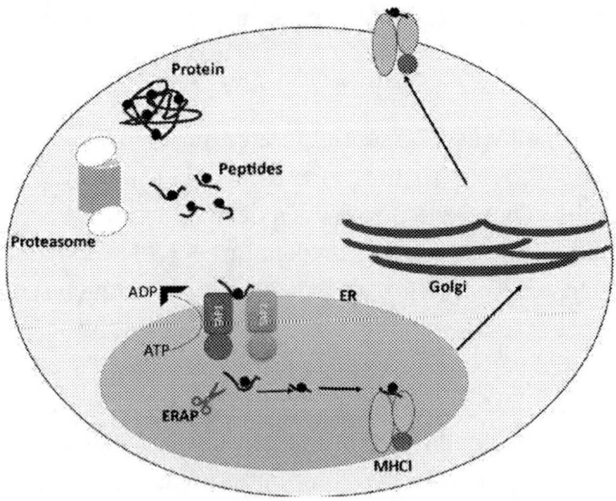

Figure 1. MHC class I pathway: In this pathway endogenous proteins are degraded into peptides by proteasome. The resulting peptides are transported inside ER where it is loaded on MHC class I molecule. The peptide-MHCI complex is translocated on the cell surface where it will be recognized by $CD8^+$ T cells.

Presenting the Exogenous Antigens- MHC Class II Pathway

The MHC class II pathway is essential in generating $CD4^+$ T cell immune response. Antigen presenting cells (APCs) such as macrophages and dendritic cells internalize antigens by phagocytosis and receptor mediated endocytosis [4]. B cells are very efficient in internalizing antigens through receptor mediated endocytosis [4]. They utilize their antigen specific membrane bound immunoglobulin molecule as the receptor [4].

Once the antigen is internalized by the APCs, it is processed into small peptides in the endocytic compartments [4]. The internalized antigen progresses through early endosome (pH ~ 6.0), late endosome (pH ~ 5.0) and lysosome (pH ~ 4.5), in these compartments it encounters different hydrolytic enzymes and lower pH [4]. In the endocytic compartments the antigen is degraded into 13-15 amino acid long peptides [4]. In the late endosome the peptides encounter and bind to MHC II molecule [4]. When MHC II molecule is synthesized in RER, it is associated with an invariant chain (Ii, CD74). The main purpose of this invariant chain is to prevent binding of peptides destined for MHC I molecule to MHC II [4]. Also, this invariant chain contains sorting signals which directs the MHC II through Golgi complex to the endocytic compartments [4].

As MHC II progresses through endocytic compartments the invariant chain is degraded in to a small fragment known as CLIP (class II–associated invariant chain peptide) [4]. This degradation occurs due to increase in the proteolytic activity of each endocytic compartment. CLIP is bound to the peptide binding groove of MHC II molecule, HLA-DM a non-classical MHC II molecule aids in the exchange of CLIP with the antigenic peptide [4].

Once the peptide is bound to MHC II molecule the peptide-MHCII complex progresses toward the cell membrane. Because of the neutral pH at the cell membrane, the complex adopts a compact form which makes it difficult to replace the already bound peptide with any other peptide [4].

Antigen specific CD4$^+$ T cell recognize and bind to the peptide-MHCII complex which results in the activation and differentiation of CD4$^+$T cells into helper T cell subsets. These subsets orchestrate the immune response further to achieve the desired response.

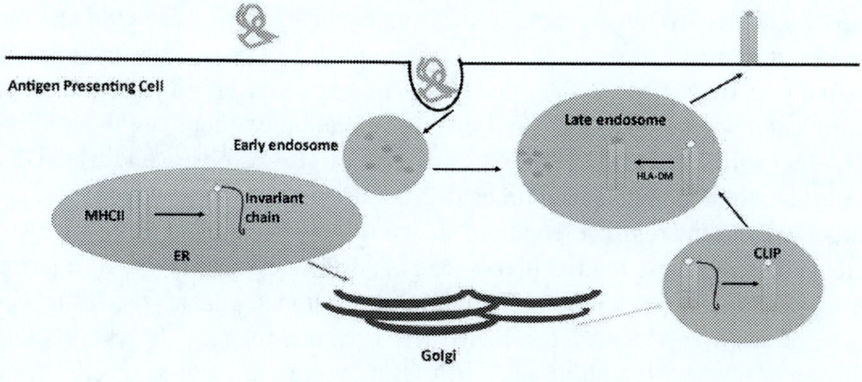

Figure 2. MHC class II pathway: In this pathway exogenous antigens are taken up by the cell through endocytosis. The antigen is degraded into peptide by various hydrolytic enzymes. In the late endosome peptide bind to MHC class II molecule. This peptide-MHC II complex is translocated to the cell surface. The peptide-MHCII complex will be recognized by CD4$^+$T cell.

Distinct but Indispensable - Antigen Cross-Presentation

Usually, the internalized antigen is processed through MHC class II pathway; however, in some instances internalized antigen is processed and presented on MHC I molecule [4]. This process is known as cross-presentation. Cross-presentation is important in eliciting immune response after immunization, against intracellular pathogens and tumour antigens [5]. Antigen cross-presentation leads to induction of cytotoxic T lymphocytes responses, which is termed as cross- priming. If it induces tolerance in $CD8^+$ T cells it is known as cross-tolerance [4]. Antigen presenting cells such as macrophages, B cells and Dendritic cells are capable of cross-presentation. [4, 5]. However, dendritic cells (DCs) are the most prominent cell type that has been shown to cross-present antigens to $CD8^+$ T cells [4, 5].

Dominant Cross-Presenting DC Subset

DCs are heterogenous population of cells. Based on phenotypic markers, murine splenic DCs are divided into two subsets i) $CD11b^{low}$ $CD8\alpha^+$ DEC-205^+ ii) $CD11b^{high}$ CD8 α^- DEC 205^-[6] It is thought that $CD8\alpha^+$ DCs belong to lymphoid lineage and CD8 α^- DCs belong to myeloid lineage [6]. A study was conducted by Bevan et al., to determine which subset of splenic DC was involved in cross-presentation. To first establish that cross-presentation was actually taking place, OVA loaded β2m-/- cells were injected through intravenous route to B6 mice and then proliferation of adoptively transferred OT-I cells, labelled with CFSE was assessed. [6] It was observed that OT-I T cells had divided up to seven times [6]. To further investigate which cell type was involved in cross-presentation, spleen cells were isolated from mice 14h after injecting OVA loaded β2m-/- cells and flow cytometry-based analysis was performed [6]. Based on the flow cytometry analysis, four cell populations were identified which included macrophages, B and T cells, myeloid DCs and lymphoid DCs [6]. Using antibodies and magnetic beads specific subsets were depleted from spleen cells and remaining cells were incubated with OT-I cells *in-vitro* to evaluate their capacity to stimulate T cells [6]. Specific depletion of $CD8\alpha^+$ lymphoid DCs resulted in significant loss of stimulatory activity [6]. This finding was further confirmed by sorting the splenocytes into myeloid and lymphoid DCs [6]. The sorted myeloid and lymphoid DCs were used as APCs to assess OT-I proliferation *in-vitro*. It was observed that sorted myeloid DCs lacked stimulatory activity whereas lymphoid DCs showed stronger stimulatory activity [6]. This finding suggested that lymphoid DC is the main cross-presenting subset [6]. Although $CD8\alpha^+$ DCs are the dominant cross-presenting subset, emerging evidence indicate that other DCs can cross-present antigens but it depends on various factors including antigen type, inflammatory signal, and the location of DCs [7].

Why CD8α+ DCs Are More Efficient as Cross-Presenters Than Other Subsets?

DCs have various mechanisms which make them very efficient at antigen cross presentation [5, 8, 9]. DCs in general have reduced capacity of endosome maturation and acidification as compared to other cells [5, 8, 9] Also, murine $CD8\alpha^+$ DCs express lower levels of lysosomal protease as compared to $CD8^-$ DCs [8, 9].

Specifically, in $CD8\alpha^+$ DCs a small GTPase Rac2 assembles the NADPH oxidase complex (NOX2) in phagosome whereas in $CD8^-$DCs it assembles NOX2 at the plasma membrane [8, 9]. Presence of NOX2 in phagosome results in generation of reactive oxygen species (ROS) resulting in alkalization of the phagosome [8, 9]. The alkalization inhibits the activity of lysosomal protease which have optimum activity at low pH. Moreover, ROS generation leads to reversible oxidation of some lysosomal protease [8, 9].

These mechanisms prevent rapid degradation of internalized antigen, as a result epitopes are not destroyed before they are loaded on MHC I molecule [5]. Further limited antigen degradation generates an antigen depot inside the cell from where continuously antigen is available for processing and presentation on MHC I molecule [5]. This allows increased life span of peptide-MHCI on the surface of DCs which enables activation of T cells [5].

It has been hypothesized that transport of antigens from endosome to cytosol is required for cross-presentation *in vivo* [5, 8, 9]. As compared to $CD8^-$DCs, $CD8\alpha^+$ DCs are more efficient in transporting antigens to cytosol. [5, 8, 9].

These are some of the reasons why $CD8\alpha^+$ DCs are more efficient at cross-presentation than DC other subsets.

Pathways Reported for Antigen Cross-Presentation

Vacuolar Pathway

The Vacuolar pathway does not require cytosolic proteasomal machinery and hence is resistant to proteasome inhibitor [5, 9] Antigens that are taken up by phagocytosis or endocytosis are degraded in phagosome or endosome [5, 9]. The peptides generated after degradation are loaded onto MHC I molecule.

Cytosolic Pathway

The Cytosolic pathway is sensitive to proteasome inhibitor suggesting that antigens internalized in phagosome or endosome move to the cytoplasm for proteasomal degradation [5, 9]. The resulting peptides are translocated back into endosome/phagosome through endosomal TAP or into ER through TAP [5, 9]. The peptides are trimmed to optimum size either by ERAP or IRAP (Insulin regulated endopeptidase) [5, 9]. Finally, the peptide is loaded on MHC I molecule.

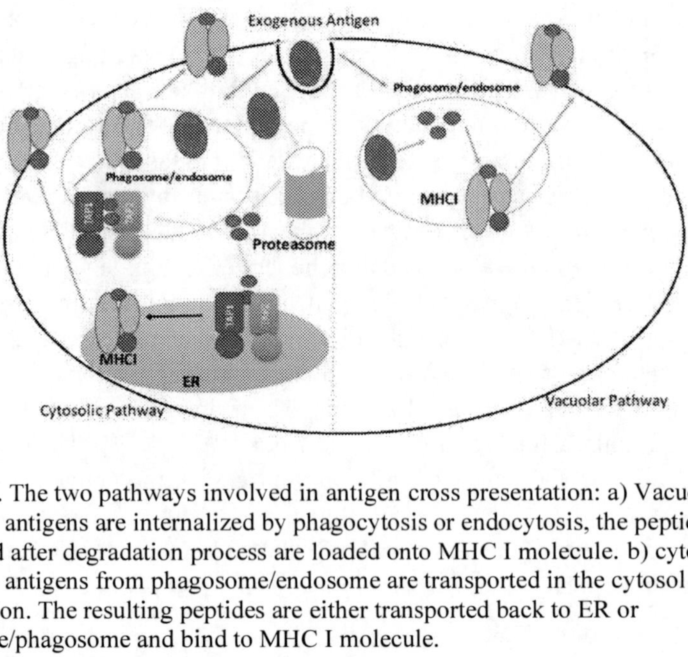

Figure 3. The two pathways involved in antigen cross presentation: a) Vacuolar pathway: antigens are internalized by phagocytosis or endocytosis, the peptides generated after degradation process are loaded onto MHC I molecule. b) cytosolic pathway: antigens from phagosome/endosome are transported in the cytosol for degradation. The resulting peptides are either transported back to ER or endosome/phagosome and bind to MHC I molecule.

But, how does MHC I molecules reach endosome? There are two ways through which this can occur. First, the newly synthesized MHC I molecule is transported from ER to endosome and is used for peptide loading [5, 9]. Second, the MHC I molecule bound to peptide, present on cell membrane is internalized through endocytosis and transported to ER [5, 9]. But, how does peptide exchange occur on recycled MHC molecule is still unclear [5].

Many components of MHC I peptide loading are present in ER which needs to be translocated to endosome/phagosome for efficient cross-presentation. The transport occurs through ERGIC (ER-Golgi intermediate compartment) [5, 9]. For, example TAP is transported from ER to endosome through membrane fusion mediated by ER soluble N-ethylmaleimide-sensitive factor attachment receptor (SNARE) Sec22b with syntaxin 4 present on endosome/phagosome [5, 9]. Further investigation is required to understand the complete molecular mechanism of these pathways.

Receptor Mediated Endocytosis - Its role in Cross-Presentation

Receptor mediated endocytosis is a mechanism through which specific molecules or antigens are taken up by cell after they bind to cell surface receptor. Different subsets of dendritic cells express different endocytic receptors that bind the antigen for uptake and cross-presentation by DCs. For example, in mice, $CD8^+$ DCs, dermal DCs, Langerhan DCs express DEC-205 that bind to dying cells and aid in presentation of cell derived antigens [10]. Similarly, another receptor CLEC9A expressed on $CD8^+$DCs, $CD103^+$ DCs and plasmacytoid DCs in mice binds to actin cytoskeleton exposed on dying cells (apoptotic or necrotic) and facilitates cross-presentation of cell associated antigens [10].

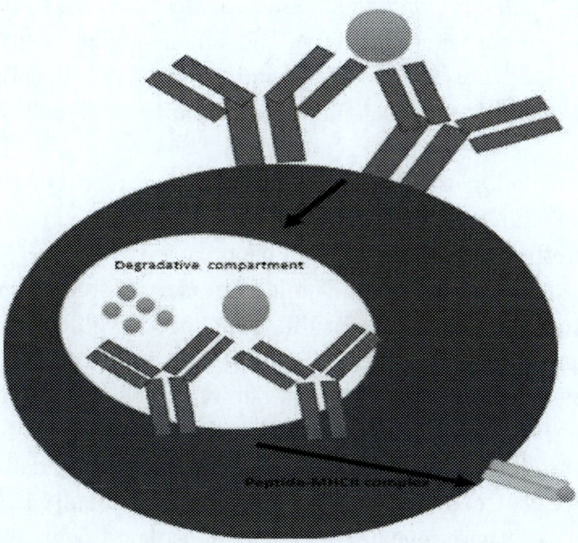

Figure 4. Antigen uptake by B cells: After binding of antigen to BCR, BCR-antigen complex is taken up and fused with the late endosome(degradative compartment). Further, the antigen is processed and presented as peptide-MHCII complex on the surface of B cells.

Expression of these endocytic receptors on dendritic cells allow them to take up antigens effectively. B cells internalize antigens very effectively through receptor mediated endocytosis as they express antigen specific membrane bound immunoglobulin on their surface which they utilize as receptors. Furthermore, B cells express MHC molecules and can express costimulatory molecules such as CD80/86 upon activation [11]. Hence, B cells can be considered as alternative APCs to DCs. In fact, few studies have highlighted the importance of B cells in activation of T cells which is discussed in the next section.

Role of B Cells in Activation of CD8+T Cells

As mentioned earlier, DCs are the primary cell type that carry out cross-presentation, however, there are evidences that suggest the role of B cells in antigen cross-presentation.

An elegant study by de Wit et al., demonstrated that B cells infected with *Salmonella* elicit $CD8^+$ T cell response which is *Salmonella* specific [12]. Further investigations revealed that B cells could activate central and effector memory $CD8^+$T cells but were unable to activate naïve $CD8^+$ T cells whereas DCs were able to activate both naïve and effector $CD8^+$ T cells [12]. This indicates that DCs are needed for priming of $CD8^+$ T cells and during reinfection B cells function as APCs to initiate a $CD8^+$ T cell recall response [12].

Cross-presentation of antigens by B cells has also been implicated in autoimmune conditions such as type I diabetes [13]. Type I diabetes develops as a result of destruction of beta cells of pancreas by cytotoxic T lymphocytes. $CD8^+$ T cells isolated from NOD mice (model to study type I diabetes) recognize autoantigens such as insulin, islet-specific glucose-6-phosphatase catalytic subunit-related protein (IGRP) [13]. In a study conducted by Mariño E., et al., using NOD mice it has been demonstrated that expansion of self-reactive $CD8^+$ T cells is diminished in B cell deficient mice (NOD μMT) as compared to the NOD mice [13]. This suggested a role of B cells in activating self-reactive $CD8^+$ T cells and subsequent onset of diabetes [13].

In this study it was revealed that B cells isolated from NOD mice and loaded with IGRP peptide can induce proliferation of IGRP-reactive $CD8^+$ T cells [13]. Further, B cells were found to promote differentiation of self-reactive $CD8^+$ T cells into effector T cells [13]. In addition to this, in NOD B cell chimeras, in which B cells lack expression of MHC I molecule the frequency of self-reactive $CD8^+$ T cells is greatly reduced [13]. This indicates that B cells present autoantigens as peptide-MHCI complex to self-reactive $CD8^+$ T cells which further drive their proliferation and differentiation into effector cells which have the capacity to kill the beta cells of pancreas [13].

The contribution of B cells in augmenting anti-tumour $CD8^+$ T cell response have also been studied. In one such study by Lee-Chang C,et al., a subset of B cell expressing 4-1BBL, a costimulatory marker has been used to boost anti-tumour $CD8^+$T cell response in glioblastoma [14]. Since 4-1BBL$^+$ B cells have the capacity of enhancing antitumor cytotoxic response this subset was used in this study [14]. These B cells were isolated from glioma-bearing mice and stimulated with CD40 and IFNγ-R ligation, which together up-regulated the expression of CD86, an important co-stimulatory molecule required for $CD8^+$T cell activation [14].

Moreover, these activated B cells showed up-regulation of APC markers such as MHC class I and MHC class II on their surface [14]. Further, their ability to cross-present exogenous antigens to $CD8^+$T cells was evaluated [14]. The activated B cells were pulsed with SIINFEKL peptide($CD8^+$ T cell epitope of OVA) or OVA protein and incubated with OT-I CD8T cells [14]. The B cells not only induced proliferation of OT-I cells but also up-regulated Granzyme B [14].

Findings from our lab also highlight the role of B cells in cross-presentation of antigens to $CD8^+$ T cells. We have observed that with each dose of RAS (radiation attenuated sporozoites) vaccination the frequency of $CD8^+$ T cells increases and protection against infectious challenge of sporozoites was achieved after 2 or 3 booster doses of RAS. It is reported that antibody response improves with multiple immunization, however, only B cell response is not sufficient to provide protection. It might be possible that B cells induced during immunization would contribute to generating $CD8^+$T cells, which are required to achieve sterile protection.

Although dendritic cells are the most efficient cross-presenting cells, the ability of B cells to cross-present antigens to $CD8^+$T cells cannot be overlooked. However, further investigation is required to answer a few questions: when do B cells take up the role of cross-presenters, is it during primary or secondary infection? Just like DCs is there any specific B cell subset that plays a dominant role in cross-presentation?

Generation of CD8+ T Cells during Infection

Antigen specific $CD8^+$T cells during infection can be primed by cross-presentation of exogenous antigens by DCs and also by direct presentation of antigens bound to MHC-I on the surface of infected DCs [8]. The relative contribution of these two pathways in activation of $CD8^+$ T cells depends on the type of pathogen and the ability of pathogen to infect a specific organ or set of organs [8]. For example, in *vaccina* virus infection $CD8^+$ T cells are activated through direct antigen presentation whereas in *toxoplasma gondii* infection uninfected DCs cross-present parasite derived antigens to $CD8^+$ T cells [8]. During blood borne infections such as *P.berghei* blood stage infection cross-presentation is carried out exclusively by $CD8^+$DCs [8].

During the influenza A infection, viral antigens present in various forms such as infective virions, defective particle, soluble antigen, intracellular antigens and dead infected cells are internalized by DCs and presented to $CD8^+$ T cells [15]. Studies from Influenza A infected mice suggest that $CD103^+$ and $CD11b^{hi}$ respiratory DCs are the primary APCs that activate naïve Influenza A virus specific $CD8^+$ T cells but with different capacities [16]. $CD103^+$ DCs are the most potent APCs and can efficiently process and present Influenza A virus antigens to $CD8^+$ T cells [16]. As $CD103^+$ DCs express CLEC9A, a receptor involved in cross-priming, they are superior in their cross-presentation ability and this might be one of the reasons for their potency as APCs [16].

Respiratory dendritic cells acquire antigens through any of the three given mechanisms: a) through direct infection of respiratory dendritic cells, b) phagocytosis of dead infected cells or virions, c) trogocytosis- where a cell acquires a portion of another cell through direct cell-cell contact [16]. After the uptake of the antigen, CD103+ carry the antigen to lung draining lymph nodes, here they encounter naïve T cells [16]. They process the antigen and present the peptide on MHC-I molecule to $CD8^+$ T cells. Following antigen recognition $CD8^+$ T cells get activated, proliferate and differentiate into effector T cells that are capable of recognizing and killing virus infected cells.

Generation of CD8+ T Cells during Vaccination

It is generally assumed that the mediators that govern the generation of $CD8^+$ T cells during infection are same as that in immunization, but it is not always true [17]. For example, following adjuvant administration maximal T cell expansion requires cytokines such as IL-27, IL-15, and transcription factors such as Tbet, Eomes which are not required for T cells responding to an infection [17].

In a recent study by Klarquist J et al., it was shown that in response to subunit vaccine, B cells enhance primary and memory CD8$^+$ T cell response [17]. Following subunit vaccination (OVA+ different adjuvants) CD8$^+$ T cell response was substantially reduced in B cell deficient mice (MD4) compared to wild-type(WT) controls. [17]. However, when CD8$^+$ T cell response was assessed in WT and B cell deficient mice following infection with LM-OVA(OVA expressing *Listera monocytogenes*) no difference was observed, suggesting a unique role of B cells in promoting CD8$^+$ T cell response to subunit vaccination [17]. Moreover, in WT mice the CD8$^+$ T cells were skewed towards memory phenotype and in MD4 mice CD8$^+$ T cells exhibited effector phenotype [17]. Further investigation was carried out to understand how B cells were promoting CD8$^+$ T cell response to subunit vaccine [17]. It was observed that B cells can cross-present antigens to CD8$^+$ T cells but this was not required to induce CD8$^+$ T cell response. Instead production of IL-27 by B cells contributed towards primary CD8$^+$ T cell response but not CD8 memory response [17].

The study unveiled a new mechanism of B cells through which they can generate CD8$^+$ T cells [17]. Generation of CD8$^+$ T cells following vaccination can greatly improve the efficacy of vaccines, but it is challenging to induce CD8$^+$ T cells in adequate numbers. Further studies are required to find out and understand various ways through which B cells can elicit robust CD8$^+$ T cell responses and specific B cells can be targeted to induce adequate numbers of CD8$^+$ T cells.

Antigen Targeting to Cross-Presenters: A Budding Vaccine Strategy

Aim of vaccination is to provide durable and long-term immunity. There are many vaccines that induce production of neutralizing antibodies; however, T cell immunity is required against infections like HIV/AIDS, tuberculosis, malaria, leishmaniasis and cancer [18]. Although neutralizing antibodies are protective, depending upon the behaviour of pathogen T cell immunity would be more desirable. For example, as new strains of influenza emerge, the neutralizing antibodies might become ineffective whereas CD8$^+$ T cell response will still be effective as the CD8$^+$T cell epitopes are conserved among strains [18].

Live attenuated vaccines induce both humoral and cell mediated response, but they pose certain risks such as reverting to virulence form, adverse reactions, etc. [18]. On the contrary, subunit vaccines are safer and stable but they usually elicit humoral immune response and not cell-mediated response [18]. Hence it is important to develop strategies that can induce cellular responses in response to subunit vaccines.

One strategy to specifically induce CD8$^+$T cells in response to extracellular vaccine molecule is to enhance the efficiency of cross-presentation. An elegant study done by Fossum et al., highlights the above-mentioned strategy [19]. In their study they targeted Xcr1, the chemokine receptor on cross-presenting DCs in murine (CD8α^+DCs) [19]. The chemokine receptor Xcr1 is uniquely expressed on CD8α^+DCs, they targeted this receptor by fusing the ligand of the receptor i.e., Xcl1 to OVA antigen [19]. The fusion protein induced proliferation of OT-I (CD8$^+$ T cells recognizing OVA peptide residues 257-264) and OT-II cells (CD4$^+$ T cells recognizing OVA peptide residues 323-339) *in-vivo* [19]. It also conferred protection against lethal challenge with influenza A virus in mice by inducing strong CD8$^+$ T cell response [19]. Hence, the study provided a novel and promising strategy to induce effective and protective CD8+ T cell response.

A similar strategy could be adopted for targeting B cell subset that is capable of presenting exogenous antigens to CD8+ T cells. Possibly this strategy would be more advantageous as it would take care of both humoral and cell mediated immune responses.

Conclusion

APCs activate T cells to induce the desired immune response. Three distinct surface proteins on APCs are involved in the activation of T cells. First, is the MHC protein that binds and present the antigenic peptide to T cells. Second, co-stimulatory molecules that bind to their corresponding receptors on T cells. Third, cell adhesion molecules that allow stable association of APCs and T cells. Although DCs are the most potent APCs, other APCs such as B cells are also efficient in activating T cells. Owing to their receptor specificity, B cells can recognize specific antigens even at low antigen concentrations. Also, they constitutively express MHC molecules on their surface.

Since antigen presentation by APCs is a pre-requisite for activation of T cells, antigen targeting to APCs offers a promising strategy to induce specific T cells in adequate numbers. For example, targeting influenza matrix protein to DC immunoreceptor (DCIR) that can mediate antigen cross-presentation, results in induction of antigen specific human $CD8^{+}T$ cell response[20]. Similarly, such receptors can be identified on B cells and can be targeted to induce antigen specific $CD8^{+}$ T cell responses. Although adopting similar strategies for both DCs and B cells can create a redundancy it might ensure that host immune system does not get hijacked by a pathogen at a single target.

Acknowledgment

The Authors would like to acknowledge Department of Biotechnology (DBT) New Delhi, India for providing funding to SKD Lab.

References

[1] Aristizábal B., González Á. Innate immune system. In: Anaya J. M., Shoenfeld Y., Rojas-Villarraga A., Roger A. Levy, Ricard Cervera. editors. *Autoimmunity: From Bench to Bedside* [Internet]. Bogota (Colombia): El Rosario University Press; 2013 Jul 18. Chapter 2.

[2] InformedHealth.org [Internet]. Cologne, Germany: Institute for Quality and Efficiency in Health Care (IQWiG); 2006-. *The innate and adaptive immune systems.*

[3] Alberts B., Johnson A., Lewis J., Martin Raff, Keith Roberts, and Peter Walter. *Molecular Biology of the Cell*. 4th edition. New York: Garland Science; 2002. Innate Immunity.

[4] Owen J. A., Punt, J., Stranford, S. A., Jones, P. P., & Kuby, J. (2013). *Kuby immunology* (7th ed.). New York: W.H. Freeman.

[5] Embgenbroich M. and Burgdorf S. (2018) Current Concepts of Antigen Cross-Presentation. *Front. Immunol.* 9:1643. doi: https://doi.org/10.3389/fimmu.2018.01643.

[6] Joke M. M. den Haan, Sophie M. Lehar, Michael J. Bevan; Cd8+ but Not Cd8− Dendritic Cells Cross-Prime Cytotoxic T Cells *in Vivo*. *J Exp Med* 18 December 2000; 192 (12): 1685-1696. doi: https://doi.org/10.1084/jem.192.12.1685.

[7] Nierkens S., Tel J., Janssen E., Adema G. J. Antigen cross-presentation by dendritic cell subsets: one general or all sergeants? *Trends Immunol.* 2013 Aug;34(8):361-70. doi: https://doi.org/10.1016/j.it.2013.02.007. Epub 2013 Mar 27. PMID: 23540650; PMCID: PMC4351710.

[8] Blander J. M. Regulation of the Cell Biology of Antigen Cross-Presentation. *Annu Rev Immunol.* 2018 Apr 26;36:717-753. doi: https://doi.org/10.1146/annurev-immunol-041015-055523. Epub 2018 Feb 28. PMID: 29490164; PMCID: PMC6430635.

[9] Joffre, O., Segura, E., Savina, A. and Sebastian Amigorena Cross-presentation by dendritic cells. *Nat Rev Immunol* 12, 557-569 (2012). https://doi.org/10.1038/nri3254.

[10] Macri C., Dumont C., Johnston A. P., Mintern J. D. Targeting dendritic cells: a promising strategy to improve vaccine effectiveness. *Clin Transl Immunology.* 2016 Mar 18;5(3):e66. doi: https://doi.org/10.1038/cti.2016.6. PMID: 27217957; PMCID: PMC4815026.

[11] Manzoor A. M., Chapter 1 - Introduction to Costimulation and Costimulatory Molecules, Developing Costimulatory Molecules for Immunotherapy of Diseases, *Academic Press*, 2015, Pages 1-43, ISBN 9780128025857 https://doi.org/10.1016/B978-0-12-802585-7.00001-7.

[12] de Wit J., Souwer Y., Jorritsma T., Klaasse Bos H., ten Brinke A., Neefjes J., S. Marieke van Ham (2010) Antigen-Specific B Cells Reactivate an Effective Cytotoxic T Cell Response against Phagocytosed Salmonella through Cross-Presentation. *PLoS One* 5(9): e13016. https://doi.org/10.1371/journal.pone.0013016.

[13] Mariño E., Tan B., Binge L., Mackay C. R., Grey S. T. B-cell cross-presentation of autologous antigen precipitates diabetes. *Diabetes.* 2012 Nov;61(11):2893-905. doi: https://doi.org/10.2337/db12-0006. Epub 2012 Jul 24. Erratum in: Diabetes. 2013 Feb;62(2):662. PMID: 22829452; PMCID: PMC3478535.

[14] Lee-Chang C., Jason Miska, David Hou, Aida Rashidi, Peng Zhang, Rachel A. Burga, Ignacio Jusué-Torres, Ting Xiao, Victor A. Arrieta, Daniel Y. Zhang, Aurora Lopez-Rosas, Yu Han, Adam M. Sonabend, Craig M. Horbinski, Roger Stupp, Irina V. Balyasnikova, Maciej S. Lesniak; Activation of 4-1BBL+ B cells with CD40 agonism and IFNγ elicits potent immunity against glioblastoma. *J Exp Med* 4 January 2021; 218 (1): e20200913. doi: https://doi.org/10.1084/jem.20200913.

[15] Korenkov D., Irina Isakova-Sivak & Larisa Rudenko (2018) Basics of CD8 T-cell immune responses after influenza infection and vaccination with inactivated or live attenuated influenza vaccine, *Expert Review of Vaccines,* 17:11, 977-987, doi: https://doi.org/10.1080/14760584.2018.1541407.

[16] Hufford M. M., Kim T. S., Sun J., Braciale T. J. The effector T cell response to influenza infection. *Curr Top Microbiol Immunol.* 2015;386:423-55. doi: https://doi.org/10.1007/82_2014_397. PMID: 25033753; PMCID: PMC4224975.

[17] Klarquist J. Eric W. Cross, Scott B. Thompson, Benjamin Willett, Daniel L. Aldridge, Alayna K. Caffrey-Carr, Zhenming Xu, Christopher A. Hunter, Andrew Getahun, Ross M. Kedl, B cells promote CD8 T cell primary and memory responses to subunit vaccines, *Cell Reports,* Volume 36, Issue 8,2021,109591, ISSN 2211-1247, https://doi.org/10.1016/j.celrep.2021.109591.

[18] Burchill M. A., Tamburini B. A., Pennock N. D., White J. T., Kurche J. S., Kedl R. M. T cell vaccinology: exploring the known unknowns. *Vaccine.* 2013 Jan

2;31(2):297-305. doi: https://doi.org/10.1016/j.vaccine.2012.10.096. Epub 2012 Nov 6. PMID: 23137843; PMCID: PMC3529748.

[19] Fossum E., Grødeland, G., Terhorst, D., Tveita, A. A., Vikse, E., Mjaaland, S., Henri, S., Malissen, B. and Bogen, B. (2015), Vaccine molecules targeting Xcr1 on cross-presenting DCs induce protective CD8+ T-cell responses against influenza virus. *Eur. J. Immunol.,* 45: 624-635. https://doi.org/10.1002/eji.201445080.

[20] Eynav Klechevsky, Anne-Laure Flamar, Yanying Cao, Jean-Philippe Blanck, Maochang Liu, Amy O'Bar, Olivier Agouna-Deciat, Peter Klucar, LuAnn Thompson-Snipes, Sandra Zurawski, Yoram Reiter, A. Karolina Palucka, Gerard Zurawski, Jacques Banchereau; Cross-priming CD8+ T cells by targeting antigens to human dendritic cells through DCIR. *Blood* 2010; 116 (10): 1685–1697. doi: https://doi.org/10.1182/blood-2010-01-264960.

Chapter 6

Categorization of Prophage Genes in *Bacillus subtilis* 168 and Assessing Their Relative Importance through RNA-Seq Gene Expression Analysis

Wenfa Ng*
Department of Chemical and Biomolecular Engineering,
National University of Singapore, Singapore

Abstract

Bacteriophage evolves to control the population of fast-growing bacterial cells, without which explosion in bacterial population may induce unimaginable harm to diverse ecosystems. But, bacteriophage also "hide" in bacterial genomes when nutritional and environmental circumstances are unfavourable. This involves the integration of phage genome into the host genome at appropriate genomic loci in a process known as lysogeny. This work sought to delineate the prophages present in the annotated genome of *Bacillus subtilis* 168, and assess their relative importance through RNA-seq expression analysis. Firstly, examination of the annotated genome of the model Gram-positive bacterium revealed five distinct prophage regions: SPBeta, prophage 6, PBSX, prophage 3 region, and prophage 1 region. All prophage regions contain host genes, which suggests that host transposase activity have swapped in host genes for phage genes in the prophage genome. Given the significant number of phage genes that have been swapped into each of the prophage genome, all prophage regions are deemed to be defective. BLAST analysis further highlighted that many of the prophages in *B. subtilis* are

* Corresponding Author's Email: ngwenfa771@hotmail.com.

In: Advances in Biology. Volume 2
Editor: Charles D. Grant
ISBN: 979-8-88697-422-5

extinct given that they do not have ancestral or daughter brethren. However, RNA-seq transcriptome analysis of *B. subtilis* turned out an interesting paradox indicative of the important role that host transposase have in swapping in host promoters for prophage genes. Specifically, a significant number of prophage genes are highly expressed, which is implausible given that phage genes should be transcriptionally silent. The result and phenomenon further suggests the relative facile nature in which host promoters could be swapped in for phage genes, which is indicative of presence of genomic motifs in prophage genome recognizable by host transposase. Existence of such sequence motifs is thus indicative of possible co-evolution of transposase and phages where transposases were originally a part of the phage genome, which latter "jumped out" into the host genome to aid the swapping in of host genes into the prophage genome for augmenting prophage genetic repertoire in the face of changing environmental conditions. Overall, it is not uncommon for bacterial species to harbour multiple prophages. But, lysogeny may not be a viable option for long-term preservation of prophage genetic repertoire given that host transposase would inevitable swap in host genes at random locations in the prophage genome.

Keywords: *Bacillus subtilis*, prophage, lysogeny, transposase, co-evolution

Subject Areas: biochemistry, microbiology, evolutionary biology, virology, bioinformatics

Introduction

Bacteriophages play a hugely important role in microbial ecology as they help check the population of bacterial species in diverse habitats. By constraining the population of bacterial species, phages prevent the excessive growth of particular bacterial species that may threaten the survival of other microbial species. This then ensures ecosystem homeostasis which is significant as it safeguards the ecological functions that microbial species bring to Earth.

Although it is known that bacteriophages infect specific bacterial species through receptor mediated binding, there remains significant challenges in identifying them and studying them in bacterial populations. One key challenge is in timing the lytic cycle of a prophage [1]. Typically, bacteriophages enter the lysogenic mode when circumstances in the environment are unfavourable for their survival [2, 3]. This phase of the phage life cycle entails the integration of the entire phage genome into the bacterial genome [4]. Bacteriophages that have activated lysogeny and whose genome has been integrated into the host genome are known as prophages.

Prophages can be viewed as bacteriophage in quiescent mode as they remain fully capable of entering the lytic phase where the whole phage genome is excised from the host genome, activated with phage genes transcribed, and help in the assembly of full phages able to leave and infect other bacterial cells of the same species. But, lysogeny also presents risk to the phage. Chief of which is the irresistible effect of evolutionary pressure on the integrity of the phage genes. In this scenario, phage genes in the host genome would remain transcriptionally silent, and thus given that they do not encode functional protein of use to the host cell, there is no driver that could guard against random genetic drift resulting from mutations incurred during DNA replication of the host genome. Hence, phage genes integrated into host genome may suffer from genetic drift, and through a relatively short span of evolutionary time, lose their functionality. If sufficient number of phage genes suffer from this fate, the prophage may lose its ability to leave the host cell, and is thus inactivated.

But, a more pressing challenge to the survival of a prophage in lysogeny mode is the threat from transposon. These small genetic elements activated by transposase could copy a functional gene from the host into genomic loci occupied by phage genes. Such random insertion events thus disrupt phage genes, and would remain retained by the host cells, as loss of phage genes do not affect the fitness of the host cell. Hence, cells with high level of transposase and transposons do not present favourable environment for the long-term survival of prophages, with the end result that these bacterial species would usually have prophage with truncated genome or which carry genes that belongs to the host rather than the phage itself.

Traditionally, research in bacteriophage genetics have relied on Sanger sequencing for analyzing each gene in turn, which are subsequently cloned into an expression vector for introduction to a recombinant host to produce the target protein. What follows next is biochemical assays for assessing and ascribing function to the target protein. But, this mode of meticulous and tedious phage research has increasingly been replaced by modern genome sequencing efforts, that through automated annotation pipelines, help identify many conserved hypothetical phage genes in genomes of bacterial species. Such approaches significantly streamlined the phage identification exercise, as functions of many genes in the sequenced genome of the bacterial species could be automatically annotated. Finding phage genes would then involve a relatively simple keyword search for "phage" in the Excel file containing the annotated genome of the bacterial species.

This work sought to use the above described approach in understanding the prophages present in the annotated genome of *Bacillus subtilis* 168. Specifically, an in-house MATLAB software was used in parsing the annotated genome of *B. subtilis* 168, and helped in the construction of a gene database of the microorganism. Information included in the gene database include gene identifier, gene function, and gene sequence. This database was subsequently searched with the keyword "phage" to yield a list of prophage genes in the genome of *B. subtilis* 168. Manual curation was then employed to categorized different sets of phage genes as those belonging to different prophages. Further refinement was also made to identify different functional genes such as those involved in DNA replication, transcription, repair and structural roles in each set of prophage genes. Finally, RNA-seq transcriptome data for *B. subtilis* 168 under different nutritional and environmental conditions was analyzed to yield a picture of the expression level of different phage genes under different conditions.

Results revealed the presence of five prophages in *B. subtilis* 168 genome. Many of the prophages are functionally inactivated given that many of their genes have been disrupted by transposons through either swapping in of host genes or insertion of insertion sequences (IS). More importantly, some of the prophages only retained a truncated genome, which attest to the power of evolutionary forces in obliterating many of the phage genes through time. Prophages with such truncated genome are thus also likely to have been the oldest in the *B. subtilis* genome. Functional chacterization of phage genes in each prophage genome turns out an incomplete list of genes for each function. Importantly, few prophages carried the genes belonging to the full complement of functions needed for a full functional prophage. This again provides additional evidence that many of the prophages are functionally inactive and could not enter the lytic cycle. Finally, expression analysis of phage genes revealed high expression level of phage genes that have been swapped in with the host genes. In turn, these genes behave as host genes armed with their own promoters. Further analysis of phage genes with no or low expression level likely highlights the remaining true prophage genes in the genome of the host cell.

Materials and Methods

Reading of Annotated Genome File from Genbank

Annotated genome file for *Bacillus subtilis* 168 was downloaded from Genbank (NCBI Reference Sequence: NC_000964.3). An in-house MATLAB software was used to read and parse information in the annotated genome file into an Excel gene database comprising gene name, gene function, and gene sequence for each gene in the genome of the bacterium.

Identification and Categorization of Phage Genes in *Bacillus subtilis*

A keyword search for "phage" was performed using the in-built Excel search function. All genes whose gene function contain the keyword "phage" would be identified, and collected as an ensemble of phage genes in *B. subtilis* in a separate database.

More importantly, the obtained ensemble of phage genes in *B. subtilis* was further curated manually to yield separate list of genes for individual prophages residing in the genome of the bacterium. This meant that a list of phage genes for each prophage in *B. subtilis* 168 is available in the new database obtained by this work.

Functional Categorization of Phage Genes in Each Prophage

To further understand the functional roles of each phage gene in the genome of individual prophage, functional categorization was performed based on the description of each phage gene to yield a dataset comprising genes belonging to different categories such as DNA replication, transcription, structural genes, and miscellaneous.

RNA-Seq Transcriptome Analysis of Phage Genes in *B. subtilis*

RNA-seq transcriptome of *B. subtilis* under different environmental and nutritional conditions were used to assess whether growth medium or culture conditions influenced the expression level of phage genes. Briefly, RNA-seq transcriptome of *B. subtilis* was downloaded from Array Express and processed by an in-house MATLAB RNA-seq transcriptome processing software to yield expression count for each gene in the genome of *B. subtilis* under specific nutritional and environmental conditions. During transcriptome processing, 5 bases were truncated from the 5' and 3' end of each sequenced read. This preprocessed sequenced read was subsequently aligned to each gene in the genome of *B. subtilis*. If the alignment was successful, the expression count of the gene would be increased by 1. This process was repeated iteratively for a total of 1.2 million sequenced reads to yield a readout of the transcriptional landscape of *B. subtilis*. The Excel file containing the output comprises gene name, gene function, gene sequence, and expression count. This file was searched with the keyword "phage" in Excel to yield a set of phage genes with expression count, which was compiled as a separate database.

Result and Discussion

Genetic Repertoire of SPBeta Prophage in *Bacillus subtilis* 168

Phage SPBeta comprises 23 genes including a serine-type phage integrase (sprA). Many of the phage genes are conserved hypothetic protein with unknown functions. Of interest is that some of the phage genes are likely to have been swapped by host genes through the action of transposon. Examples of this include: N-acetyl transferase (yokL), and aminoglycoside N3' acetyltransferase (aacD).

Interestingly, phage SPBeta comes with calcium dependent DNA nuclease, RNase, and phage toxin. Overall, lack of understanding of the functions of most of the genes associated with SPBeta precludes gaining a further appreciation of the mode of operation and function of this prophage in *B. subtilis*.

Functional categorization of the annotated genes in phage SPBeta reveals only two integral genes important to phage DNA replication (uvrX) and phage integration into the host genome (sprA) (Table 2). Interestingly, the analysis also reveals the swapping of host genes into the prophage genomes. Examples of such genes are type 1 toxin-antitoxin system (bsrG), putative N-acetyltransferase (yokL), putative Rnase (yokI), putative transposase (yoyK), a calcium-dependent DNA nuclease lipoprotein (nukF), and an aminoglycoside N3'-acetyltransferase (aacD). Overall, many of the genes in prophage SPBeta have been swapped with host genes likely due to transposase activity.

Table 1. Genetic repertoire of SPBeta prophage in *Bacillus subtilis* 168

Gene	Gene function
sunA	sublancin 168 lantibiotic antimicrobial precursor peptide
uvrX	lesion bypass phage DNA polymerase
yolD	conserved hypothetical protein
yolC	conserved phage protein of unknown function
yolB	conserved protein of unknown function
yolA	conserved exported protein of unknown function
bsrG	phage toxin; type I toxin-antitoxin system
yokL	putative N-acetyltransferase
yokK	conserved protein of unknown function
yokJ	conserved protein of unknown function
yokI	putative Rnase
yokH	conserved protein of unknown function
yoyK	fragment of putative transposase
yokG	conserved protein of unknown function
nukF	calcium-dependent DNA nuclease, lipoprotein
yokE	conserved protein of unknown function
aacD	aminoglycoside N3'-acetyltransferase
	hypothetical protein
	hypothetical protein
yokC	conserved hypothetical protein
yokB	hypothetical protein
sprA	serine-type phage integrase

Table 2. Categorization of function of annotated genes in phage SPBeta in *Bacillus subtilis*

Function	DNA replication
Gene	Gene function
uvrX	lesion bypass phage DNA polymerase
sprA	serine-type phage integrase
Function	Miscellaneous
Gene	Gene function
bsrG	phage toxin; type I toxin-antitoxin system
yokL	putative N-acetyltransferase
yokI	putative Rnase
yoyK	fragment of putative transposase
nukF	calcium-dependent DNA nuclease, lipoprotein
aacD	aminoglycoside N3'-acetyltransferase

Given the loss of many prophage genes, it is likely that prophage SPBeta is defective, and could not enter the lytic cycle.

Genetic Repertoire of Prophage 6 in the Genome of *Bacillus subtilis*

Similar to prophage SPBeta, prophage 6 in *Bacillus subtilis* comprises a collection of 21 genes (Table 3). Although many of the genes are putative prophage genes or prophage related, there exists instances in which prophage genes have been swapped by host genes through the action of transposase. Example of such swapped-in host genes is the putative SOS response associated protein (yoaM). But such examples are rare, and the genome of prophage 6 consist of mainly phage related genes. This suggests that the genome of prophage 6 lacks genomic sites that could be recognized by transposase for the insertion of transposons, and thus, the genome of prophage 6 remains largely intact.

Table 3. Genetic repertoire of prophage 6 in *Bacillus subtilis* 168

Gene	Gene function
yoaM	putative SOS response associated protein
yozS	putative permease, phage-related
yozT	conserved hypothetical protein (putative phage origin)
yobB	putative transcriptional regulator from prophage
yozV	putative phage protein
yobD	transcriptional regulator (phage-related, Xre family)
yozH	hypothetical protein; putative defective prophage 6
yozI	conserved hypothetical protein
yobE	putative SOS response associated phage protein
yozW	hypothetical protein
yozX	putative phage protein
yozY	putative transcriptional regulator from bacteriophage
yobHc	fragment of putative DNA phage repair protein
yobHm	putative phage DNA repair protein fragment
yozL	conserved hypothetical protein of phage origin
yozM	putative bacteriophage protein
yobI	putative phage repair NTPase with transmembrane helix
yoyA	putative fragment of phage protein
yobJ	conserved protein of unknown function
rttL	phage toxin ribonuclease
yobM	putative phage protein
yobO	putative phage-related pre-neck appendage

Functional categorization of prophage genes in prophage 6 of *B. subtilis* reveals that the prophage is endowed with important genes and functions in DNA replication, transcription, and other miscellaneous functions (Table 4). For example, in the area of DNA replication, prophage 6 has three DNA repair protein genes, which suggest that the genome of prophage 6 may be vulnerable to increased rate of mutations due to an error-prone DNA replication machinery.

Table 4. Categorization of function of annotated genes in prophage 6 of *Bacillus subtilis*

Function	DNA replication
Gene	Gene function
yobIIc	fragment of putative DNA phage repair protein
yobHm	putative phage DNA repair protein fragment
yobI	putative phage repair NTPase
Function	Transcription
Gene	Gene function
yobB	putative transcriptional regulator
yobD	transcriptional regulator (phage-related)
yozY	putative transcriptional regulator
Function	Miscellaneous
Gene	Gene function
yoaM	putative SOS response associated protein
Function	DNA replication
Gene	Gene function
yozS	putative permease, phage-related
yobE	putative SOS response associated phage protein
rttL	phage toxin ribonuclease
yobO	putative phage-related pre-neck appendage

At the level of transcription, prophage 6 is also endowed with three transcriptional regulator genes which indicates that the prophage plays an active role in regulating the transcription of its genes, thereby, providing it with a fitness advantage in timing its replication and exit from the cell after the lysogeny period. Besides DNA replication and transcription, prophage 6 is also endowed with two genes associated with DNA damage SOS response, which provides tantalizing evidence that bacteriophage also uses a mechanism similar to SOS to maintain the integrity of its genome. Finally, other phage genes with important functions include the phage related permease (yozS), phage toxin ribonuclease (rttL), and phage-related pre-neck appendage (yobO).

Genetic Repertoire of PBSX Prophage in the Genome of *Bacillus subtilis*

Table 5 shows the genetic repertoire of PBSX prophage in *Bacillus subtilis* 168. Specifically, the prophage genome comprises 41 genes, which is larger than the genome of prophage SPBeta and prophage 6 in *B. subtilis*. Classified as a defective prophage, PBSX prophage nevertheless still retain many prophage structural genes such as RNA polymerase, capsid portal protein, prophage terminase, and phage chromosome binding protein. Similar to other prophages in *B. subtilis* genome, prophage PBSX has many conserved hypothetical proteins whose function await elucidation.

Table 5. Genetic repertoire of PBSX prophage in *Bacillus subtilis* 168

Gene	Gene function
xlyB	phage PBSX; N-acetylmuramoyl-L-alanine amidase
yjqA	phage PBSX; conserved hypothetical protein
pghB	phage PBSX; gamma-polyglutamate hydrolase
xpdC	phage PBSX; manganese-containing peroxidase
xkdA	phage PBSX; putative peptidase
xre	phage PBSX transcriptional regulator
yjzJ	phage PBSX; conserved hypothetical protein
xkdB	phage PBSX; conserved hypothetical protein
xkdC	phage PBSX; putative DNA binding protein
ykzK	phage PBSX; conserved hypothetical protein
xkdD	phage PBSX; conserved hypothetical protein
xtrA	phage PBSX; conserved protein of unknown function
xpf	phage PBSX; putative RNA polymerase PBSX sigma factor-like
xtmA	phage PBSX; prophage terminase (small subunit)
xtmB	phage PBSX; prophage terminase (large subunit)

Table 5. (Continued)

Gene	Gene function
xkdE	phage PBSX; phage capsid portal protein
xkdF	phage PBSX; conserved hypothetical protein
xkdG	phage PBSX; putative phage chromosome binding protein
ykzL	phage PBSX; conserved hypothetical protein
xkdH	phage PBSX; conserved hypothetical protein
xkdI	phage PBSX; conserved hypothetical protein
xkdJ	phage PBSX; conserved hypothetical protein
ykzM	phage PBSX; conserved hypothetical protein
xkdK	phage PBSX; conserved hypothetical protein
xkdM	phage PBSX; conserved hypothetical protein
xkdN	phage PBSX; conserved hypothetical protein
xkzB	phage PBSX; conserved hypothetical protein
xkdO	phage PBSX; putative tape measure protein
xkdP	phage PBSX; putative murein binding protein
xkdQ	phage PBSX; conserved hypothetical protein
xkdR	phage PBSX; conserved hypothetical protein
xkdS	phage PBSX; conserved hypothetical protein
xkdT	phage PBSX; putative base plate assembly protein
xkdU	phage PBSX; conserved hypothetical protein
xkzA	phage PBSX; conserved hypothetical protein
xkdV	phage PBSX; conserved hypothetical protein
xkdW	phage PBSX; conserved hypothetical protein
xkdX	phage PBSX; conserved hypothetical protein
xepA	phage PBSX; lytic exoenzyme
xhlA	phage PBSX; putative enzyme
xhlB	phage PBSX; holin
xlyA	phage PBSX; N-acetylmuramoyl-L-alanine amidase

Categorization of genes in the PBSX prophage based on function reveals distinct functional categories in DNA replication, transcription, structural, and lytic phase (Table 6). Although PBSX is known to be a defective prophage in the *B. subtilis* virology community, it nevertheless still retains key genes in various important aspects of a prophage life cycle. For example, in the area of DNA replication, PBSX still retains DNA binding protein and phage chromosome binding protein. While these two genes may not be sufficient for orchestrating the entire DNA replication process, it indicates that host transposase and transposon activity are unable to completely destroy the prophage's ability to replicate. Next, PBSX still retain at least two genes regulating transcription. These genes are critical as a prophage must possess the ability to regulate transcriptional activity at a fine balance in order not to put undue stress on the expression of host genes, while ensuring that sufficient phage genes are expressed and proteins produced. In PBSX, the two transcription related phage genes that are retained are transcriptional regulator and putative RNA polymerase PBSX sigma factor, but, it is important to note that this is likely to be an incomplete list of prophage genes involved in transcriptional regulation.

Structural genes are important for enabling prophages to self-assemble into virions. In this area, PBSX prophage still retain important genes encoding proteins such as prophage terminase (small and large subunit), phage capsid portal protein, putative tape measure protein, and phage base plate assembly protein. Similar to the case for DNA replication and transcription regulatory protein, PBSX is unlikely to retain the full complement of structural genes that defines the structure of the phage virion. For a prophage to enter the lytic phase of the phage life cycle, genes encoding proteins important to performing tasks during the lytic phase are important in addition to the key excisionase enzyme. For PBSX, it still retains lytic exoenzyme and holin as genes important to the lytic phase of the phage life cycle. Although incomplete, the genetic repertoire for lytic phase genes in PBSX prophage shows that transposase and transposon activity in the host genome is indeed random, and not specifically targeted at particular groups of genes. Finally, the genetic repertoire of PBSX prophage also contain a couple of host genes swapped in possibly by the action of transposase and transposons. These genes are categorized as miscellaneous in the functional analysis of the genetic complement of PBSX. A cursory analysis of the function of the swapped-in host genes suggests that they belong to different classes of enzymes and proteins, and therefore reaffirmed the random nature in which transposase and transposon swapped-in host genes for prophage genes.

Table 6. Functional categorization of genes in PBSX prophage

Function	DNA replication
Gene	Gene function
xkdC	phage PBSX; putative DNA binding protein
xkdG	phage PBSX; putative phage chromosome binding protein
Function	Transcription
Gene	Gene function
xre	phage PBSX transcriptional regulator
xpf	phage PBSX; putative RNA polymerase PBSX sigma factor-like
Function	Structural
Gene	Gene function
xtmA	phage PBSX; prophage terminase (small subunit)
xtmB	phage PBSX; prophage terminase (large subunit)
Function	DNA replication
Gene	Gene function
xkdE	phage PBSX; phage capsid portal protein
xkdO	phage PBSX; putative tape measure protein
xkdT	phage PBSX; putative base plate assembly protein
Function	Lytic phase
Gene	Gene function
xepA	phage PBSX; lytic exoenzyme
xhlB	phage PBSX; holin
Function	Miscellaneous
Gene	Gene function
xlyB	phage PBSX; N-acetylmuramoyl-L-alanine amidase
pghB	phage PBSX; gamma-polyglutamate hydrolase
xpdC	phage PBSX; manganese-containing peroxidase
xkdA	phage PBSX; putative peptidase
xkdP	phage PBSX; putative murein binding protein
xlyA	phage PBSX; N-acetylmuramoyl-L-alanine amidase

Prophage 1 and Prophage 3 No Longer Qualify as Prophages

Besides SPBeta, prophage 6 and PBSX prophage, the genome of *B. subtilis* 168 also contains two other genomic regions that suggest previous integration of prophage genes. These are prophage region 3 and prophage region 1. However, these prophage regions can no longer be characterized as prophages as too many of the prophage genes have been swapped out by host genes. Comparing the relatively intact nature of prophage genes in SPBeta, prophage 6 and PBSX prophage, the results suggest that the genomic region in which prophage 3 and prophage 1 integrated into may be favourable for integration of transposon or possess specific sequence recognition motifs for transposases.

Expression Level of Prophage Genes in *B. subtilis*

Table 7 shows a tabulation of the most highly expressed prophage genes in *B. subtilis* 168. A quick glance of the data suggests an anomaly in the data that should first be dissected. Specifically, prophage genes should be transcriptionally silent in the *B. subtilis* genome. But, the RNA-seq transcriptome data of the organism shows that many of the prophage genes are transcriptionally active at high level. Another 215 prophage genes register a transcription count of 5 or less, and 82 prophage genes were not expressed. From a theoretical perspective, prophage genes with no expression level may be true phage genes controlled by phage promoters, and therefore could not be expressed in *B. subtilis*.

Presence of sizeable number of prophage genes with high expression activity suggests that these genes are controlled by promoters endogenous to *B. subtilis*. This could happen by way of transposase swapping in host promoters in place of prophage promoters in the prophage genome. Significant number of highly expressed prophage genes in *B. subtilis*, thus, further cements the role of transposase in recognizing unique genomic loci in prophage genome for splicing in of host genes and endogenous promoters. More importantly, many of the highly expressed prophage genes are putative phage genes, and not endogenous host genes swapped-in by transposase activity. The result highlights an evolutionary dance between prophage genome motif, and transposase activity endogenous to the host, whose provenance maybe possible close evolutionary relatedness and co-evolution between prophage and transposase and transposon outside of the context of the bacterial cell that serves as current host for the prophage and transposase.

Table 7. Expression level of highly expressed prophage genes in *B. subtilis* 168

Gene	Gene function	Count
ybcL	putative efflux transporter; prophage 1 region	311
ybdO	putative phage protein; prophage region 1	257
ndhF	putative NADH dehydrogenase; prophage 1 region	240
ybdN	putative phage protein; prophage region 1	207
xkdO	phage PBSX; putative tape measure protein	161
yqaQ	putative phage DNA-binding protein; skin	159
adaB	O6-methylguanine-DNA methyltransferase; prophage	158
yomG	putative DNA wielding protein; SPbeta phage	138
yjcO	putative DNA binding protein; phage island	106
xkdE	phage PBSX; phage capsid portal protein	96
alkA	DNA-3-methyladenine glycosylase; prophage 1	91
yonO	conserved protein of unknown function; phage	80
gutP	H+-glucitol symporter; prophage region 3	79
yjcM	conserved hypothetical protein; phage island	65
cwlP	lytic transglycosylase; SPbeta phage protein	64
ydiM	hypothetical protein; prophage 3 region	63
yobO	putative phage-related pre-neck appendage	60
ybcF	putative enzyme; prophage 1 region	58
yopR	putative DNA breaking-rejoining enzyme; phage	49
yonI	conserved hypothetical protein; phage SPbeta	44

Such evolutionary relationships remain difficult to disentangle without the aid of further molecular evidence at the sequence level. But, one possible reason may be that transposase in *B. subtilis* might have "jumped out" of the prophage genome to reside at a separate location in the *B. subtilis* genome. Functional role of the transposase in the context of phages may be to aid the reshuffling of host genes into the phage genome to introduce mutations and augment the phage genetic repertoire for gaining a survival advantage against a multitude of restriction-modification, and CRISPR systems in bacterial system.

Analysis of the distribution of highly expressed prophage genes in different prophages further reveals that a significant percentage of the high expression genes are in prophage 1 region of *B. subtilis* (Table 8). Distribution of highly expressed prophage genes in other prophages are SPBeta prophage (30%), prophage 3 region (10%), and PBSX prophage (10%). Such a result confirms that prophage region 1 may harbour transposase recognition motifs that afford the swapping in of host genes or promoters, which, in the latter case, resulted in higher expression level of genes. But, this phenomenon is not restricted to prophage 1 region, and is also present in other prophage genome, but to a smaller extent.

Table 8. Distribution of highly expressed prophage genes in different prophages

Phage category	Number	Percentage abundance (%)
Prophage 1 region	6	30
SPBeta prophage	3	15
Prophage 3 region	2	10
PBSX prophage	2	10

Overall, RNA-seq transcriptome analysis reveals the paradoxical result that many prophage genes have high expression level. But, the result reveals that transposase in *B. subtilis* likely aid in swapping in of host promoters and genes into prophage genome, which helps expression of prophage genes. Evidence collected herein also suggests that transposase and transposons may be co-evolutionary mates with bacteriophages where transposase's function is to aid the swapping of host genes into the phage genome to engender variation in the phage genetic repertoire that helps the phage gain a survival advantage in a harsh cellular environment policed by multiple bacterial restriction enzymes and CRISPR systems.

Conclusion

Throughout evolution, bacteriophage entered an evolutionary dance with their host bacterial species. Such co-evolution led to the development of lysogeny where bacteriophage integrated their genome into the host bacterial genome during times of environmental and nutritional stress. Analysis of the repertoire of prophage genes in the Gram-positive model bacterium, *B. subtilis* 168, reveals five distinct prophage regions: SPBeta, Prophage 6, PBSX, Prophage 3 region, and Prophage 1 region. The latter two prophages have been almost totally destroyed by transposase activity that swapped in host genes for phage genes. The remaining three distinct prophages are defective and likely extinct in the environment given that many of their key structural genes have been replaced by host genes likely due to transposase activity, and the prophage lost the ability to replicate its genome and exit the bacterial cell. RNA-seq transcriptome analysis further reveals the paradoxical result that many prophage genes enjoy high expression level. This points to the swapping in of host promoters into prophage genome to aid the expression of prophage genes as the only mechanism explaining the observations. Such a mechanism, in turn, highlights the evolutionary dance between host transposase and prophage, where transposase may be once a part of the prophage genome whose activity is to help swap in host genes to augment the genetic repertoire and diversity to help improve the survivability of the prophage against bacterial adaptive immune system and restriction-modification system.

Conflict of Interest

The author declares no conflicts of interest.

Funding

No funding was used in this work.

References

[1] Wang I-N. Lysis Timing and Bacteriophage Fitness. *Genetics,* 172, 17-26 (2006).

[2] Howard-Varona C, Hargreaves KR, Abedon ST and Sullivan MB. Lysogeny in nature: Mechanisms, impact and ecology of temperate phages. *ISME J,* 11, 1511-1520 (2017).

[3] Benler S and Koonin EV. Phage lysis-lysogeny switches and programmed cell death: Danse macabre. *Bio Essays,* 42, 2000114 (2020).

[4] Garriss G and Henriques-Normark B. Lysogeny in Streptococcus pneumoniae. *Microorganisms,* 8, 1546 (2020).

Chapter 7

Optimization of Naturally Occurring Compounds through Click Chemistry Approach a Powerful Synthetic Toolbox

Idris Arslan*
Zonguldak Bülent Ecevit University, Biomedical Engineering, Zonguldak, Turkey

Abstract

In this rapidly evolving era, naturally occurring product-driven discovery of hybrid molecules by click chemistry have recently revolutionized promising results. Molecular optimization depending on modification of pharmacophoric moiety is often necessary for improvement on target specificity, toxicity, potency and stability. Doubtlessly, click chemistry is one the most efficient synthetic approaches which gives rise to molecular diversity. This review focuses on optimization of naturally occurring products by click chemistry protocol which is a powerful synthetic toolbox and their promising mode of actions compared to parental molecules.

Keywords: click chemistry, naturally occurring products, hybridization, CuAAC, SPAAC, SuFEx

* Corresponding Author's Emails: idris.arslan@beun.edu.tr; idris.arslan@yahoo.com.

In: Advances in Biology. Volume 2
Editor: Charles D. Grant
ISBN: 979-8-88697-422-5

Introduction

Natural products are secondary metabolites produced by living organisms in an attempt to optimize nature. Natural product-driven drug discovery dates back to the identification of morphine isolated from opium by German pharmacist Friedrich Sertürner in 1804. It is usually acknowledged as being the first pharmacologically active and naturally occurring product from any plant species (Luch 2009).

To date, a great number of natural product-based clinical drugs have been discovered for use as biological probe and/or therapeutic agents, including anticancer (e.g., taxol), multiple sclerosis (e.g., fingolimod) immuno-suppressants (e.g., cyclosporine and doxorubicin), cardiovascular disease (e.g., statins) (Newman and Cragg, 2016) and searches continue for novel drugs. Approximately 65% of the 1211 drugs and small-molecule new chemical entities approved for marketing between 1981-2014 are directly or indirectly orginate from naturally occurring compounds (Newman and Cragg 2016). Today, natural products still constitute a large reservoir for pharmacologically active molecules, facilitating the drug design (Arslan et al. 2021; Arslan and Cenzano 2021; Arslan and Cenzano 2020).

Because natural products are highly complex metabolites with distinct stereo (chiral) centers, sp^3 C atoms, rigid skeleton, and unstable functional groups, the *de novo* synthesis is often used in an attempt to optimize naturally occurring compounds. The structural complexity of natural products gives rise to challenges for chemical synthesis and hybridization which causes a high attrition rate in the course of drug design (Pelaez 2006).

Molecules that cross cell membranes easily are often required in drug discovery, thereby their cell permeability, molecular size, solubility (polarity) are critical factors for drug design. Recently, many reports on structural modification of natural products are available in an attempt to increase the desired biological properties such as better efficacy, lesser toxicity to normal cells (Majeed et al. 2013), as well as more selective. In the recent past, structural modification of natural products have been discovered to improve target specificity, potency and stability.

In 2001, Professor K. Barry Sharpless termed 'click chemistry' as the most powerful synthetic toolbox for the development of a set of strong, remarkable reliable and selective reactions for the rapid synthesis of beneficial novel analogues and combinatorial libraries (Kolb et al. 2001). In this sense, click chemistry can be defined shortly as prototypic 1,3-dipolar cycloaddition of an azide-alkyne in an attempt to generate 1, 2, 3-triazole hybrids, which has been used for a broad spectrum of practices owing to itse the simple synthesis, quickly creating novel analogues.

1, 2, 3-triazole, a highly stable heterocyclic structural unit, has a high dipole moment and can make up hydrogen bonds with drug targets. Furthermore, the introduction of 1, 2, 3-triazole moiety into naturally occurring products to form the promising potent biological derivatives has demonstrated an impressive mode of actions. Molecular hybridization mediated by the modification of pharmacophoric moieties has appeared as a powerful strategy for the new drug design which could act as multitarget ligands (Viegas Júnior et al. 2013).

Recently, several click chemistry models such as strain-promoted azide-alkyne cycloaddition (SPAAC) and Sulfur (VI) Fuoride Exchange chemistry (SuFEx) have been developed. In this context, we herein describe the modification of natural products of varied classes by employing modern click chemistry strategies such as CuAAP, SPAAC, and SuFEx and their promising biomedical properties (Figure 1).

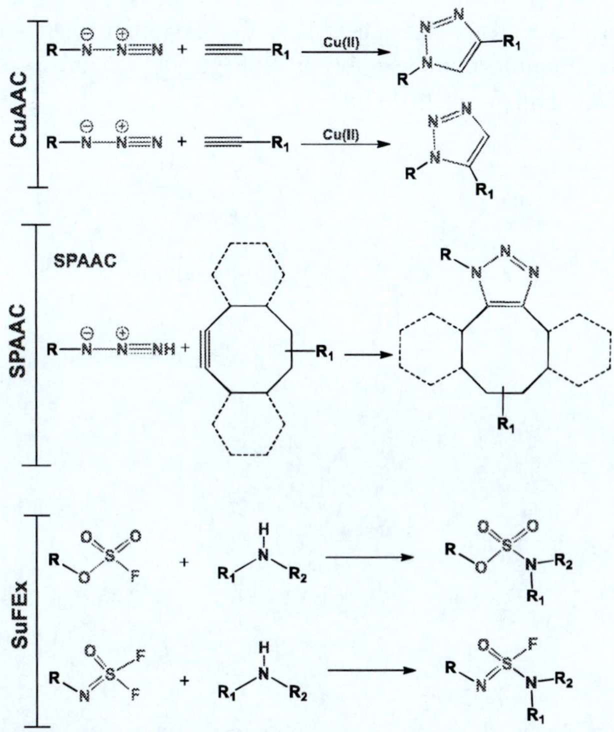

Figure 1. Representative click chemistries.

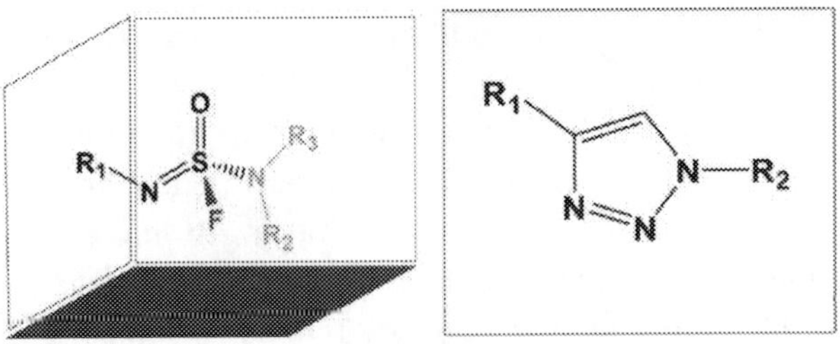

Figure 2. Molecular assembly of SuFEx (tetrahedral) and CuAAC (plenar) click chemistries

In the SuFEx click chemistry, aryl fluorosulfates ($ArOSO_2F$) and iminosulfur oxydifluorides ($RN{=}S(O)F_2$) are initially synthesized through two connective oxyfluoride gases, sulfuryl fluoride (SO_2F_2) and thionyl tetrafluoride ($O{=}SF_4$), respectively (Li et al. 2017). These two SVI−F motifs have been successfully used as connective linkers in polymer synthesis and for construction of various functional molecules (Dong et al. 2014). SO_2F_2 and $ArOSO_2F$ functionalities were introduced into a great number of naturally occurring products (Kitamura et al. 2020). CuAAC click chemistry is a planar molecule, while molecular assembly of SuFEx click chemistry is tetrahedral (Figure 2) (Kitamura et al. 2015).

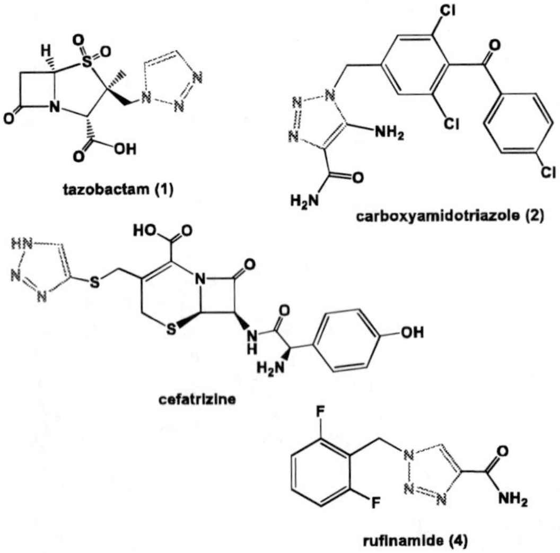

On the other hand, several drug molecules containing 1, 2, 3-triazole are commercially available including tazobactam (**1**) (Xiao et al. 2016), carboxyamidotriazole (**2**) (Guo et al. 2015), cefatrizine (**3**) (Choi et al. 2004), and rufinamide (**4**) (Panebianco et al. 2018).

Click Chemistry-Driven Hybridization of Naturally Occurring Compounds

Terpenoids are the largest class of plant-based naturally occurring metabolites, and account for approximately more than 60% of identified secondary metabolites. Most diterpenoids and triterpenoids equipped with astonishing structural characteristics including chiral centers which cause optically active, rigid and/or cyclic skeletons, sp^3 C atoms, distinct functionalities (Zhang et al. 2021). Pentacyclic triterpene class of compounds generally have oxygen-containing functionalities and carry privileged moieties for chemical hybridization. Many terpenoid derivatives have been synthesized up to now by arching azide or alkyne (CuAAC) at the appropriate position of terpenoid compounds, and then reactions with different building blocks contain alkyne or azide.

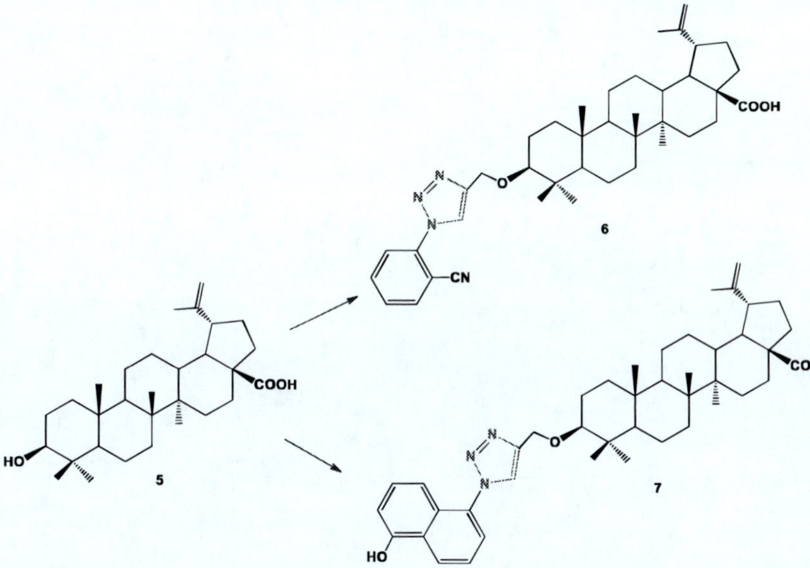

A lupane triterpenoid betulinic acid (**5**) is reported to be cytotoxic towards a variety of cancer cells and involves in modulation of Bcl-2 family, cell cycle key proteins and nuclear factor kappa B (NF-kB) (Yang and Dou 2010). Chemical modification of betulinic acid using a water-soluble 1,2,3-triazole at C-3 position in ring A moiety *via* click chemistry strategy displayed that its 1,2,3-triazoles had better potency than molecule itself. Two derivatives i.e., 3[1*N*(2-cyanophenyl)-1*H*-1,2,3-triazol-4 yl] methyloxy betulinic acid (**6**) and 3[1*N*(5-hydroxy-naphth-1yl)-1*H*-1,2,3-triazol-4 yl] methyloxy betulinic acid (**7**) exhibited effective IC_{50} values (2.5 and 3.5 mM, respectively) towards leukemia cell line HL-60. Research results displayed better capacity of two semi-synthetic esters (**6**) and (**7**) along with prominent arguments for their apoptotic efficiency against THP-1 and HL-60 cancer cell lines. Moreover, the stimulation of apoptosis by (**6**) and (**7**) were also proven through the loss in mitochondrial membrane potential (ΔΨm) and remarkable DNA fragmentation. Triggering the intrinsic pathway via destruction of the mitochondrial membrane potential and DNA fragmentation seems to be the mode of action (Majeed et al. 2013).

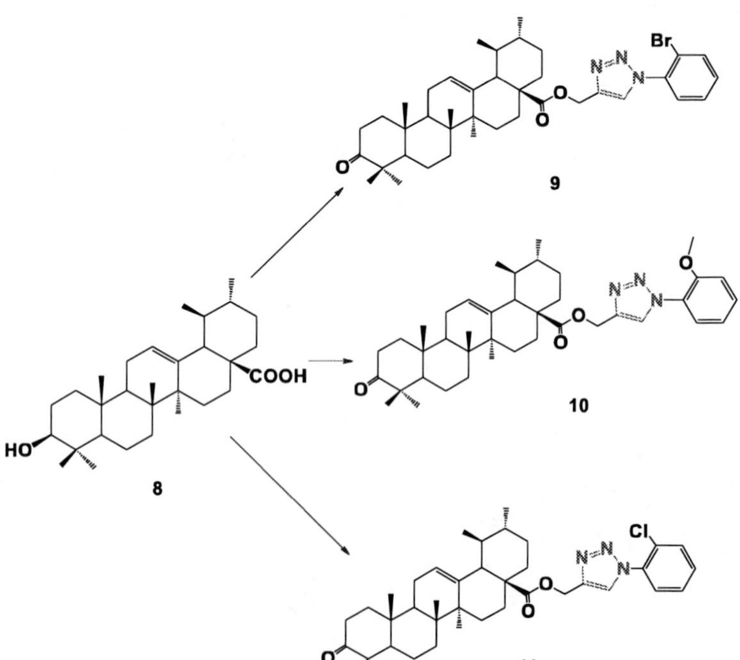

Ursolic acid (**8**) is a naturally occurring ursane triterpenoid that is reported to display anticancer activity by blocking the cell cycle progression in the G_1 phase, resulting in the stimulation of apoptosis (Hsu et al. 2004). Rashid and co-workers (2013) synthesized a series of ursolic acid-1-phenyl-1*H*-[1,2,3]triazol-4 yl methylester derivatives by using Cu(I) catalyzed Huisgen 1,3-dipolar cycloaddition reaction of terminal alkyne congeners of ursolic acid with diverse aromatic azides for boosting their potent antitumor agents. 1D/2D ^{1}H and ^{13}C NMR spectroscopy techniques were employed for structural elucidation of all the compounds synthesized. Anti-cancer potential of newly synthesized derivatives were assayed towards an array of human cancer cells including A-549, MCF-7, HCT-116 and THP-1 through cell density-based sulforhodamine-B (SRB) assay. Research results showed that most of the ursolic acid derivatives displayed more preferred anti-cancer activity towards cancer cells analyzed compared to ursolic acid. Congeners (**9**), (**10**), and (**11**), with *o*-bromo, *o*-methoxy and *o*-chloro substituents were observed to be the most encouraging compounds (Rashid et al. 2013).

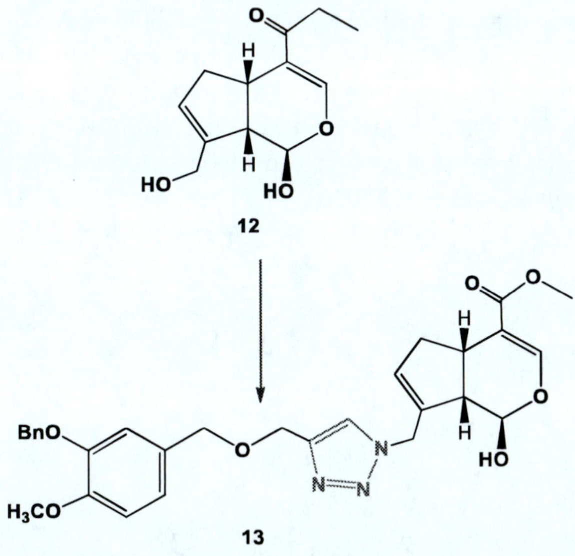

Genipin (**12**) is a secoiridoid originated from *Genipa americana* (Rubiaceae) and in a minor proportion of *Gardenia jasminoides* (Rubiaceae) (Ramos-de-la-Peña et al. 2014). Genipin is an iridoid compound that consist of a cyclopentanoid fused with a dihydropyan ring with OH moiety at C-1 position. It is commonly used in the pharmaceutical practices for cross-linking in an attempt to synthesize the biopolymers as a drug delivery agent. An amphiphilic genipin molecule is able to react easily with primary amine groups of amino acids in any protein to generate dark-blue pigments that enable it to be served as a food colorant (Ueda et al. 1991). Genipin was reported to possess various biological properties including hepatoprotective, anti-inflammatory, anticancer, antithrombotic, antioxidant, gastritis curative, antidiabetic, neuroprotective. Additionally, the anti-inflammatory potency of genipin molecule might be the result from the inhibition of inducible nitric oxide synthase (iNOS), cyclooxygenase-2 (COX-2), interleukin (IL)-6, IL-1β tumor necrosis factor alpha (TNF-α) expression through the down-regulation of nuclear transcription factor kappa-B (NF-κB) (Wang et al. 2012). Genipin with added triazole moiety has been shown to form analogues with enhanced anti-cancer activity. A series of 10-triazolyl-1-methoxygenipin derivatives was synthesized in four-step reactions including (1) methoxylation at C-1 (2) mesylation at C-10 (3) azidation of mesylate moiety with NaN_3 (4) Huisgen 1, 3-dipolar cycloaddition reaction with some substituted alkynes, resulting in twenty seven derivatives.

Benzylether-triazole genipin synthetic analogue (**13**), showed the best anti-cancer activity against P-388 (murine lymphocytic leukemia cells) and A-549 (human lung carcinoma) cancer cell lines with IC_{50} values of 2.54 and 4.53 μM, respectively. Results showed that substituted groups at C-10 of triazole ring displayed a pivotal role on their cytotoxicity than that of the parental genipin (Silalai et al., 2020).

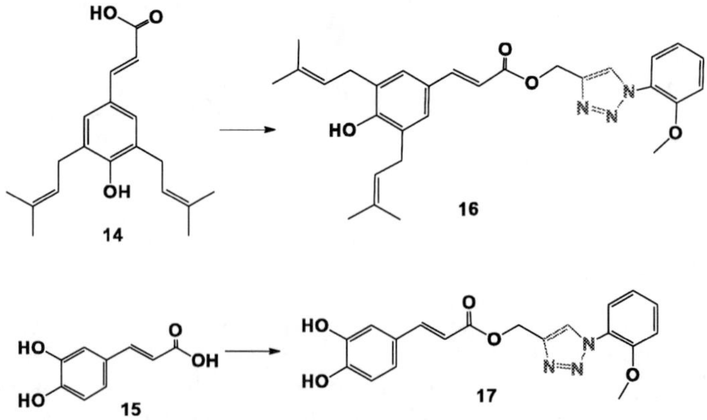

Artepillin C (**14**) and caffeic acid (**15**) are among the major PAK1-inhibiting components of bee-product propolis. On the other hand, primarily owing to their negatively-charged terminal carboxyl group, membrane-permeability of these natural products are rather restricted. Thereby, in an attempt to enhance both membrane-permeability and anti-cancer activity without any considerable loss of their polarity (water-solubility), artepillin C and caffeic acid were successfully esterized with the water-soluble 1,2,3-triazolyl alcohol via Click Chemistry. From the data, it was obvious that both 15A (**16**), and 15C (**17**), derivatives displayed better membrane permeability and anti-cancer activity compared to compounds themselves (Takahashi et al. 2017).

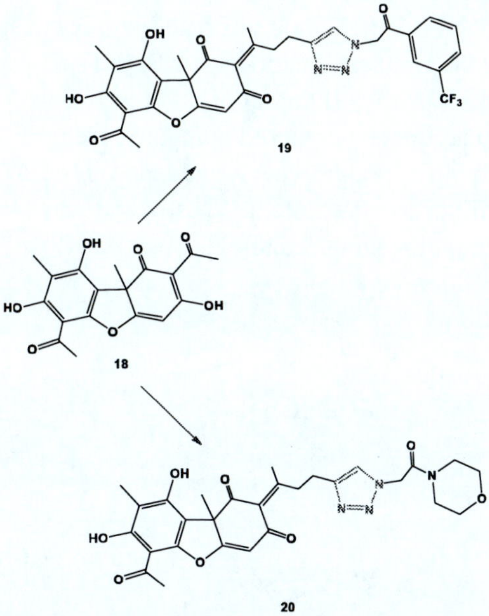

Usnic acid (**18**) is a naturally occurring dibenzofuran analogue present in vario lichen sources such as *Usnia, Ramalina, Cladonia* genera (Ingolfsdottir 2002). Interestingly, both the (+) and (-) enantiomers of usnic acid are influential towards a wide range of Gram-positive (G+) bacterial strains, including strains from clinical isolates, irrespective of their resistant phenotype. (Cocchietto et al. 2002). Usnic acid enaminone-conjugated 1, 2, 3-triazoles as antimycobacterial agents have been synthesized using a 1,3-dipolar cycloaddition strategy. Among them, saccharin synthetic analogue 36 proved to be the most promising derivative, inhibiting *Mycobacterium tuberculosis* (MIC value of 2.5 μM), the newly synthesized derivatives were effective on *Bacillus subtilis* (MTCC 441) with analogue 3-(trifluoromethyl) phenacyl (**19**), and *N*-acylmorpholinyl (**20**), showing inhibitory concentrations of 41 and 90.7 μM, respectively, while they were inactive on the other microorganisms tested (Bangalore et al., 2020).

Oridonin (**21**) is an *ent*-kaurane diterpenoid secondary metabolite used in East Asian herbal medicine isolated from *Rabdosia rubescens*. Over the past decade, considerable advancements have been performed in structural optimization and mode of action efforts of oridonin in order to improve its effectiveness. Recent studies have clearly displayed mode of actions by which oridonin could induce autophagy, increase phagocytosis, suppress cell cycle progression, and trigger apoptosis through regulation of relevant signaling pathways regarding of the modulation of reactive oxygen species (ROS), Bcl-2/Bax, p53/p21, JNK, nuclear factor-kappa B (NF-κB), MAPK, PI3K, and fatty acid synthesis pathways (Ding et al. 2016; Owona et al. 2015; Li et al. 2011). Lately, a Lianyungang-centered company, Hengrui Medicine Co. Ltd. (China), was now given approval to advance HAO472 (L-alanine-(14-oridonin) ester trifluoroacetate into a Phase I human clinical trial for the treatment of acute myelogenous leukemia (Sun et al. 2014).

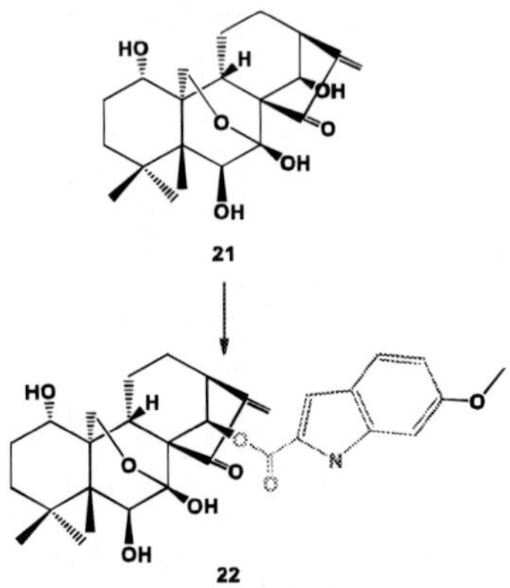

An array of oridonin-triazole analogs containing distinct substituents on the 14-OH position were designed and synthesized to optimize the lead molecule. Most of the analogs exhibited enhanced anti-proliferative activity more than that of original compound oridonin and the positive control drug 5-fluorouracil. Among hybrids, compound (**22**) displayed the strongest anti-proliferative effect against HCT116 (human colorectal carcinoma) cancer cells with IC_{50} value of 0.16μM (Shen et al., 2019). Similarly, Hou and co-workers (2019) synthesized the oridonin analogues via Heck reaction and copper-catalyzed alkyne-azide cycloaddition techniques and biologically evaluated their apoptotic and cell cycle arrest effect of derivatives. The overall results revealed that C-14(2-triazole) acetoxyl-moiety could retain or enhance the anti-cancer activity, while the introduction of C-17 phenyl group displayed the negative effect.

SpeB is a cysteine protease secreted by bacterial pathogen *Streptococcus pyogenes* bacteria involved in host immune system evasion and bacterial virulence (Spaulding et al. 2013). Wang and co-workers (2015) identified a benzyl (cyanomethyl) carbamate-based inhibitor termed SpeB 2477 (**23**) by the hydrolysis of a fluorogenic peptide substrate Ac-AIK-AMC. Most recently, a high-throughput screening hit SpeB 2477 has been modified by SuFEx click chemistry and diversified into 480 analogues in overnight reactions. The existence of analogue 7 (**24**) reduced the viability of wild-type *Streptococcus pyogenes* in a dose-dependent manner, whereas no similar mode of action of (**24**) observed in the ΔSpeB mutant bacterium (Kitamura et al. 2020).

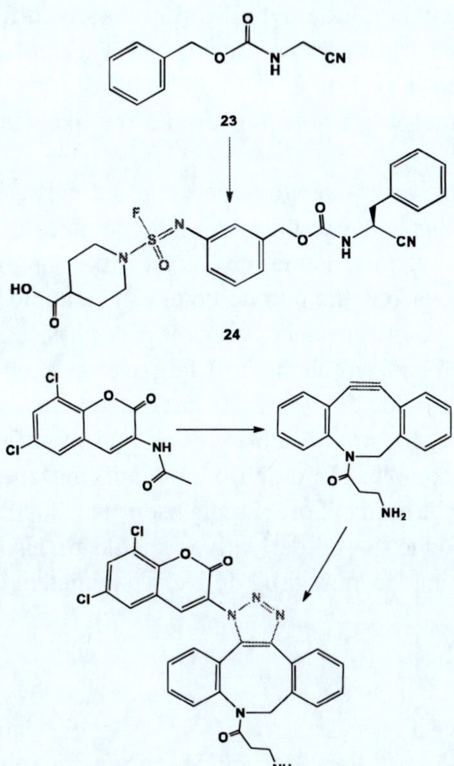

Lately, a fluorogenic version of 1, 3-dipolar cycloaddition between azides and alkynes (DBCO) has been introduced to synthesize the triazolyl coumarin based fluorescent scaffold (**25**) through strain-promoted alkyne azide cycloaddition (SPAAC) chemistry (Sinha et al. 2016).

Conclusion

One of the overall aims of click chemistry is to accelerate the drug discovery with beneficial features. Hybrid molecules synthesized by fusing or conjugating distinct active molecular probes are reported to synergistically prevent different signal pathways that play a role pathogenesis of metabolic disease. In this rapidly evolving era, natural product-mediated hybrid molecules by click chemistry have now revolutionized promising results.

Even though naturally occurring products are generally complex metabolites that contain a little hybridization moiety, different hybridization strategies contribute to alternative analogues. The molecular hybridization protocols regarding novel natural product-based analogues would open horizons for pharmacological searches and are likely to be the center of attraction.

Natural product-based analogues were substantially assayed by cell-based phenotypic screenings (~ %91), thus their target molecules are unclear (Zhang et al. 2021). Similarly, most natural product triazole analogues were analyzed in *in-vitro* conditions but their toxicity profiles should be presented with appropriate assays. As for instance, a recent paper has suggested that an anti-cancer drug termed mubritinib has off-target effects on the compartments within cells which provide the cell with most of their energy, the mitochondria. Chemical derivatives of mubritinib revealed that modifying the 1*H*-1, 2 3-triazole changed complex I inhibition, identifying the heterocyclic 1,3-nitrogen moiety as the toxicophore (Stephenson et al. 2020). Considering these aspects, natural product-based derivatives should be also screened with the protocols that take into account drug delivery parameters.

References

Arslan, I., Akgul, H., and Kara, M. (2021). Saporin, a polynucleotide–adenosine nucleosidase, may be an efficacious therapeutic agent for SARS-CoV-2 Infection. *SLAS Discovery*, 26, 330–335.

Arslan, I., and Cenzano, A. M., (2021). Characterization of Triterpene Saponins from Gypsophila arrostii by Electrospray Ionization Ion Trap Multiple-Stage Mass Spectrometry. *Revista Brasileira de Farmacognosia.*, 31, 107–111.

Arslan, I., and Cenzano, A. M. (2020). N-triterpene saponins in cancer therapy: A review of mode of action. *Revista Brasileira de Farmacognosia*, 2020, 30(1), 1–6.

Bangalore, P. K., Vagolu, S. K., Bollikanda, R. K, Veeragoni, D. K., Choudante, P. C., Misra, S., Sriram D., Sridhar, B., and Kantevari, S. (2020). Usnic acid enaminone-coupled 1, 2, 3-triazoles as antibacterial and antitubercular agents. *Journal of Natural Products*, 83(1), 26-35.

Choi, H. G., Jun, H. W., Kim, D. D., Sah, H., Yoo, B. K., and Yong, C. S. (2004). Simultaneous determination of cefatrizine and clavulanic acid in dog plasma by HPLC. *Journal of pharmaceutical and biomedical analysis*, 35(1), 221-231.

Cocchietto, M., Skert, N., Nimis, P., and Sava, G. (2002). A review on usnic acid, an interesting natural compound. *Naturwissenschaften*, 89(4), 137-146.

Ding, Y., Ding, C., Ye, N., Liu, Z., Wold, E. A., Chen, H., Wilda, C., Shen, Q., and Zhou, J. (2016). Discovery and development of natural product oridonin-inspired anticancer agents. *European Journal of Medicinal Chemistry*, 122, 102-117.

Dong, J., Sharpless, K. B., Kwisnek, .L, Oakdale, J. S., and Fokin, V. V. (2014). SuFEx-based synthesis of polysulfates. Angew. *Chem. Int. Ed.*, 53 (36), 9466-9470.

Guo L, Luo L, Ju R, Chen C, Zhu L, Li J,Yu X, Ye C, and Zhang, D. (2015). Carboxyamidotriazole: a novel inhibitor of both cAMP-phosphodiesterases and cGMP-phosphodiesterases. *European Journal of Pharmacology*, 746, 14-21.

Hou, W., Fan, Q., Su, L., and Xu, H. (2019). Synthesis of Oridonin Derivatives via Mizoroki-Heck Reaction and Click Chemistry for Cytotoxic Activity. *Acamc*, 19, 935–947.

Hsu, Y. L., Kuo, P. L., and Lin, C. C. (2004). Proliferative inhibition, cell-cycle dysregulation, and induction of apoptosis by ursolic acid in human non-small cell lung cancer A549 cells. *Life sciences*, 75(19), 2303-2316.

Ingolfsdottir K. (2002). Usnic acid. *Phytochemistry*, 61(7), 729-736.

Kitamura, S., Zheng, Q., Woehl, J. L., Solania, A., Chen, E., Dillon, N., Hull, M. V., Kotaniguchi, M., Cappiello, J. R., Kitamura, S., Nizet V, Sharpless, K. B., and Wolan, D. W. (2020). Sulfur(VI) Fluoride Exchange (SuFEx)-Enabled High-Throughput Medicinal Chemistry. *Journal of the American Chemical Society*, 142(25), 10899–10904.

Kolb, H. C., Finn, M. G, and Sharpless, K. B. (2001). Click Chemistry: Diverse Chemical Function from a Few Good Reactions. *Angewandte Chemie* (International ed. in English), 40(11), 2004–2021.

Li, S., Wu, P., Moses, J. E., and Sharpless, K. B. (2017). Multidimensional SuFEx click chemistry: Sequential Sulfur(VI) Fluoride Exchange connections of diverse modules launched from an SOF4 hub. *Angew. Chem. Int. Ed.*, 56 (11), 2903-2908.

Li, C. Y., Wang, E. Q., Cheng, Y., and Bao, J. K. (2011). Oridonin: An active diterpenoid targeting cell cycle arrest, apoptotic and autophagic pathways for cancer therapeutics. *The International Journal of Biochemistry & Cell Biology*, 43, 701-704.

Luch, A. (2009). On the impact of the molecule structure in chemical carcinogenesis. *Molecular, clinical and Environmental Toxicology*, 151-179.

Majeed, R., Sangwan, P. L., Chinthakindi, P. K., Khan, I., Dangroo, N. A., Thota, N., Hamid, A., Sharma, P. R., Saxena, A. K., and Koul, S. (2013). Synthesis of 3-O-propargylated betulinic acid and its 1,2,3-triazoles as potential apoptotic agents, *European Journal of Medicinal Chemistry*, 63,-792.

Newman, D. J., and Cragg, G. M. (2016). Natural products as sources of new drugs from 1981 to 2014. *Journal of Natural Products*, 79(3), 629-661.

Owona, B. A., and Schluesener, H. J. (2015). Molecular insight in the multifunctional effects of oridonin. *Drugs in R&D*, 15(3), 233-244.

Panebianco, M., Prabhakar, H., and Marson, A. G. (2018). Rufinamide add-on therapy for refractory epilepsy. *Cochrane Database of Systematic Reviews*, (4).

Peláez, F. (2006). The historical delivery of antibiotics from microbial natural products-can history repeat? *Biochemical pharmacology*, 71(7), 981–990

Rashid, S., Dar, B. A., Majeed, R., Hamid, A., and Bhat, B. A. (2013). Synthesis and biological evaluation of ursolic acid-triazolyl derivatives as potential anti-cancer agents. *Eur. J. Med. Chem.*, 66, 238-245.

Ramos-De-La-Pena, A. M., Renard, C. M., Wicker, L., Montanez, J. C., Garcia-Cerda, L. A., and Contreras-Esquivel, J. C. (2014). Environmental friendly cold-mechanical/sonic enzymatic assisted extraction of genipin from genipap (Genipa americana). *Ultrasonics sonochemistry*, 21(1), 43-49.

Shen, Q. K., Deng, H., Wang, S. B., Tian, Y. S., and Quan, Z. S. (2019). Synthesis, and evaluation of *in vitro* and *in vivo* anticancer activity of 14-substituted oridonin analogs: a novel and potent cell cycle arrest and apoptosis inducer through the p53-MDM2 pathway. *European Journal of Medicinal Chemistry*, 173, 15-31.

Silalai, P., Sirion, U., Piyachaturawat, P., Chairoungdua, A., Suksen, K., and Saeeng, R. (2020). Design, Synthesis and Evaluations of New 10-Triazolyl-1-methoxygenipin Analogues for Their Cytotoxicity to Cancer Cells. *ChemistrySelect*, 5(30), 9540-9546.

Spaulding, A. R., Salgado-Pabón, W., Kohler, P. L., Horswill, A. R., Leung, D. Y., and Schlievert, P. M. (2013). Staphylococcal and Streptococcal superantigen exotoxins. *Clinical microbiology reviews*, 26(3), 422-447.

Sun, P., Wu, G., Qiu, Z., and Chen, Y. (2014). Preparation of L-alanine-(14-oridonin) Ester Trifluoroacetate for Treatment of Cancer. Patent Application CN104017000A, 3.

Stephenson, Z. A., Harvey, R. F., Pryde, K. R., Mistry, S., Hardy, R. E., Serreli, R., et al. (2020). Identification of a Novel Toxicophore in Anti-cancer Chemotherapeutics that Targets Mitochondrial Respiratory Complex I. *Elife*, 9, 1–31.

Takahashi, H., Nguyen, B. C. Q., Uto, Y., Shahinozzaman, M., Tawata, S., and Maruta, H. (2017). 2, 3-Triazolyl esterization of PAK1-blocking propolis ingredients, artepillin C (ARC) and caffeic acid (CA), for boosting their anti-cancer/anti-PAK1 activities along with cell-permeability, *Drug Discov*, 104–109.

Ueda, S., Iwahashi, Y., and Tokuda, H. (1991). Production of anti-tumor-promoting iridoid glucosides in Genipa americana and its cell cultures. *Journal of natural products*, 54(6), 1677-1680.

Viegas-Junior, C., Danuello, A., da Silva Bolzani, V., Barreiro, E. J., and Fraga, C. A. (2007). Molecular hybridization: a useful tool in the design of new drug prototypes. *Current medicinal chemistry*, 14(17), 1829–1852.

Wang, A. Y., González-Páez, G. E., and Wolan, D. W. (2015). Identification and co-complex structure of a new S. pyogenes SpeB small molecule inhibitor. *Biochemistry*, 54(28), 4365-4373.

Wang, Q. S., Xiang, Y., Cui, Y. L., Lin, K. M., and Zhang, X. F. (2012). Dietary blue pigments derived from genipin, attenuate inflammation by inhibiting LPS-induced iNOS and COX-2 expression via the NF-κB inactivation. *PloS One*, 7(3), e34122.

Xiao, A. J., Miller, B. W., Huntington, J. A., and Nicolau, D. P. (2016). Ceftolozane/tazobactam pharmacokinetic/pharmacodynamic-derived dose justification for phase 3 studies in patients with nosocomial pneumonia. *The Journal of Clinical Pharmacology*, 56(1), 56-66.

Yang, H., and Ping Dou, Q. (2010). Targeting apoptosis pathway with natural terpenoids: implications for treatment of breast and prostate cancer. *Current drug targets*, 11(6), 733-744.

Zhang, X., Zhang, S., Zhao, S., Wang, X., Liu, B., and Xu, H. (2021). Click chemistry in natural product modification. *Frontiers in Chemistry*., 9, 1-29.

Chapter 8

Molecular Docking Study of Asafoetida Phytoconstituents as 14α-Demethylase and Transpeptidase Inhibitors

Nidhi Rani[1], Randhir Singh[2,*], Rohini Agrawal[3], Rajwinder Kaur[1] and Praveen Kumar[4]

[1]Chitkara College of Pharmacy, Chitkara University, Punjab, India
[2]Department of Pharmacology, Central University of Punjab, Bathinda, Punjab, India
[3]College of Pharmacy, JSS Academy of Technical Education, Noida, India
[4]Sun Pharmaceutical Industries Ltd., Himachal Pardesh, India

Abstract

Asafoetida, an oleo gum resin obtained from *Ferula asafoetida* of medicinal and nutritional importance. It has been traditionally used for the treatment of various ailments as spice and medicine from centuries. A number of phytoconstituents are present in *Ferula asafoetida* which are responsible for pharmacological properties. To evaluate to potential of *Ferula asafoetida* in the treatment of microbial infections, molecular modeling technique was utilized. 74 phytoconstituents of asafoetida were created from ACD Chem sketch software and PDBs for 14α-demethylase and transpeptidase were retrieved from RCSB PDB database. Docking studies were carried out using Mol Dock score software. All the prepared ligands were evaluated for drug likeliness properties using Swiss ADME. Among 74 compounds, Asacoumarin A was found to be good inhibitor of enzymes under investigation and can be a potential lead molecule for treatment of bacterial and fungal infections.

* Corresponding Author's Email: nidhiprajapati8@gmail.com. Tel: +919896029234.

In: Advances in Biology. Volume 2
Editor: Charles D. Grant
ISBN: 979-8-88697-422-5

Keywords: antibacterial, antifungal, asafoetida, Asocoumarin A, 14α-demethylase, transpeptidase

Introduction

In ayurvedic system, several plants have been described to treat bacterial and fungal infections. Asafoetida, is one of the plants mentioned in Ayurveda for various therapeutic uses. It is oleo gum resin obtained from the Ferula plants, family Umbelliferone. *F. asafoetida* is one of the most prominent *Ferula* species that is prevalent to Afghanistan and Iran. It exists in two forms bitter and sweet and due to its strong, clinging and sulfur like fragrance it is popularly used as an ingredient of cuisine in India [1]. Traditionally, it is utilized for treatment of various disorders such as whooping cough, asthma, ulcer, epilepsy, stomachache, flatulence, bronchitis, intestinal parasites, antispasmodic, indigestion, and influenza [2, 3]. Recent studies have also indicated asafoetida to possess antioxidant [4, 5], antimicrobial [6-10], antiviral [11], antifungal [12-15], cancer chemopreventive [16], anti-diabetic [17], anti-carcinogenic [18, 19] antispasmodic and hypotensive [20], relaxant [21, 22], neuroprotective [23] and molluscicidal [24] activity. Phytoconstituents are the chemical constituents obtained from plant species. These phytoconstituents can be used as inhibitors of 14α-demethylase and transpeptidase having good to excellent drug likeliness properties.

In-silico methods are computer-based approaches commonly utilized in the pharmaceutical field of science to aid the discovery of inhibitors with strong protein binding capabilities, drug-likeliness features, and ADMET (Absorption, Distribution, Metabolism, Excretion, and Toxicity) elucidation. This necessitates the use of protein databases such as the RCSB PDB, Quantitative Structure Activity Analysis correlations (which predict the activity of novel compounds purely based on chemical structure), and computational molecular docking. Various inhibitors for a variety of diseases have been created using *in-silico* approaches [25, 26]. In molecular biology and computer-aided drug creation, molecular docking is a crucial technique. Ligand-protein docking is a technique for predicting the binding mechanism of a ligand and protein with known three-dimensional structures. Protein-protein docking and nucleic acid-protein docking are two other essential forms of docking [27].

In this study, an attempt was made to investigate and identify the inhibitors from Asafoetida, against the enzymes 14α-demethylase and transpeptidase, fungal and bacterial enzymes respectively using *in-silico* docking and ADME studies.

Materials and Methods

Protein Target Preparation

3-dimensional structures of 14α-demethylase and transpeptidase were retrieved from the RCSB protein data bank (RCSB PDB: Homepage) with pdb code 1EA1 and 3PTE, respectively and were saved as pdb files.

Ligand Generation

The two-dimensional structures of 74 phytoconstituents present in asafoetida [28, 29] were created from ACD/Chem sketch software. The generated structures were then saved in MDL molfile format and ligands were then imported in MVD software and optimized to 3D format.

Protein-Ligand Docking

The phytoconstituents were screened for antifungal and antibacterial properties using the software molergo virtual docker (MVD 2010.4.1.0). Average of 10 trial runs for each structure was performed by software and best five poses were obtained. The post docking tool was then studied to attain the best docking pose and energy values.

ADME Study

After molecular docking study, the compounds with best dock scores were further screened for ADME properties using Swiss ADME, an online software.

Result and Discussion

A total of 74 ligands from Asafoetida were identified and structures were generated in Chem sketch software which were saved in mol files, MVD acceptable format. The PDB was chosen and imported in software and optimized. The cavities were predicted and the best cavity having the active ligand was selected. The structures were imported and allowed to dock in the active site. The drugs with five best poses were obtained and the pose with lowest docking score and good number of interactions were selected as depicted in Table 1. Based, on the docking results, the ligand with the overall lowest binding energy was considered as the best inhibitor.

Table 1. Mol dock score of the compounds against 14α-demethylase and transpeptidase

S.No.	Compound	Mol Dock Score	
		1EA1	3PTE
1	Umbelliprenin	-162.035	-102.709
2	5-Hydroxyumbelliprenin	-160.961	-99.8259
3	8-Hydroxyumbelliprenin	-145.859	-92.1275
4	Tadshiferin	-138.473	-92.574
5	Gallbanic acid	-136.525	-63.2683
6	8-Acetoxy-5-S-hydroxyumbelliprenin	-160.559	-104.894
7	Conferol	894.183	945.507
8	Gummosin	899.584	948.556
9	Epi-Samarcandin	1891.58	1944.28
10	Epi-Samarcandin acetate	1896.24	1947.17
11	Fransiferol A	879.499	947.515
12	Fransiferol B	-120.253	-58.9402
13	Fransiferol C	1886.57	1940.06
14	Asacoumarin A	-145.648	-107.824
15	Assafoetidin	2870.13	2939.59
16	Ferocaculicin	1897.32	1945.39
17	Assafoetidinol A	3916.45	3957.56
18	Assafoetidinol B	3924.36	3950.34
19	Polyanthinin	7923.99	7970.06
20	Kamolonol	1919.18	1963.99
21	Foetdidine	878.492	941.217
22	Saradaferin	2880.8	2921.83
23	10R-Karatavicinol	135.708	-90.6955
24	Methyl galbanate	-103.58	-56.5878
25	Feselol	893.908	946.782
26	Ligupersin A	1891.64	1943.13
27	Epi-conferdione	1902.87	1943.57
28	Microlobin	2890.61	2953.2
29	Umbelliferone	-84.7961	-51.3339

S.No.	Compound	Mol Dock Score	
		1EA1	3PTE
30	2-Butyl-1-propenyl disulfide	-70.9187	-55.406
31	1-(Methylthio)propenyl disulfide	-77.0732	-54.5035
32	2-Butyl-3-(methylthio)-2-propenyldisulfide	-83.9089	-67.0737
33	2-Methyl-2-propanethiol	-51.7112	-39.3931
34	2,3-Dimethylthiirane	-49.8867	-38.4665
35	1-Methylthio(Z)-1-propene	-47.8902	-39.6116
36	1-methylthio€-1-propene	-47.5282	-37.3227
37	Dimethyl disulfide	-52.3239	-37.1513
38	S-Methylpropanethioate	-52.0051	-39.8218
39	2-(methylthio)butane	-63.1393	-47.3257
40	3,4-Dimethylthiophene	-55.7345	-42.9128
41	Methyl (Z)-1-propenyldisulfide	-46.9516	-35.6895
42	Dimethyltrisulfide	-64.7475	-48.328
43	2-Butylmethyl disulfide	-70.2365	-51.741
44	Dipropyl disulfide	-65.9204	-55.5231
45	2,3,4-Trimethylthiophene	-65.7311	-47.5383
46	2-Butylvinyl disulfide	-73.0955	-56.2606
47	2-Butyl-1-propenyl disulfide	-60.4158	-45.6725
48	Methyl-1-(methylthio)propyl disulfide	-70.786	-64.9069
49	Di-2-butyl disulfide	-59.2642	-41.4609
50	Methyl-1-(methylthio)ethyldisulfide	-77.099	-55.3387
51	1-(Methylthio)propyl propyl disulfide	-76.1646	-57.8111
52	1-(Methylthio)propyl-1-propenyl disulfide	-107.731	-71.758
53	Asadisulfide	-68.1226	-51.1198
54	2-butylmethyl trisulfide	-81.6939	-55.6658
55	Di-2-butyl trisulfide	-79.2355	-53.6935
56	Di-2-butyl tetrasulfide	-89.4315	-70.378
57	Foetisulfide A	-87.4899	-73.995
58	Foetisulfide C	938.908	974.802
59	7-Oxocallitristic acid	1919.87	1941.37
60	Picealactone	-	1986.73
61	15-Hydroxy-6-en-dehydoabietic acid	-72.4314	-47.6247
62	Vanillin	-146.981	-103.128
63	3,4-Dimethoxycinnamyl-3-(3,4-diacetoxyphenyl) acrylate	-102.746	-65.7167
64	Taraxacin	1921.82	1970.73
65	Fetidone A	1917.23	1969.17
66	Falcarinolone	-126.216	-108.623
67	Oleic acid	-152.364	-100.286
68	β-sitosterol	1901.83	1928.93
69	Galactose	-68.9823	-42.6185
70	Arabinose	-66.2757	-33.8639
71	Glucuronic acid	-82.0261	-40.5631
72	Rhamnose	-67.0273	-37.9448
73	Luteloin-7-β-D-glucopyranoside	-125.942	-66.3524
74	Ferulic acid	-95.2081	-66.5196

Among 74 compounds screened for antifungal properties, 11 compounds were found to exhibit potential antifungal potency with low docking score namely, Umbelliprenin, 5-hydroxyumbelliprenin, 8-hydroxyumbelliprenin, tadshiferin, gallbanic acid, 8-acetoxy-5-s-hydroxyumbelliprenin, fransiferol B, Asacoumarin A, 10R-Karatavicinol, vanillin, falcarinolone, oleic acid and luteolin-7-β-D-glucopyranoside. Moreover, umbelliprenin (-162.035, 2 interactions), 5-hydroxyumbelliprenin (-160.961, 2 interactions) and 8-acetoxy-5-s-hydroxyumbelliprenin (-160.559, 3 interactions) were found to be more potent as compared to other 11 compounds. It was also found that Gln 72 and Tyr 76 are important amino acids for umbelliprenin (2 interactions), Arg 96 and Phe 255 are important amino acids for 5-hydroxyumbelliprenin (2 interactions), Ile 322 and Ile 323 are important amino acids for 8-acetoxy-5-s-hydroxyumbelliprenin (2 interaction) and antifungal property of respective molecules.

These 74 compounds present in Asafoetida were further screened for antibacterial potential against transpeptidase enzyme. The molecules were docked in the active site of the enzyme using pdb 3PTE. The screening results depicted that the compounds Umbelliprenin, 8-acetoxy-5-s-hydroxy-umbelliprenin, Asacoumarin A, 3,4-dimethoxycinnamyl-3(3,4-diacetoxy-phenyl)acrylate and falcrinolone with mol dock score of -102.709, -104.894, -107.824, -103.128 and -108.623, respectively exhibited good antibacterial properties. Further, the study depicted that amino acids Arg 285, Tyr 159 and Thr 299 are vital for antibacterial property of molecules under investigation. Among these 74 compounds, compounds with best dock score were further tested for ADMET properties using Swiss ADME online software.

Thus, in the present study, an attempt was made to find a lead compound from Asafoetida which can be effectively utilized for the treatment of microbial infections such as fungus and bacteria. The study depicted that Asacoumarin A, phytoconstituent present in Asafoetida shows excellent binding by docking studies to both the drug targets. Moreover, the compound also displayed good ADME profile which can be further used for *in-vitro* and *in-vivo* study to develop effective formulation for treatment of fungal and bacterial infections.

Thus, in the present study, an attempt was made to find a lead compound from Asafoetida which can be effectively utilized for the treatment of microbial infections such as fungus and bacteria. The study depicted that Asacoumarin A, phytoconstituent present in Asafoetida shows excellent binding by docking studies to both the drug targets. Moreover, the compound also displayed good ADME profile which can be further used for *in-vitro* and *in-vivo* study to develop effective formulation for treatment of fungal and bacterial infections.

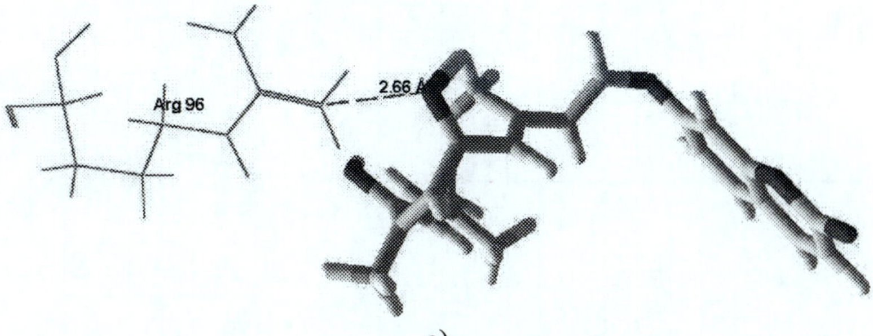

a)

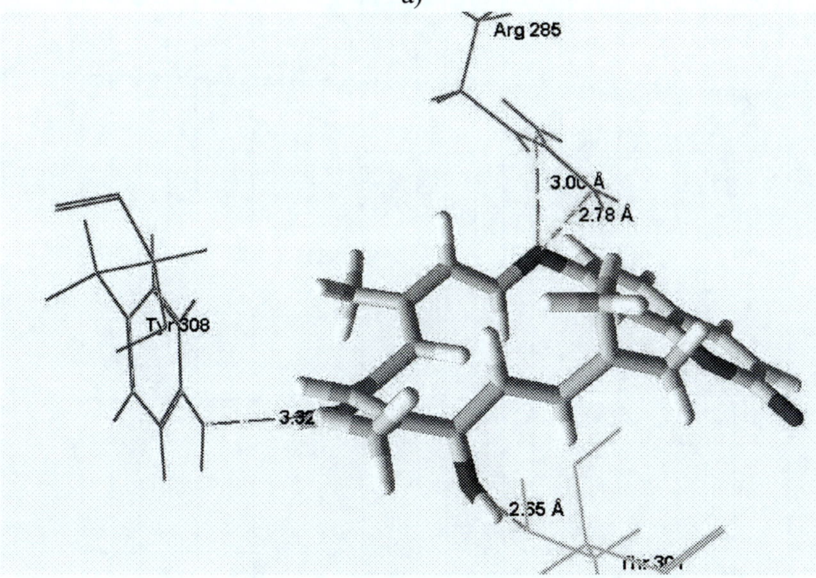

b)

Figure 1. a) Molecular Docking Pose of Asacoumarin A in 14α-demethylase pocket. b) Docking Pose of Asacoumarin in the active site of transpeptidase enzyme.

Conclusion

Increasing dependence on allopathic medicine has led to the development of resistance among the various microbes with some serious side effects. The inhibition of microbial enzymes could be used to the treatment of these infectious diseases. In this *in-silico* study, with the help of molecular docking and ADME, Asocoumarin A has been identified to inhibit the enzymes transpeptidase and 14α-demethylase. Asocoumarin may be a potential lead compound for the treatment of fungal and bacterial infections.

Table 2. Drug properties of the selected ligand

A. Molecular Property

Attributes → Compounds	Molecular Mass	Log P	H-Bond acceptor	H-Bond donor	Rotatable Bonds	TPSA	Molar Refractivity
Umbelliprenin	366.49	4.53	3	0	9	39.44	114.85
5-Hydroxyumbelliprenin	382.49	3.68	4	1	9	59.67	116.01
8-Hydroxyumbelliprenin	382.49	3.68	4	1	9	59.67	116.01
Tadshiferin	382.49	3.68	4	1	9	59.67	116.01
Gallbanic acid	398.49	3.74	5	1	6	76.74	115.2
Fransiferol B	382.49	3.76	4	1	6	59.67	114.11
Asacoumarin A	398.49	2.85	5	2	9	79.9	117.18
10R-Karatavicinol	400.51	2.93	5	2	10	79.9	117.69
Vanillin	326.39	2.82	4	1	1	63.6	90.75
Falcarinolone	242.27	2.21	3	0	0	43.37	67.47
Oleic acid	232.32	2.45	2	0	0	34.14	68.92
Luteloin-7-β-D-glucopyranoside	150.13	-2.32	5	4	0	90.15	29.77

B. Drug Likeliness

Attributes Compounds	Lipinski	Ghose	Veber	Egan	Muegge	Bioavailability Score
Umbelliprenin	1	1	0	1	1	0.55
5-Hydroxyumbelliprenin	0	0	0	0	1	0.55
8-Hydroxyumbelliprenin	0	0	0	0	1	0.55
Tadshiferin	0	0	0	0	1	0.55

Attributes Compounds	Lipinski	Ghose	Veber	Egan	Muegge	Bioavailability Score
Gallbanic acid	0	0	0	0	1	0.56
Fransiferol B	0	0	0	0	1	0.55
Asacoumarin A	0	0	0	0	0	0.55
10R-Karatavicinol	0	0	0	0	0	0.55
Vanillin	0	0	0	0	0	0.55
Falcarinolone	0	0	0	0	0	0.55
Oleic acid	0	0	0	0	0	0.55
Luteloin-7-β-D-glucopyranoside	0	3	0	0	2	0.55

C. Medicinal Chemistry

Attributes Compounds	Lead likeliness	Synthetic Accessibility
Umbelliprenin	3	3.94
5-Hydroxyumbelliprenin	3	4.42
8-Hydroxyumbelliprenin	3	4.45
Tadshiferin	3	4.44
Gallbanic acid	2	4.78
Fransiferol B	2	4.64
Asacoumarin A	3	4.76
10R-Karatavicinol	3	4.55
Vanillin	1	4.37
Falcarinolone	1	4.12
Oleic acid	1	4.19
Luteloin-7-β-D-glucopyranoside	1	3.8

^ Synthetic accessibility: 1(very easy) 10 (very difficult).

D. Pharmacokinetics

Attributes Compounds	GI absorption	BBB permeation	Pgp substrate	CYP1A2 inhibitor	CYP2C19 inhibitor	CYP2C9 inhibitor	CYP2D6 inhibitor	CYP3A4 inhibitor	Log Kp
Umbelliprenin	High	No	No	No	No	Yes	No	Yes	-3.6
5-Hydroxyumbelliprenin	High	No	No	No	Yes	Yes	No	Yes	-4.49
8-Hydroxyumbelliprenin	High	No	No	No	Yes	Yes	No	Yes	-4.49
Tadshiferin	High	No	No	No	Yes	Yes	No	Yes	-4.49
Gallbanic acid	High	No	No	No	Yes	Yes	No	Yes	-5.09
Fransiferol B	High	Yes	No	No	Yes	Yes	Yes	Yes	-4.87
Asacoumarin A	High	No	No	Yes	No	Yes	No	Yes	-5.37
10R-Karatavicinol	High	No	No	No	No	No	Yes	Yes	-5.43
Vanillin	High	Yes	Yes	No	Yes	Yes	No	Yes	-5.36
Falcarinolone	High	Yes	No	No	No	No	No	No	-7.14
Oleic acid	High	Yes	No	No	No	No	No	No	-6.32
Luteloin-7-β-D-glucopyranoside	Low	No	No	No	No	No	No	No	-9.36

GI absorption, BBB and Pgp substrate: The evaluation study depicted that GI absorption is high for all the compounds except Luteloin-7-β-D-glucopyranoside so can be administered orally. Moreover, most of the compounds do not cross BBB. Compounds except Vanillin are not Pgp substrate so oral bioavailability will be good.

CYP1A2 inhibitor: All except Asacoumarin A are non-inhibitors of CYP1A2.

CYP2C19 inhibitor: 5-Hydroxyumbelliprenin, 8-Hydroxyumbelliprenin, Gallbanic acid, Fransiferol B and Vanillin are CYP2C9 inhibitors which may lead to prolonged sedation and unconsciousness.

CYP2C9 inhibitor: Most of the compounds are CYP2C9 inhibitors therefore, increased risk of drug interaction.

CYP2D6 inhibitor: Most of the compounds except Fransiferol B and 10R-Karatavicinol are non-CYP2D6 inhibitor so chances of antidepressant action are low.

CYP3A4 inhibitor: All except Falcarinolone, oleic acid and Luteloin-7-β-D-glucopyranoside are CYP3A4 inhibitors so increases the effect of medicine.

Negative log Kp implies that compound is non-skin permeable.

Conflict of Interest

None.

References

[1] Iran Herbal Pharmacopeia Edition Committee. *Iran Herbal Pharmacopeia*. Ministry of Health & Medical Education, Food and Medicine Deputy Office Publication; 2002.

[2] Takeoka, G. Volatile constituents of Asafoetida. In: Takeoka, G. R., Guntert, M., Engel, K-H, eds. Aroma Active Compounds in Foods. Washington, DC: *American Chemical Society*; 2001:33e44.

[3] Al-Jafari, A. H., Vila, R., Freixa, B., Costa, J., Canigueral, S. Antifungal compounds from the rhizome and roots of F. hermonis. *Phytother. Res.*, 2012, http://dx.doi.org/10.1002/ptr.4806.

[4] Dehpour, A. A., Ebrahimzadeh, M. A., Fazel, N. S., Mohammad, N. S. Antioxidant activity of the methanol extract of *Ferula asafoetida* and its essential oil composition. *Grasas Aceites,* 2009, 60, 405-412.

[5] Kavoosi, G., Rowshan, V. Chemical composition, antioxidant and antimicrobial activities of essential oil obtained from *Ferula asafoetida* oleo-gum-resin: effect of collection time. *Food Chem.*, 2013, 138, 2180-2187.

[6] Shrivastava, V., Bhardwaj, U., Sharma, V., Mahajan, N., Sharma, V., Shrivastava, G. Antimicrobial activities of Asafoetida resin extracts (a potential Indian spice). *J. Pharm. Res.*, 2012, 5, 5022-5024.

[7] Divya, K., Ramalakshmi, K., Murthy, P. S., Rao, L. J. M. Volatile oils from *Ferula asafoetida* varieties and their antimicrobial activity. *LWT Food Sci. Technol.*, 2014, 59,774-779.

[8] Padhy, S., Rai, S., Lamba, N. N. H. K., Upadhyay, M. Spices as potent antibacterial agents against Staphylococcus aureus. *ARPN J. Sci. Technol.*, 2014, 4, 46-51.

[9] Patil, S. D., Shinde, S., Kandpile, P., Jain, A. S. Evaluation of antimicrobial activity of asafoetida. *Int. J. Pharm. Sci. Res.*, 2015, 6, 722-727.

[10] Bhatnager, R., Rani, R., Dang, A. S. Antibacterial activity of Ferula asafoetida: a comparison of red and white type. *J. Appl. Biol. Biotechnol.*, 2015, 3, 18-21.

[11] Mahendra, P., Bisht, S. *Ferula asafoetida*: traditional uses and pharmacological activity. *Pharmacogn. Rev.*, 2012, 6, 141-146.

[12] Kamble, V. A., Patil, S. D. Spice-derived essential oils: effective antifungal and possible therapeutic agents. *J. Herbs Spices Med. Plants.*, 2008, 14, 129-143.

[13] Rani, A., Jain, S., Dureja, P. Synergistic fungicidal efficacy of formulations of neem oil, nicotinic acid and *Ferula asafoetida* with a, b-unsaturated carbonyl compounds against *Sclerotium rolfsii* ITCC 5226 & Macrophomina phaseolina ITCC 0482. *J. Pestic. Sci.*, 2009, 34,253-258.

[14] Angelini, P., Pagiotti, R., Venanzoni, R., Granetti, B. Antifungal and allelopathic effects of asafoetida against *Trichoderma harzianum* and *Pleurotus* spp. *Allelopath. J.*, 2009, 23, 357-368.

[15] Mostafa, Z., Soheil, P., Mahdi, J., Mahmoodi, S. Antifangal effects of asafoetida seed essential oil on in vitro growth of five species of plant pathogenic fungi. *Int. Res. J. Appl. Basic Sci.*, 2013, 4, 1159-1162.

[16] Saleem, M., Alam, A., Sultana, S. Asafoetida inhibits early events of carcinogenesis: a chemopreventive study. *Life Sci.*, 2001, 68, 1913-1921.

[17] Abu-Zaiton, A. S. Anti-diabetic activity of *Ferula asafoetida* extract in normal and alloxan-induced diabetic rats. *Pak. J. Biol. Sci.*, 2010, 13, 97-100.

[18] Mallikarjuna, G. U., Dhanalakshmi, S., Raisuddin, S., Ramesha Rao, A. Chemomodulatory influence of *Ferula asafoetida* on mammary epithelial differentiation, hepatic drug metabolizing enzymes, antioxidant profiles and N-methyl-Nnitrosourea-induced mammary carcinogenesis in rats. *Breast Cancer Res. Treat.*, 2003, 81, 1-10.

[19] Fatehi, M., Farifteh, F., Fatehi-Hassanabad, Z. Antispasmodic and hypotensive effects of *Ferula asafoetida* gum extract. *J. Ethnopharmacol.*, 2004, 91, 321-324.

[20] Bagheri, S. M., Hejazian, S. H., Dashti-R, M. H. The relaxant effect of seed's essential oil and oleo-gum-resin of *Ferula asafoetida* on isolated rat's ileum. *Ann. Med. Health Sci. Res.*, 2014, 4, 238-241.

[21] Khazdair, M. R., Boskabady, M. H. The relaxant effect of *Ferula asafoetida* on smooth muscles and the possible mechanisms. *J. HerbMed. Pharmacol.*, 2015, 4, 40-44.

[22] Tayeboon, G. S., Tavakoli, F., Hassani, S., Khanavi, M., Sabzevari, O., Ostad, S.N. Effects of Cymbopogon citratus and *Ferula asafoetida* extracts on glutamate-induced neurotoxicity. *In vitro Cell Dev. Biol-Anim.*, 2013, 49, 706-715.

[23] Moghaddama M, Farhadi N. Influence of environmental and genetic factors on resin yield, essential oil content and chemical composition of *Ferula asafoetida* L. populations. *J. App. Res. Med. Aromat. Plants*. 2015;2:69e76.

[24] Kumar, P., Singh, D. K. Molluscicidal activity of Ferula asafoetida, *Syzygium aromaticum* and *Carum carvi* and their active components against the snail *Lymnaea acuminata. Chemosphere,* 2006, 63, 1568-1574.

[25] Raj, L. S. M., Jude, J., Kannan, I., Krishna, P. S., Shankar, K. A. Molecular docking study for inhibitors of *Aggregatibacter actinomycet amcomitans* toxins in treatment of aggressive perioodontitis. *Journal of Clinical and Diagnostic Research.,* 2014, 9(11), ZC48 –ZC51.

[26] Jorgensen, W. L. The many roles of computation in drug discovery. *Science,* 2004, 303(5665), 1813–1818.

[27] Cross, J. B., Thompson, D. C., Rai, B. K., Baber, J. C., Fan, K. Y., Hu, Y., Humblet, C. Comparison of several molecular docking programs pose prediction and virtual screening accuracy. *J. Chem. Inf. Model.,* 2009, 49(6), 1455–1474.

[28] Amalraj, A., Gopi, S. Biological activities and medicinal properties of Asafoetida: A review, *J. Trad. and Comp. Med.*, 2016, http://dx.doi.org/10.1016/j.jtcme. 2016.11.004.

[29] Bandyopadhyay, D., Basak, B., Chatterjee, A., Lai, T., Banerji, A., Banerji, J., Neuman, A., Prangé, T. Saradaferin, a new sesquiterpenoid coumarin from *Ferula assafoetida, Natural Product Research: Formerly Natural Product Letters*, 2006, 20(10), 961-965, DOI: 10.1080/14786410600823431.

Contents of Earlier Volume

Index

D

E

F

G

H

I

L

M

N

O

P

R